Molecular Genetics and Pathogenesis of Ehlers-Danlos Syndrome and Related Connective Tissue Disorders

Molecular Genetics and Pathogenesis of Ehlers-Danlos Syndrome and Related Connective Tissue Disorders

Special Issue Editors

Marina Colombi
Marco Ritelli

MDPI • Basel • Beijing • Wuhan • Barcelona • Belgrade • Manchester • Tokyo • Cluj • Tianjin

Special Issue Editors

Marina Colombi
University of Brescia
Italy

Marco Ritelli
University of Brescia
Italy

Editorial Office
MDPI
St. Alban-Anlage 66
4052 Basel, Switzerland

This is a reprint of articles from the Special Issue published online in the open access journal *Genes* (ISSN 2073-4425) (available at: https://www.mdpi.com/journal/genes/special_issues/Connective_Tissue_Disorders).

For citation purposes, cite each article independently as indicated on the article page online and as indicated below:

LastName, A.A.; LastName, B.B.; LastName, C.C. Article Title. *Journal Name* **Year**, *Article Number*, Page Range.

ISBN 978-3-03936-322-3 (Hbk)
ISBN 978-3-03936-323-0 (PDF)

© 2020 by the authors. Articles in this book are Open Access and distributed under the Creative Commons Attribution (CC BY) license, which allows users to download, copy and build upon published articles, as long as the author and publisher are properly credited, which ensures maximum dissemination and a wider impact of our publications.

The book as a whole is distributed by MDPI under the terms and conditions of the Creative Commons license CC BY-NC-ND.

Contents

About the Special Issue Editors . . . . **vii**

Marco Ritelli and Marina Colombi
Molecular Genetics and Pathogenesis of Ehlers–Danlos Syndrome and Related Connective Tissue Disorders
Reprinted from: *Genes* **2020**, *11*, 547, doi:10.3390/genes11050547 . . . . **1**

Marco Ritelli, Valeria Cinquina, Marina Venturini, Letizia Pezzaioli, Anna Maria Formenti, Nicola Chiarelli and Marina Colombi
Expanding the Clinical and Mutational Spectrum of Recessive *AEBP1*-Related Classical-Like Ehlers-Danlos Syndrome
Reprinted from: *Genes* **2019**, *10*, 135, doi:10.3390/genes10020135 . . . . **7**

Daisy Rymen, Marco Ritelli, Nicoletta Zoppi, Valeria Cinquina, Cecilia Giunta, Marianne Rohrbach and Marina Colombi
Clinical and Molecular Characterization of Classical-Like Ehlers-Danlos Syndrome Due to a Novel *TNXB* Variant
Reprinted from: *Genes* **2019**, *10*, 843, doi:10.3390/genes10110843 . . . . **21**

Lucia Micale, Vito Guarnieri, Bartolomeo Augello, Orazio Palumbo, Emanuele Agolini, Valentina Maria Sofia, Tommaso Mazza, Antonio Novelli, Massimo Carella and Marco Castori
Novel *TNXB* Variants in Two Italian Patients with Classical-Like Ehlers-Danlos Syndrome
Reprinted from: *Genes* **2019**, *10*, 967, doi:10.3390/genes10120967 . . . . **33**

Chloe Angwin, Angela F. Brady, Marina Colombi, David J. P. Ferguson, Rebecca Pollitt, F. Michael Pope, Marco Ritelli, Sofie Symoens, Neeti Ghali and Fleur S. van Dijk
Absence of Collagen Flowers on Electron Microscopy and Identification of (Likely) Pathogenic *COL5A1* Variants in Two Patients
Reprinted from: *Genes* **2019**, *10*, 762, doi:10.3390/genes10100762 . . . . **45**

Amanda J. Miller, Jane R. Schubart, Timothy Sheehan, Rebecca Bascom and Clair A. Francomano
Arterial Elasticity in Ehlers-Danlos Syndromes
Reprinted from: *Genes* **2020**, *11*, 55, doi:10.3390/genes11010055 . . . . **53**

Nicola Chiarelli, Marco Ritelli, Nicoletta Zoppi and Marina Colombi
Cellular and Molecular Mechanisms in the Pathogenesis of Classical, Vascular, and Hypermobile Ehlers-Danlos Syndromes
Reprinted from: *Genes* **2019**, *10*, 609, doi:10.3390/genes10080609 . . . . **63**

Pei Jin Lim, Uschi Lindert, Lennart Opitz, Ingrid Hausser, Marianne Rohrbach and Cecilia Giunta
Transcriptome Profiling of Primary Skin Fibroblasts Reveal Distinct Molecular Features Between *PLOD1*- and *FKBP14*-Kyphoscoliotic Ehlers–Danlos Syndrome
Reprinted from: *Genes* **2019**, *10*, 517, doi:10.3390/genes10070517 . . . . **85**

Stefano Giuseppe Caraffi, Ilenia Maini, Ivan Ivanovski, Marzia Pollazzon, Sara Giangiobbe, Maurizia Valli, Antonio Rossi, Silvia Sassi, Silvia Faccioli, Maja Di Rocco, Cinzia Magnani, Belinda Campos-Xavier, Sheila Unger, Andrea Superti-Furga and Livia Garavelli
Severe Peripheral Joint Laxity is a Distinctive Clinical Feature of Spondylodysplastic-Ehlers-Danlos Syndrome (EDS)-*B4GALT7* and Spondylodysplastic-EDS-*B3GALT6*
Reprinted from: *Genes* **2019**, *10*, 799, doi:10.3390/genes10100799 103

Marco Ritelli, Valeria Cinquina, Edoardo Giacopuzzi, Marina Venturini, Nicola Chiarelli and Marina Colombi
Further Defining the Phenotypic Spectrum of *B3GAT3* Mutations and Literature Review on Linkeropathy Syndromes
Reprinted from: *Genes* **2019**, *10*, 631, doi:10.3390/genes10090631 125

Camille Kumps, Belinda Campos-Xavier, Yvonne Hilhorst-Hofstee, Carlo Marcelis, Marius Kraenzlin, Nicole Fleischer, Sheila Unger and Andrea Superti-Furga
The Connective Tissue Disorder Associated with Recessive Variants in the *SLC39A13* Zinc Transporter Gene (Spondylo-Dysplastic Ehlers–Danlos Syndrome Type 3): Insights from Four Novel Patients and Follow-Up on Two Original Cases
Reprinted from: *Genes* **2020**, *11*, 420, doi:10.3390/genes11040420 145

Tomoki Kosho, Shuji Mizumoto, Takafumi Watanabe, Takahiro Yoshizawa, Noriko Miyake and Shuhei Yamada
Recent Advances in the Pathophysiology of Musculocontractural Ehlers-Danlos Syndrome
Reprinted from: *Genes* **2020**, *11*, 43, doi:10.3390/genes11010043 155

Aude Beyens, Kyaran Van Meensel, Lore Pottie, Riet De Rycke, Michiel De Bruyne, Femke Baeke, Piet Hoebeke, Frank Plasschaert, Bart Loeys, Sofie De Schepper, Sofie Symoens and Bert Callewaert
Defining the Clinical, Molecular and Ultrastructural Characteristics in Occipital Horn Syndrome: Two New Cases and Review of the Literature
Reprinted from: *Genes* **2019**, *10*, 528, doi:10.3390/genes10070528 169

Carmela Fusco, Silvia Morlino, Lucia Micale, Alessandro Ferraris, Paola Grammatico and Marco Castori
Characterization of Two Novel Intronic Variants Affecting *Splicing* in *FBN1*-Related Disorders
Reprinted from: *Genes* **2019**, *10*, 442, doi:/10.3390/genes10060442 185

Letizia Camerota, Marco Ritelli, Anita Wischmeijer, Silvia Majore, Valeria Cinquina, Paola Fortugno, Rosanna Monetta, Laura Gigante, Marfan Syndrome Study Group Tor Vergata University Hospital, Federica Carla Sangiuolo, Giuseppe Novelli, Marina Colombi and Francesco Brancati
Genotypic Categorization of Loeys-Dietz Syndrome Based on 24 Novel Families and Literature Data
Reprinted from: *Genes* **2019**, *10*, 764, doi:10.3390/genes10100764 197

About the Special Issue Editors

Marina Colombi, PhD, is a clinical and molecular geneticist who specializes in heritable disorders of connective tissue. She graduated with honors in Biological Sciences, option Genetics, at the University of Pavia, Italy, and she completed a PhD in genetic engineering at the University of Helsinki, Finland. She was research assistant professor of Applied Biology at the School of Medicine, University of Brescia, Italy, where she became full professor of Medical Genetics. Prof Colombi is the Director of the Division of Biology and Genetics, Department of Molecular and Translational Medicine and of the Postgraduate School of Medical Genetics. She is the director of the Ehlers–Danlos Syndrome and Inherited Connective Tissue Disorders Outpatient Clinic (CESED) at the Spedali Civili University Hospital of Brescia and of the Observatory on Connective Tissue Disorders (OCE) at the University of Brescia. She has various responsibilities at the University of Brescia, with a focus on the clinical genetics of rare diseases and genetic laboratory testing. Her major diagnostic and research interests include the clinical and molecular characterization of patients affected with several connective tissue disorders, and the definition of associated pathogenetic mechanisms. Most recently, her group characterized large cohorts of patients affected with classical and vascular Ehlers-Danlos syndrome as well as with ultrarare Ehlers-Danlos syndromes, arterial tortuosity syndrome, Loeys-Dietz syndrome, Marfan syndrome, cutis laxa, pseudoxanthoma elasticum, osteogenesis imperfecta, and dystrophic epidermolysis bullosa. The work of her group contributed to the definition of disease mechanisms involved in the pathogenesis of hypermobile, vascular, and classical Ehlers-Danlos syndrome, and arterial tortuosity syndrome by studying transcriptome profiling, extracellular matrix alterations, and aberrant signal transduction pathways in skin fibroblasts from patients affected with these disorders. She authored/co-authored 160 articles in international journals and various book chapters and books.

Marco Ritelli, PhD, is a clinical molecular geneticist who specializes in heritable disorders of connective tissue. He graduated with honors in Biological Sciences, option Genetics, at the University of Padua, Italy, and, at the Division of Biology and Genetics, Department of Molecular and Translational Medicine, of the University of Brescia, he performed specific training in human and molecular genetics as a research fellow, mainly focusing on the molecular characterization of patients affected with Ehlers-Danlos syndrome and related heritable connective tissue disorders and on the study of the pathomechanisms of these rare diseases. After obtaining his residency in Medical Genetics and the qualification as research assistant professor at the School of Medicine, University of Brescia, Italy, he currently has full-time involvement in diagnostic and research activity at the same institution within the group of Prof. Marina Colombi. Overall, his activities resulted in 80 peer-reviewed publications, most of which are on clinical and molecular research on Ehlers-Danlos syndrome and related disorders.

Editorial

Molecular Genetics and Pathogenesis of Ehlers–Danlos Syndrome and Related Connective Tissue Disorders

Marco Ritelli and Marina Colombi *

Division of Biology and Genetics, Department of Molecular and Translational Medicine, University of Brescia, 25123 Brescia, Italy; marco.ritelli@unibs.it
* Correspondence: marina.colombi@unibs.it; Tel.: +39-03-0371-7240

Received: 8 May 2020; Accepted: 11 May 2020; Published: 13 May 2020

Abstract: Ehlers–Danlos syndromes (EDS) are a group of heritable connective tissue disorders (HCTDs) characterized by a variable degree of skin hyperextensibility, joint hypermobility and tissue fragility. The current EDS classification distinguishes 13 subtypes and 19 different causal genes mainly involved in collagen and extracellular matrix synthesis and maintenance. EDS need to be differentiated from other HCTDs with a variable clinical overlap including Marfan syndrome and related disorders, some types of skeletal dysplasia and cutis laxa. Clinical recognition of EDS is not always straightforward and for a definite diagnosis, molecular testing can be of great assistance, especially in patients with an uncertain phenotype. Currently, the major challenging task in EDS is to unravel the molecular basis of the hypermobile EDS that is the most frequent form, and for which the diagnosis is only clinical in the absence of any definite laboratory test. This EDS subtype, as well as other EDS-reminiscent phenotypes, are currently investigated worldwide to unravel the primary genetic defect and related pathomechanisms. The research articles, case report, and reviews published in this Special Issue focus on different clinical, genetic and molecular aspects of several EDS subtypes and some related disorders, offering novel findings and future research and nosological perspectives.

Keywords: Ehlers–Danlos syndrome; heritable connective tissue disorders; differential diagnosis; next generation sequencing (NGS); transcriptomics; integrated omics approaches

---

Ehlers–Danlos syndromes (EDS), with an estimated prevalence of about 1/5000, belong to the large group of heritable connective tissue disorders (HCTDs) and are characterized by a variable degree of skin hyperextensibility, joint hypermobility (JHM), and tissue fragility. The clinical and genetic heterogeneity of these conditions has long been recognized, but the subjective interpretation of some semiquantitative clinical signs, such as skin hyperextensibility, skin texture, JHM, tissue fragility and bruising, led to diagnostic ambiguity and confusion regarding the type of EDS and the inclusion of similar phenotypes under the broad diagnosis of EDS. With more systematic research on clinical data and with the clarification of the molecular basis and associated pathomechanisms of several of these EDS phenotypes, different classification systems have been formulated in the past 50 years. A first classification with five main types was introduced in 1970 by Beighton [1], followed by the Berlin classification with 11 types [2], and the Villefranche nosology of 1997, in which six EDS types were included [3]. The rapid development of genetic techniques has allowed the recognition of many distinct disorders that, while dissimilar from the initially-described classic EDS types, have been given the umbrella term of EDS as an image of the presence of generalized connective tissue fragility. Hence, the last version of the EDS classification published in 2017, which recognized 13 types with 19 different causal genes mainly involved in collagen and extracellular matrix (ECM) synthesis and maintenance, has been expanded to include a wide range of clinically heterogenous disorders [4].

It should be noted that in an important percentage of EDS patients, no pathogenic variant in any of the known EDS-associated genes is identified. Therefore, it is expected that by taking advantage of next generation sequencing (NGS) technologies, further EDS types will be molecularly defined, thus demanding updating of the existing classification.

It always has been, and still is, a challenge to classify single patients in one of the existing EDS subtypes because the currently defined clinical criteria remain relatively unspecific. Often it is not possible to reach a clinical diagnosis and, therefore, the identification on molecular genetic testing of a clear pathogenic variant in a specific gene can be of great assistance, especially in patients with a clinical presentation that does not completely fit into one of the existing subtypes. The difficulty of clinical diagnosis is, among other reasons, due to the clinical overlap not only between many of the EDS subtypes but also with other HCTDs, such as Marfan, Loeys–Dietz, and arterial tortuosity syndromes, as well as some types of skeletal dysplasia and cutis laxa [4–10].

The major challenging task in EDS today is to unravel the molecular basis of the most frequent EDS type, namely hypermobile EDS (hEDS), the diagnosis of which remains reliant on clinical findings for the absence of any definite laboratory test. This EDS subtype as well as other EDS-reminiscent phenotypes are currently investigated worldwide to unravel the primary genetic defect and related pathomechanisms. In 2018, the groundbreaking "Hypermobile Ehlers Danlos Genetic Evaluation" (HEDGE) was launched by the International EDS Society (https://www.ehlers-danlos.com). There has never been such a worldwide collaborative effort before dedicated to discovering the underlying genetic markers for hEDS. Until the end of 2020, the HEDGE study aims to recruit, screen and undertake NGS on 1000 individuals who have been diagnosed with hEDS by the most recent clinical criteria established in 2017, which are stricter than the Villefranche criteria, in order to form homogeneous cohorts for research purposes [4]. Understanding the genetic causes of hypermobile EDS is undeniably central to the EDS community, since it will allow us to make unequivocal diagnoses for a huge number of patients. Furthermore, understanding the genetic pathways and etiopathomechanisms leading to hEDS will advise the search for possible therapeutic approaches for this disorder.

The original research articles, case report and reviews published in this Special Issue focus on different clinical, genetic, biological and molecular aspects of several EDS subtypes and some related disorders. When the first edition of McKusick's book entitled "Heritable disorders of connective tissue" was published in 1956, less than 100 manuscripts had been dedicated to EDS; they were mainly case reports [11]. Nowadays, the search term "Ehlers–Danlos syndrome" in PubMed yields more than 4000 papers including the 14 contributions of this Special Issue, demonstrating how significant and fertile scientific research is for rare genetic diseases such as EDS and related HCTDs.

The relevance of clinical and genetic research in the continuous definition of new EDS types is exemplified in the original research article by Ritelli et al. [12], who describe a novel patient with the classical-like EDS type 2 (clEDS 2) that is caused by recessive variants in *AEBP1*, and review the clinical and molecular findings of the few patients reported to date. This rare EDS type in differential diagnosis with the more frequent classical EDS (cEDS) is not yet included in the current EDS classification, since the first description was noted in 2018 [13], but certainly it will be incorporated in the forthcoming revision of the EDS nosology. Two original research articles, respectively by Rymen et al. [14] and Micale et al. [15], further define the phenotype of the other classical-like EDS type (*TNXB* deficiency) by reporting three novel patients and performing a literature review. The authors highlight that clEDS 1 is likely underdiagnosed due to the complex structure of the *TNXB* locus which complicates diagnostic molecular testing. Rymen et al. [14] also provide an in vitro characterization of the clEDS 1 cellular phenotype, demonstrating the disorganization of the type I, III and V collagen ECMs in patient's fibroblasts. The case report by Angwin et al. [16] underscores the importance of molecular analysis for a definite cEDS diagnosis by showing that patients with pathogenic *COL5A1* variants can have an absence of collagen flowers on skin biopsy transmission electron microscopy (TEM) analysis, which for many years has been recommended as a first line of investigation to confirm or exclude a cEDS diagnosis. The original research article of Miller et al. [17] provides novel clinical and instrumental findings on hEDS, cEDS, and vascular EDS (vEDS) by performing the assessment of pulse wave velocity measurement in a large patient cohort, which is recognized as a gold standard for determining the stiffness of arteries. The authors evidenced an increased arterial elasticity in all EDS subtypes that was associated with lower supine and seated systolic and diastolic blood pressure, thus likely contributing to the orthostatic symptoms frequently encountered in EDS, especially hEDS.

The paper by Chiarelli et al. [18] offers a wide overview on molecular mechanisms likely involved in cEDS, vEDS, and hEDS that could direct future studies to possible therapeutic strategies. The authors review their previous transcriptome and protein studies on patient dermal fibroblasts, emphasizing that these cells, despite sharing a common ECM remodeling, show differences in the underlying pathomechanisms. In cEDS and vEDS fibroblasts, key processes such as collagen biosynthesis/processing, protein folding quality control, endoplasmic reticulum (ER) homeostasis, autophagy, and wound healing emerged as perturbed. In hEDS cells, gene expression changes related to cell–matrix interactions, inflammatory/pain responses, and the acquisition of an in vitro pro-inflammatory myofibroblast-like phenotype seem to contribute to the complex pathogenesis of this molecularly unsolved EDS type. The evidence that the application of untargeted general omics approaches may serve as a valuable tool to identify novel proteins or pathways involved in the pathogenesis of the different EDS types is also documented in the original research article by Lim et al. [19] that focuses on the very rare kyphoscoliotic EDS (kEDS) type, which groups two clinically indistinguishable disorders caused by biallelic variants in either *PLOD1* or *FKBP14*. This article also proves that nowadays it is possible to perform a high-profile scientific investigation for very rare genetic disorders which have been neglected for too long. The authors performed transcriptome profiling by RNA sequencing of kEDS patient-derived skin fibroblasts that revealed the differential expression of genes encoding ECM components that are unique between *PLOD1*-kEDS and *FKBP14*-kEDS, as well as genes involved in inner ear development, vascular remodeling, ER stress and protein trafficking that were differentially expressed in patient cells compared to controls, addressing possible pharmacological targets to improve disease symptoms.

We recommend reading two papers of the Special Issue together, namely first the research article by Caraffi et al. [20] and then the paper by Ritelli et al. [21], since they present stimulating results offering nosological viewpoints concerning the so-called linkeropathies (LKs), which are caused by defects in genes involved in the glycosaminoglycan (GAG) biosynthesis. Specifically, LK genes encode for enzymes that add GAG chains onto proteoglycans via a common tetrasaccharide linker region. LKs include two different subtypes of the spondylodysplastic EDS (spEDS type 1 and 2) and further related disorders that are characterized by a variable mixed phenotype with signs of EDS and skeletal dysplasia. Of note, some of these conditions are in fact included either in the 2017 EDS classification [4] or in the 2019 nosology of skeletal dysplasia [10]. In the original research article by Caraffi et al. [20], the clinical and molecular findings of three spEDS patients are reported. Through the description of one patient with *B4GALT7*- and two patients with *B3GALT6*-spEDS and a review of previous literature reports, the authors contribute to a more accurate definition of the clinical features associated with these rare conditions. Ritelli et al. [21] report on a patient fulfilling the diagnostic criteria for spEDS according to the 2017 nosology, in whom, however, NGS identified compound heterozygosity for two pathogenic variants in *B3GAT3* that is not recognized as an EDS-causing gene. The authors review the spectrum of *B3GAT3*-related disorders and provide a comparison of all LK patients reported at the time of writing, corroborating the notion that LKs are a phenotypic continuum bridging EDS and skeletal disorders. Following these papers, we suggest reading the research article by Kumps et al. [22] that further accentuates the existing nosological confusion concerning these very rare syndromes with huge clinical overlap. Indeed, the authors describe four patients with recessive variants in *SLC39A13* that are associated with spEDS type 3, even if the encoded gene product, namely a putative zinc transporter (contrariwise to the proteins encoded by *B4GALT7* and *B3GALT6*), is not involved in GAG biosynthesis. Given that the clinical presentation of this condition in childhood consists mainly of short stature and characteristic facial features, the authors propose that the differential diagnosis is not necessarily that of a connective tissue disorder and that *SLC39A13* should be included in gene panels designed to address dysmorphism and short stature. In the outstanding review by Kosho et al. [23], the authors discuss recent advances in the pathophysiology of the rare musculocontractural EDS (mcEDS), which is caused by biallelic variants in *CHST14* and *DSE* (which are also involved in GAG synthesis). By describing novel glycobiological, pathological, and animal model-based findings, the authors highlight the critical roles of dermatan sulfate (DS) and DS-proteoglycans in the multisystem development and maintenance of connective tissues and provide fundamental evidence to support future etiology-based therapies.

Finally, three papers deal with different HCTDs in differential diagnosis with EDS. Beyens et al. [24] describe the clinical and molecular characteristics of two novel and 32 previously reported patients with occipital horn syndrome (OHS), previously known as EDS type IX or X-linked cutis laxa [2], caused by pathogenic variants in *ATP7A* (encoding a copper transporter). The main clinical features of OHS, such as cutis laxa, bony exostoses and bladder diverticula, are attributed to defective ATP7A trafficking and decreased activity of lysyl oxidase, a cupro-enzyme involved in collagen crosslinking, in line with a pathogenetic scheme shared with many EDS types. The authors explored the pathomechanisms of OHS by performing TEM analysis on skin biopsies and collagen biochemical analysis on fibroblast cultures that showed increased collagen diameter, elastic fiber abnormalities and multiple autophagolysosomes. Fusco et al. [25] report two unrelated individuals with Marfan syndrome (MFS) and Mitral valve–Aorta–Skeleton–Skin (MASS) syndrome, respectively, which were associated with different intronic variants in *FBN1*, pointing out the importance of intronic sequence analysis and the need for integrative functional studies in the diagnosis of MFS and related disorders. Camerota et al. [26] report on a cohort of 34 patients with Loeys–Dietz syndrome (LDS) with a defined molecular defect either in *TGFBR1*, *TGFBR2*, *SMAD3*, or *TGFB2*. The study broadens the clinical and molecular spectrum of LDS, corroborates and expands previously delineated genotype–phenotype correlations, and shows that a phenotypic continuum emerges as more patients are described, paving the way for a gene-based classification of the different disease subtypes.

In conclusion, this Special Issue, by offering novel findings and future research perspectives, will be of interest not only to a wide range of investigators but also to patients with EDS and related disorders, as well as to all healthcare practitioners who may encounter such syndromes during their work and we hope they will enjoy reading it.

Author Contributions: M.R. and M.C. equally contributed to the managing of the Special Issue and wrote this editorial. All authors have read and agreed to the published version of the manuscript.

Funding: No funding was active on this project.

Acknowledgments: M.R. and M.C. would like to thank the Editor-in-Chief and the Section Editors of Genes for the fruitful collaboration and all the authors and reviewers for their contributions to this Special Issue.

Conflicts of Interest: The authors declare that there is no conflict of interest concerning this work.

References

1. Beighton, P. Ehlers–Danlos syndrome. *Ann. Rheum. Dis.* **1970**, *29*, 332–333. [CrossRef] [PubMed]
2. Beighton, P.; De Paepe, A.; Danks, D.; Finidori, G.; Gedde-Dahl, T.; Goodman, R.; Hall, J.G.; Hillister, D.W.; Horton, W.; McKusick, V.A.; et al. International nosology of heritable disorders of connective tissue, Berlin, 1986. *Am. J. Med. Genet.* **1988**, *29*, 581–594. [CrossRef] [PubMed]
3. Beighton, P.; De Paepe, A.; Steinmann, B.; Tsipouras, P.; Wenstrup, R.J. Ehlers–Danlos syndromes: Revised nosology, Villefranche, 1997. Ehlers–Danlos National Foundation (USA) and Ehlers–Danlos Support Group (UK). *Am. J. Med. Genet.* **1998**, *77*, 31–37. [CrossRef]
4. Malfait, F.; Francomano, C.; Byers, P.; Belmont, J.; Berglund, B.; Black, J.; Bloom, L.; Bowen, J.M.; Brady, A.F.; Burrows, N.P.; et al. The 2017 international classification of the Ehlers–Danlos syndromes. *Am. J. Med. Genet. Part C Semin. Med. Genet.* **2017**, *175*, 8–26. [CrossRef]

5. Colombi, M.; Dordoni, C.; Chiarelli, N.; Ritelli, M. Differential diagnosis and diagnostic flow chart of joint hypermobility syndrome/Ehlers–Danlos syndrome hypermobility type compared to other heritable connective tissue disorders. *Am. J. Med. Genet. Part C Semin. Med. Genet.* **2015**, *169*, 6–22. [CrossRef] [PubMed]
6. Meester, J.A.N.; Verstraeten, A.; Schepers, D.; Alaerts, M.; Van Laer, L.; Loeys, B.L. Differences in manifestations of Marfan syndrome, Ehlers–Danlos syndrome, and Loeys-Dietz syndrome. *Ann. Cardiothorac. Surg.* **2017**, *6*, 582–594. [CrossRef]
7. Dietz, H. *Marfan Syndrome 1993–2020*; Adam, M.P., Ardinger, H.H., Pagon, R.A., Eds.; University of Washington: Seattle, WA, USA, 2001.
8. Loeys, B.L.; Dietz, H.C. *Loeys-Dietz Syndrome 1993–2020*; Adam, M.P., Ardinger, H.H., Pagon, R.A., Eds.; University of Washington: Seattle, WA, USA, 2008.
9. Callewaert, B.; De Paepe, A.; Coucke, P. *Arterial Tortuosity Syndrome 1993–2020*; Adam, M.P., Ardinger, H.H., Pagon, R.A., Eds.; University of Washington: Seattle, WA, USA, 2014.
10. Mortier, G.R.; Cohn, D.H.; Cormier-Daire, V.; Hall, C.; Krakow, D.; Mundlos, S.; Nishimura, G.; Robertson, S.; Sangiorgi, L.; Savarirayan, R.; et al. Nosology and classification of genetic skeletal disorders: 2019 revision. *Am. J. Med. Genet. Part A* **2019**, *179*, 2393–2419. [CrossRef] [PubMed]
11. McKusick, V.A. Heritable disorders of connective tissue: IV. The Ehlers–Danlos syndrome. *J. Chronic Dis.* **1956**, *3*, 2–24. [CrossRef]
12. Ritelli, M.; Cinquina, V.; Venturini, M.; Pezzaioli, L.; Formenti, A.M.; Chiarelli, N.; Colombi, M. Expanding the clinical and mutational spectrum of recessive *AEBP1*-related classical-like Ehlers–Danlos syndrome. *Genes* **2019**, *10*, 135. [CrossRef] [PubMed]
13. Blackburn, P.R.; Xu, Z.; Tumelty, K.E.; Zhao, R.W.; Monis, W.J.; Harris, K.G.; Gass, J.M.; Cousin, M.A.; Boczek, N.J.; Mitkov, M.V.; et al. Bi-allelic alterations in *AEBP1* lead to defective collagen assembly and connective tissue structure resulting in a variant of Ehlers–Danlos syndrome. *Am. J. Hum. Genet.* **2018**, *102*, 696–705. [CrossRef]
14. Rymen, D.; Ritelli, M.; Zoppi, N.; Cinquina, V.; Giunta, C.; Rohrbach, M.; Colombi, M. Clinical and molecular characterization of classical-like Ehlers–Danlos syndrome due to a novel *TNXB* variant. *Genes* **2019**, *10*, 843. [CrossRef] [PubMed]
15. Micale, L.; Guarnieri, V.; Augello, B.; Palumbo, O.; Agolini, E.; Sofia, V.M.; Mazza, T.; Novelli, A.; Carella, M.; Castori, M. Novel *TNXB* variants in two Italian patients with classical-like Ehlers–Danlos syndrome. *Genes* **2019**, *10*, 967. [CrossRef] [PubMed]
16. Angwin, C.; Brady, A.F.; Colombi, M.; Ferguson, D.J.P.; Pollitt, R.; Pope, F.M.; Ritelli, M.; Symoens, S.; Ghali, N.; Van Dijk, F.S. Absence of collagen flowers on electron microscopy and identification of (Likely) pathogenic *COL5A1* variants in two patients. *Genes* **2019**, *10*, 762. [CrossRef] [PubMed]
17. Miller, A.J.; Schubart, J.R.; Sheehan, T.; Bascom, R.; Francomano, C.A. Arterial elasticity in Ehlers–Danlos syndromes. *Genes* **2020**, *11*, 55. [CrossRef]
18. Chiarelli, N.; Ritelli, M.; Zoppi, N.; Colombi, M. Cellular and molecular mechanisms in the pathogenesis of classical, vascular, and hypermobile ehlers-danlos syndromes. *Genes* **2019**, *10*, 609. [CrossRef]
19. Lim, P.J.; Lindert, U.; Opitz, L.; Hausser, I.; Rohrbach, M.; Giunta, C. Transcriptome profiling of primary skin fibroblasts reveal distinct molecular features between *PLOD1*-and *FKBP14*-kyphoscoliotic Ehlers–Danlos syndrome. *Genes* **2019**, *10*, 517. [CrossRef]
20. Caraffi, S.G.; Maini, I.; Ivanovski, I.; Pollazzon, M.; Giangiobbe, S.; Valli, M.; Rossi, A.; Sassi, S.; Faccioli, S.; Di Rocco, M.; et al. Severe peripheral joint laxity is a distinctive clinical feature of spondylodysplastic-Ehlers–Danlos syndrome (EDS)-*B4GALT7* and spondylodysplastic-EDS-*B3GALT6*. *Genes* **2019**, *10*, 799. [CrossRef]
21. Ritelli, M.; Cinquina, V.; Giacopuzzi, E.; Venturini, M.; Chiarelli, N.; Colombi, M. Further defining the phenotypic spectrum of *B3GAT3* mutations and literature review on linkeropathy syndromes. *Genes* **2019**, *10*, 631. [CrossRef]
22. Kumps, C.; Campos-Xavier, B.; Hilhorst-Hofstee, Y.; Marcelis, C.; Kraenzlin, M.; Fleischer, N.; Unger, S.; Superti-Furga, A. The connective tissue disorder associated with recessive variants in the SLC39A13 zinc transporter gene (spondylodysplastic Ehlers–Danlos syndrome type 3): Insights from four novel patients and follow-up on two original cases. *Genes* **2020**, *11*, 420. [CrossRef] [PubMed]

23. Kosho, T.; Mizumoto, S.; Watanabe, T.; Yoshizawa, T.; Miyake, N.; Yamada, S. Recent advances in the pathophysiology of musculocontractural Ehlers–Danlos syndrome. *Genes* **2020**, *11*, 43. [CrossRef]
24. Beyens, A.; Van Meensel, K.; Pottie, L.; De Rycke, R.; De Bruyne, M.; Baeke, F.; Hoebeke, P.; Plasschaert, F.; Loeys, B.; De Schepper, S.; et al. Defining the clinical, molecular and ultrastructural characteristics in occipital horn syndrome: Two new cases and review of the literature. *Genes* **2019**, *10*, 528. [CrossRef] [PubMed]
25. Fusco, C.; Morlino, S.; Micale, L.; Ferraris, A.; Grammatico, P.; Castori, M. Characterization of two novel intronic variants affecting splicing in *FBN1*-related disorders. *Genes* **2019**, *10*, 442. [CrossRef] [PubMed]
26. Camerota, L.; Ritelli, M.; Wischmeijer, A.; Majore, S.; Cinquina, V.; Fortugno, P.; Monetta, R.; Gigante, L.; Hospital, M.S.S.G.T.V.U.; Sangiuolo, F.C.; et al. Genotypic categorization of Loeys-Dietz syndrome based on 24 novel families and literature data. *Genes* **2019**, *10*, 764. [CrossRef] [PubMed]

© 2020 by the authors. Licensee MDPI, Basel, Switzerland. This article is an open access article distributed under the terms and conditions of the Creative Commons Attribution (CC BY) license (http://creativecommons.org/licenses/by/4.0/).

Article

Expanding the Clinical and Mutational Spectrum of Recessive *AEBP1*-Related Classical-Like Ehlers-Danlos Syndrome

Marco Ritelli [1], Valeria Cinquina [1], Marina Venturini [2], Letizia Pezzaioli [1,3], Anna Maria Formenti [4], Nicola Chiarelli [1] and Marina Colombi [1,*]

[1] Division of Biology and Genetics, Department of Molecular and Translational Medicine, University of Brescia, 25123 Brescia, Italy; marco.ritelli@unibs.it (M.R.); v.cinquina@studenti.unibs.it (V.C.); l.pezzaioli001@studenti.unibs.it (L.P.); nicola.chiarelli@unibs.it (N.C.)
[2] Division of Dermatology, Department of Clinical and Experimental Sciences, Spedali Civili University Hospital, 25123 Brescia, Italy; marina.venturini@unibs.it
[3] Spedali Civili of Brescia, 25123 Brescia, Italy
[4] IRCCS Istituto Ortopedico Galeazzi, 20161 Milano, Italy; annaformenti@live.it
* Correspondence: marina.colombi@unibs.it; Tel.: +39-030-3717-240; Fax +39-030-371-7241

Received: 10 January 2019; Accepted: 8 February 2019; Published: 12 February 2019

Abstract: Ehlers-Danlos syndrome (EDS) comprises clinically heterogeneous connective tissue disorders with diverse molecular etiologies. The 2017 International Classification for EDS recognized 13 distinct subtypes caused by pathogenic variants in 19 genes mainly encoding fibrillar collagens and collagen-modifying or processing proteins. Recently, a new EDS subtype, i.e., classical-like EDS type 2, was defined after the identification, in six patients with clinical findings reminiscent of EDS, of recessive alterations in *AEBP1*, which encodes the aortic carboxypeptidase–like protein associating with collagens in the extracellular matrix. Herein, we report on a 53-year-old patient, born from healthy second-cousins, who fitted the diagnostic criteria for classical EDS (cEDS) for the presence of hyperextensible skin with multiple atrophic scars, generalized joint hypermobility, and other minor criteria. Molecular analyses of cEDS genes did not identify any causal variant. Therefore, *AEBP1* sequencing was performed that revealed homozygosity for the rare c.1925T>C p.(Leu642Pro) variant classified as likely pathogenetic (class 4) according to the American College of Medical Genetics and Genomics (ACMG) guidelines. The comparison of the patient's features with those of the other patients reported up to now and the identification of the first missense variant likely associated with the condition offer future perspectives for EDS nosology and research in this field.

Keywords: classical Ehlers-Danlos syndrome; classical-like Ehlers-Danlos syndrome type 2; *AEBP1*; aortic carboxypeptidase-like protein; differential diagnosis; high-frequency ultrasonography; reflectance confocal microscopy

---

1. Introduction

Ehlers-Danlos syndrome (EDS), with an estimated prevalence of 1/5000, comprises a group of clinically heterogeneous heritable connective tissue disorders (HCTDs) with diverse molecular etiologies. The 2017 revised EDS classification recognized 13 distinct subtypes caused by pathogenic variants in 19 genes mainly encoding fibrillar collagens, collagen-modifying proteins, or processing enzymes [1]. Classical EDS (cEDS) (MIM #130000), with an estimated prevalence of 1/20,000, is an autosomal dominant disorder primarily characterized by cutaneous and articular involvement. Indeed, cEDS is suggested by skin hyperextensibility plus atrophic scarring that must be present together with the other major criterion, i.e., generalized joint hypermobility (gJHM) evaluated according to the Beighton score (BS ≥5/9), and/or with at least three of the minor criteria among easy bruising, soft, doughy skin, skin fragility, molluscoid pseudotumors, subcutaneous spheroids, hernia (or a history of thereof), epicanthal folds, JHM complications (e.g., sprains, luxation/subluxation, pain, flexible flatfoot), and family history of a first-degree relative who meets clinical criteria [1–3]. Furthermore, cEDS patients may present distinctive facial features, premature rupture of fetal membranes, scoliosis, osteoporosis, gastroesophageal reflux, and cardiac and blood vessel fragility [2,4–8]. Skin is hyperextensible if it can be stretched over a standardized cut off in the following areas: 1.5 cm for the distal part of the forearms and the dorsum of the hands; 3 cm for neck, elbow and knees; 1 cm on the volar surface of the hand (palm) [1,2,5]. Atrophic scarring can range in severity; however, most cEDS patients have wide atrophic scars in different body areas that can variably assume a cigarette paper, papyraceous, or hemosiderotic appearance [1,2,5].

Point mutations or intragenic rearrangements of the *COL5A1* and *COL5A2* genes encoding type V collagen are recognized in over 90% of patients [4,9], the recurrent heterozygous *COL1A1* c.934C>T (p.Arg312Cys) substitution is rarely found [4,10,11]. Negative molecular testing does not exclude the diagnosis, as specific types of mutations (e.g., deep intronic variants) may go undetected by standard diagnostic molecular techniques. Nevertheless, alternative diagnoses should be taken into account in the absence of a *COL5A1*, *COL5A2*, and *COL1A1* mutation [1].

Recognition of cEDS is straightforward in the patient with the typical cutaneous signs and BS ≥5. However, intra- and interfamilial variability tells a much broader clinical presentation and significant overlap with other EDS types and HCTDs [1,2,4–6,12]. Differential diagnosis of cEDS should include the hypermobile EDS (hEDS), particularly in patients without a striking cutaneous involvement [1,12]. Indeed, hEDS shares with cEDS gJHM and many mucocutaneous signs, but generally a lower grade of skin hyperextensibility and only few small atrophic or post-surgical enlarged scars are observed [13,14]. In case of a family history compatible with autosomal recessive transmission, differential diagnosis comprises the rare classical-like EDS type 1 (MIM #606408) due to biallelic *TNXB* mutations. These patients show marked skin hyperextensibility, easy bruising, and joint laxity, but unlike cEDS patients, they do not have atrophic scarring or poor wound healing. Furthermore, minor criteria such as foot deformities, edema in the legs, mild proximal and distal muscle weakness, axonal polyneuropathy, and atrophy of muscles in hands and feet facilitates the differential [1,11]. Severe progressive cardiac-valvular problems distinguish the cardiac-valvular EDS type (*COL1A2*) from cEDS, severe skin fragility and unusual craniofacial features discriminates the dermatosparaxis EDS (*ADAMTS2*), whereas (congenital) kyphoscoliosis and muscle hypotonia differentiates the kyphoscoliotic EDS (*PLOD1*, *FKBP14*), which are other rare recessive EDS types. Bilateral congenital hip dislocation differentiates the autosomal dominant arthrochalasia EDS (*COL1A1*, *COL1A2*) [1,11,12].

Recently, in six individuals from four unrelated families who presented with a constellation of clinical findings reminiscent of cEDS such as gJHM, redundant and hyperextensible skin with poor wound healing and abnormal scarring [15–17], and recessive alterations in the *AEBP1* gene, which encodes the aortic carboxypeptidase-like protein (ACLP) associating with collagens in the extracellular matrix, were recognized, thus defining a new EDS form labelled as classical-like EDS type 2 (MIM #618000).

Herein, we describe an additional patient with a homozygous missense *AEBP1* causative variant and compare her clinical features with those of the other patients reported so far, offering future perspectives for EDS nosology and research in this field.

2. Patient and Methods

2.1. Molecular Analyses

The patient was evaluated at the specialized outpatient clinic for the diagnosis of EDS and related connective tissue disorders, i.e., the Ehlers-Danlos Syndrome and Inherited Connective Tissue Disorders Clinic (CESED), at the University Hospital Spedali Civili of Brescia. Molecular analysis was achieved in compliance with the Italian legislation on genetic diagnostic tests and the patient provided written informed consent for publication of clinical data and photographs according to the Italian bioethics laws. Since this report is based on data obtained through routine clinical care and is not considered research at the involved institutions; formal ethics review was not obtained. Genomic DNA was extracted from peripheral blood leukocytes using standard procedures; the exons and intron-flanking regions of COL5A1, COL5A2, and exon 14 of COL1A1 (c.934C>T (p.Arg312Cys)) were amplified by PCR and directly sequenced using an ABI PRISM® 3130XL Genetic Analyzer (Life Technologies, Carlsbad, CA, USA), as previously reported [4]. For the multiplex ligation-dependent probe amplification (MLPA), the commercially available SALSA MLPA kits P331 and P332 for COL5A1 gene were used, according to the manufacturer's recommendations (MRC-Holland, Amsterdam, The Netherlands), as previously described [4]. The primers for AEBP1 Sanger sequencing (Supplementary Table 1) were designed for all coding exons, including the intron-exon boundaries, and primer sequences were analyzed for the absence of known variants using the GnomAD database [18]. The sequences were analyzed with the Sequencher 5.0 software and variants were annotated according to the Human Genome Variation Society (HGVS) nomenclature by using the Alamut Visual software version 2.11. To evaluate the putative pathogenicity of the AEBP1 missense variant, which was submitted to the LOVD Ehlers–Danlos Syndrome Variant Database [19], we used the following mutation prediction programs: Mutation Assessor [20], PhD-SNP [21], Align GVD [22], SIFT [23], Mutation Taster [24], PolyPhen2 [25], PROVEAN [26], MutPred [27], M-CAP [28], CADD [29], DANN [30], Fathmm-MKL [31], and VEST [32]. The nucleotide and protein accession numbers correspond to the AEBP1 (NM_001129.4, NP_001120.3) reference sequences.

2.2. High-Frequency Ultrasonography and In Vivo Reflectance Confocal Microscopy

To investigate patient's skin by a non-invasive approach, we performed high-frequency ultrasonography (HF-USG) and *in vivo* reflectance confocal microscopy (RCM) as previously described [33–35].

Briefly, HF-USG was performed on the dorsal and volar side of the forearm of the patient and 10 age- and gender-matched healthy individuals with the same skin phototype and similar sun exposure history by digital 50-MHz ultrasonography B mode scanning (DUB-USB Skin Scanner, Taberna Pro Medicum Company, Lueneburg, Germany). For ultrasound transmission, water was employed as a coupling medium between the transducer and the skin surface. The usable depth of signal penetration was 4 mm, and the gain was 40 dB. Ultrasonography images were collected under standard conditions (environmental temperature was 20–23 °C and the patient remained in a lying position for at least 10 min before examination). Acquired images were exported into a dedicated database and were evaluated using specific image-analysis software to assess epidermal and dermal thickness (µM) and lesional echogenicity.

RCM investigation on the same sides of the forearm was achieved with a Vivascope 1500® microscope (MAVIG GmbH, Lucid Technologies, Henrietta, NY, USA) to visualize *in vivo* the horizontal optical skin sectioning at cellular-level resolution (lateral resolution = 0.5–1 µM, axial resolution = 3–5 µM) from the epidermis to the papillary dermis (200–250 µM in depth). The system uses a laser source with a wavelength of 830 nm and a power <35 mW at the tissue level. The microscope objective is attached to the skin through an adhesive ring to diminish motion artefacts during investigation. Water was used between the adhesive window and the skin, and ultrasound gel (Aquasonic 100 Gel; Parker Laboratories Inc., Fairfield, NJ, USA) was used between the adhesive window and the lens as detailed in [35]. VivaScan 7.0, Viva Stack™ and Viva Block™ software (Lucid Technologies, Henrietta, NY, USA) was employed to acquire blocks of 4 × 4 mm horizontal optical sections, obtained from 64 individual horizontal optical sections (500 × 500 µm images). The system saves images in bitmap format with digital resolution of 1000 × 1000 pixels and 256 levels of grey.

3. Results

3.1. Clinical Findings

The proband (LOVD ID AN_006205) was an Italian 53-year-old woman, born from healthy second-cousins parents, and had two healthy brothers. Clinical history was remarkable for premature birth at 30 weeks (height 44 cm, weight 1.2 kg) associated with perinatal respiratory distress. Neonatal severe hypotonia and delayed motor development, i.e., delays in walking (she took her first steps at four years of age) and acquisition of fine motor skills, were also reported. Medical history further included propensity to develop ecchymoses either spontaneously or upon minimal trauma often occurring for motor clumsiness, surgically treated umbilical hernia in infancy, myopia and astigmatism since childhood, and complete dental loss due to unspecified periodontitis at 14 years old. At age 18, a clinical diagnosis of unspecified EDS was given for gJHM, skin hyperextensibility, delayed wound healing, and easy bruising; genetic analyses were not performed. The patient suffered from recurrent dislocations of knees and occasionally of shoulders and elbows since the age of 10; the objective patellar instability was surgically treated by capsuloplasty and transposition of the insertion of the common patellar tendon by tibial tuberosity transplantation followed by skin plastic surgery at the age of 29 leading to a wide atrophic post-surgical scar (Figure 1A). At 21 and 23 years old, respectively, she underwent bilateral saphenectomy for symptomatic varicosities with pain, fatigability, heaviness, and recurrent superficial thrombophlebitis and surgical removal of nodules on vocal cords. At age 41, the patient was subjected to operative treatment of rotator cuff disease in the setting of weakness and substantial functional disability. Since age 42, she suffered from Achilles tendinopathy with pain and stiffness, especially at the back of the ankle, treated with on-demand NSAIDs use, conservative physical therapy, and orthotic insoles for severe pes planus. At 43 years of age, metatarsal osteotomy on the 3rd toe of the right foot for metatarsalgia and aggravating Achilles tendinopathy was performed. In the same period, she developed disabling bilateral gonarthrosis, treated with arthroscopic abrasion, epitrochleitis, and subacromial shoulder impingement associated with night pain. Hypotrophy of the scapular girdle and weak osteotendinous reflexes were observed at age 50, when she also experienced the dislocation of the left ankle with soft tissue effusion without reabsorption.

On examination, at 52 years of age, she presented with a height of 150 cm (genetic target 157 cm, arm span/height ratio 1.03, normal value <1.05), a weight of 52 kg, hyperextensible, soft, doughy, fragile and redundant skin, with an old-aging appearance of face and extremities, and multiple atrophic papyraceous scars, especially on knees, defective wound healing, easy bruising, spheroids on the elbows, and BS 5/9 (Figure 1A). She also showed multiple papules with some follicular prominence that looked like a diffuse poikiloderma of Civatte (PoC-like dermatitis) more pronounced in photo-exposed sites, androgenetic alopecia, high palate, elongated uvula, scoliotic attitude, mobile patellae and flat feet (even though surgical intervention and orthotics, respectively), hallux valgus, bilateral piezogenic papules, peripheral artery disease (i.e., intermittent claudication, peripheral cyanosis, and cold skin), and varicose veins (Figure 1A). The patient reported persistent lumbar back pain and sporadic pain of hips, knees, left ankle, elbows, shoulders, and feet. Multidimensional fatigue inventory (MFI) questionnaire was suggestive for chronic fatigue (total score 69, higher score in the questions investigating physical fatigue). Cognitive development and mentation were normal. Heart ultrasound detected normal cardiac/valve morphology and function. Dual-energy X-ray absorptiometry (DXA) disclosed femoral osteopenia (T-score left femoral neck −1.5 SD, T-score total hip −1.6 SD); lumbar BMD was normal (T score −0.9 DS) in the presence of marked degenerative arthritis. Nevertheless, we found a mild dorsal vertebral deformity (T10) in the presence of a low TBS value (1.23). The patient also presented mild scoliosis and lumbar spine rectilinization. Due to hypovitaminosis D, Cholecalciferol 50,000 UI monthly was commenced. Other bone metabolism blood and urinary samples and markers of bone remodeling were normal.

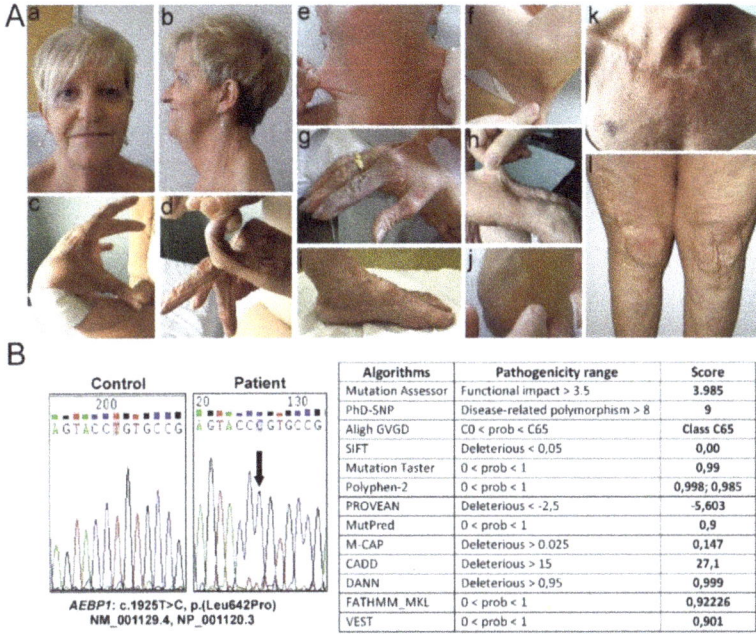

Figure 1. Clinical and molecular findings of the patient. (**A**) Old-aging appearance of face and androgenetic alopecia (a,b); laxity of the thumb (c), laxity of the fifth finger (d); hyperextensible skin in different body areas: neck (e), elbow (f), dorsum of the hand (g) and forearm (h); flat feet and piezogenic papules (i); subcutaneous spheroid on elbow (j); diffuse PoC-like dermatitis and easy bruising (k); skin redundancy, atrophic papyraceous scars on knees, postsurgical enlarged scar after right knee capsuloplasty and skin plastic surgery (l). (**B**) Sequence chromatograms showing the position of the c.1925T>C p.(Leu642Pro) variant (arrow) identified in homozygosity in exon 16 of the *AEBP1* gene (seq. Ref.: NM_001129.4, NP_001120.3) and *in silico* prediction of the pathogenicity of the p.(Leu642Pro) missense substitution by using 13 different algorithms [20–32].

3.2. Molecular Findings

The patient's phenotype was suggestive for cEDS, since she fulfilled both major (skin hyperextensibility plus atrophic scarring and gJHM) and 6 minor criteria according to the 2017 EDS nosology, i.e., easy bruising, soft, doughy skin, skin fragility, subcutaneous spheroids, a history of hernia, and JHM complications. Therefore, after written informed consent was obtained, we performed Sanger sequencing of *COL5A1*, *COL5A2*, and of exon 14 of *COL1A1* (p.Arg312Cys), integrated by MLPA analysis of *COL5A1*, which did not identify any pathogenic variant. Although negative molecular testing, a clinical diagnosis of cEDS was maintained, since the other EDS types in differential diagnosis with cEDS (including periodontal EDS) were excluded clinically. Following the discovery of *AEBP1* biallelic variants [15,16], Sanger sequencing of this gene was achieved, which revealed the homozygosity for the rare c.1925T>C p.(Leu642Pro) variant in exon 16 (Figure 1B), leading to the substitution of a highly conserved leucine residue with a proline at position 642 within the metallocarboxypeptidase-like domain of the protein. This variant has been observed in 3 individuals in GnomAD (rs753531562, 3/282140, no homozygotes, total MAF: C = 0.00001063). Its putative pathogenicity was estimated through an array of 13 different *in silico* prediction algorithms that agreed to define p.(Leu642Pro) as high impacting variant. Given that the variant is located in a critical and well-established functional domain without benign variation, the extremely low frequency in publicly available population databases, the multiple lines of computational evidence supporting a deleterious effect on the gene product, and the patient's phenotype highly suggestive for a disease with a single genetic etiology, the p.(Leu642Pro) missense variant is classified as likely pathogenic (class 4) according to the guidelines of the ACMG. Samples of the healthy parents or brother were not available for molecular analyses.

3.3. Instrumental Findings on Patient's Skin

In order to investigate the skin by a non-invasive approach, HF-USG and *in vivo* RCM were performed on selected skin areas, i.e., dorsal and volar side of the forearm, showing clinically significant differences between our patient and 10 healthy individuals (Table 1 and Figure 2). Digital 50-MHz ultrasonography scanning demonstrated an increase in epidermal entrance echo (highly echogenic band produced by the differences of the acoustic impedance between gel and skin) corresponding to increased epidermal thickness, but a decrease in dermal thickness compared to control skin of age- and gender-matched healthy individuals with the same skin phototype II and similar sun exposure history. The patient's epidermis (dorsal thickness = 172 µM; volar thickness = 141 µM) was thicker than that of healthy controls (dorsal thickness (mean ± standard deviation, SD) = 121 ± 22 µM; volar thickness (mean ± SD) = 102 ± 12 µM), likely due to the multiple and diffuse papules (Table 1). The increased thickness was more evident on the dorsal side of the forearm that is chronically more photoexposed compared to the volar side. Contrariwise, the patient's dermis appeared thinner (dorsal dermal thickness = 570 µM; volar epidermal thickness = 289 µM) compared to healthy controls (dorsal dermal thickness (mean ± SD) = 1108 ± 320 µM; volar epidermal thickness (mean ± SD) = 983 ± 205 µM) (Table 1). Moreover, the considerable hypoechogenicity of the dermal layer suggests disruption of collagen fibers and accumulation of elastotic material that is typical of chronological and photoinduced skin aging (Figure 2A). This ultrastructural pattern is known as subepidermal low echogenic band (SLEB) and derives from skin elastosis and accumulation of glycosaminoglycans that have increased water-binding capacity [33]. *In vivo* RCM investigation demonstrated loss of the typical honey-comb pattern (corresponding to alteration of epidermal thickness), irregularity of the dermal-epidermal junction and the disarray of the dermis, which was characterized by coarse and fragmented collagen fibers both on the dorsal and volar side of patient's forearm (Figure 2B). These alterations are independent of sun exposure, given that they are present both on dorsal and volar side of the forearm, suggesting a pronounced and diffuse skin aging due to *AEBP1*-defect.

Table 1. Epidermal and dermal thickness of patient's forearm evaluated by high-frequency ultrasonography (HF-USG) compared to 10 healthy individuals.

	Dorsal Forearm		Volar Forearm	
	Patient	Controls (mean ± SD)	Patient	Controls (mean ± SD)
Epidermal thickness (µM)	172	121 ± 22	141	102 ± 12
Dermal thickness (µM)	570	1108 ± 320	289	983 ± 205

4. Discussion

Recently, taking advantage from NGS, a new, autosomal recessive type of EDS has been discovered due to variants in the *AEBP1* gene. This EDS type is very rare and, so far, found in only seven individuals (including the present patient) from four unrelated families (Table 2). The International Consortium on EDS and Related Disorders has not yet classified and named this type, but in OMIM it is labeled as classical-like EDS type 2 (MIM #618000). Indeed, the few patients reported hitherto (Table 2) share many similarities representative of the classical type as much as they all fulfill the cEDS diagnostic criteria of the 2017 nosology [1,2] for the presence of the pathognomonic cutaneous involvement, i.e., soft, doughy and very hyperextensible skin, delayed wound healing with abnormal atrophic scarring, JHM, and other minor criteria such as easy bruising, subcutaneous spheroids (observed only in our patient), and JHM complications such as dislocations/subluxations (shoulders, knees, hips, ankles, elbows, clavicula, wrist, mandibular and distal radioulnar joints, in some cases requiring surgical treatment), sprain, pain, and flexible flatfoot (Table 2)

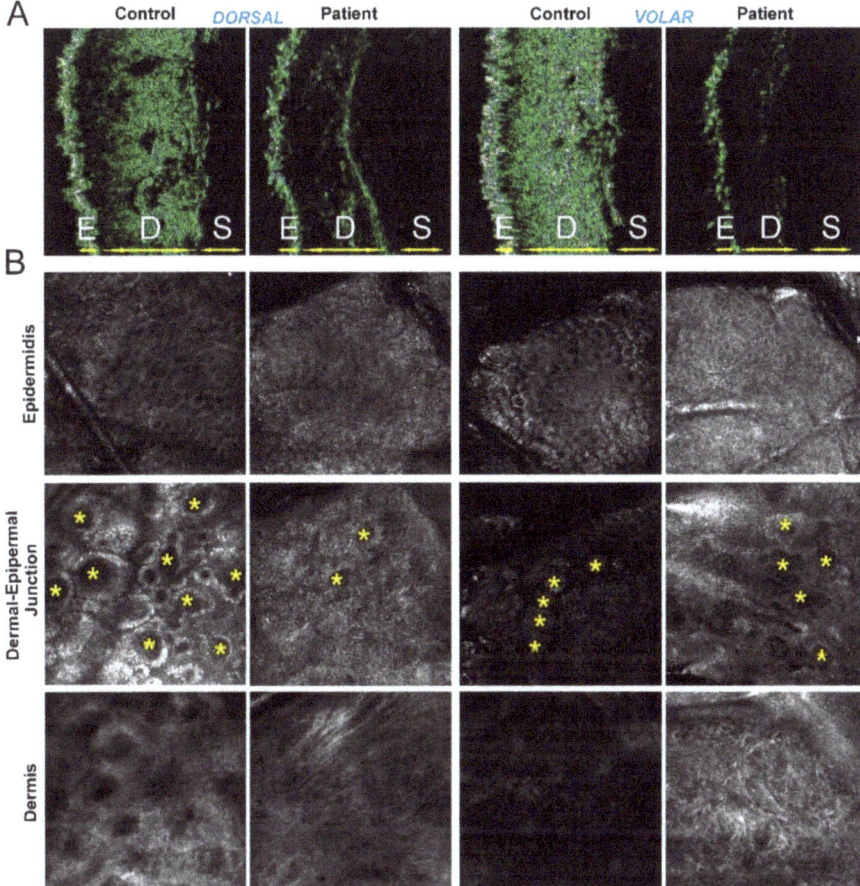

Figure 2. Instrumental findings on patient's skin. (**A**) Ultrasonography (50 MHz) images of the forearm skin from the patient and a representative age- and gender-matched healthy individual (control). E, epidermis, D, dermis, S, subcutaneous adipose tissue (depth of imaging: 4 mm). Disorganization of collagen fibers and elastosis in patient's skin appears as a significant thinning and hypoechogenicity of the dermal layer both on dorsal (left) and volar side (right) compared to control (**B**) Reflectance confocal microscopy images of the forearm skin from patient and control (magnification: 500 × 500 µm). Epidermis: typical honey-comb pattern on dorsal (left) and volar (right) side in healthy skin are not detectable in patient's skin. Dermal-epidermal junction: regular edge papillae [rings of basal keratinocytes surrounding dark circular structures corresponding to dermal papillae (*)] on dorsal and volar side of control skin are reduced both in number and definition in patient's skin. Dermis: Irregular and fragmented collagen fibers that appear bright and coarse on both dorsal and volar side of patient's skin compared to control. An increased brightness of all skin structures, corresponding to chronological and photoinduced skin aging, is present on the dorsal side of healthy skin but not on the volar side that usually is not photoexposed, whereas in the patient this pronounced skin aging is present at both sides.

Table 2. Summary of clinical features of individuals with autosomal recessive variants in AEBP1

Citation	Present patient	P1*	P2*	P3*	P4*	P5*	P6*
Sex	female	male	male	female	male	female	male
Ethnicity	white	white	white	Middle Eastern	Middle Eastern	white	white
Age at evaluation	53y	35y	33y	12y	24y	39y	38y
AEBP1 variant(s) (NM_001129.4)	c.1925T>C homozygous	c.1470del, c.1743C>A compound heterozygous	c.1320_1326del homozygous	c.1630+1G>A (r.1609_1630del) homozygous	c.1630+1G>A (r.1609_1630del) homozygous	c.917dup homozygous	c.917dup homozygous
Protein change (NP_001120.3)	p.(Leu642Pro)	p.(Asn490_Met495delins40), p.(Cys581*)	p.(Arg440Serfs*3)	p.(Val537Leufs*31)	p.(Val537Leufs*31)	p.(Tyr306*)	p.(Tyr306*)
Joint hypermobility (BS)	+ (5/9)	+ (8/9)	+ (8/9)	+ (8/9)	+ (NA)	+ (6/9)	+ (2/9)
Dislocations/Subluxations	left ankle, knees, shoulders, elbows	hip, right distal radioulnar joint	hip (congenital), shoulders	hip, knees, ankles shoulders, interphalangeal joints	hips, knees and ankles	wrist, mandibular and distal radioulnar joints	ankles, knees, clavicula
Foot deformities	pes planus, hallux valgus	pes planus, hallux valgus, hammer toes	pes planus, hallux valgus, hammer toes	pes planus, hallux valgus, hammer toes	pes planus, hallux valgus, toe deformities	pes planus, hallux valgus, sandal gap	hindfoot deformity, sandal gap
Extensive skin hyperextensibility	+	+	+	+	+	+	+
Delayed wound healing (abnormal scarring)	+ (widened atrophic scars)	+ (widened atrophic scars)	+ (widened atrophic scars, keloids)	+ (widened atrophic scars, keloids)	+ (widened atrophic scars)	+ (widened atrophic scars)	+ (widened atrophic scars)
Redundant skin	+ old-aging appearance	+ old-aging appearance	+	+	+	+ old-aging appearance	+ old-aging appearance
Easy bruising	+	+	+	+	NA	+	+
Prominent chest superficial veins	-	NA	+	NA	NA	NA	NA
Hernia	umbilical surgically treated	-	large ventral surgical hernia	umbilical, ventral, inguinal	NA	-	-
Genitourinary abnormalities	-	cryptorchidism surgically corrected	-	-	-	-	cryptorchidism surgically corrected
Gastrointestinal abnormalities	-	motility issues	bowel rupture	-	-	-	-
Vascular abnormalities	peripheral artery disease, varicose veins	MVP	MVP, mildly dilated aortic root, bilateral carotids stenosis, aortic dilation requiring surgery	-	-	MVP, circular pericardial effusion	varicose veins
Dentition	Pyorrhea, complete dental loss at age 14	retention of a single baby tooth	-	abnormal dental alignment	abnormal dental alignment	-	-
Citation	Present patient	P1*	P2*	P3*	P4*	P5*	P6*
Sex	female	male	male	female	male	female	male
Ethnicity	white	white	white	Middle Eastern	Middle Eastern	white	white
Age at evaluation	53y	35y	33y	12y	24y	39y	38y

Table 2. Cont.

Citation	Present patient	P1*	P2*	P3*	P4*	P5*	P6*
Facial dysmorphisms	high palate, elongated uvula	-	micrognathia	bilateral ptosis webbed neck, sagged cheeks large ears, narrow palate	bilateral ptosis webbed neck, sagged cheeks large ears, narrow palate	-	-
Skeletal anomalies (MRI findings)	femoral osteopenia, T10 vertebral deformity, scoliosis, lumbar spine rectilinization with marked degenerative arthritis	severe osteopenia of hips (mild disc bulging at the C4-5 and C7-T1 levels)	hip replacement for severe osteopenia, upper thoracic scoliosis with degenerative disease and facet arthrosis of spine (empty sella)	skull with 'copper beaten' appearance, severe osteopenia, narrowing of the interpedicular distance of the lumbar spine distally, short and squared iliac bones, remodeled long bones of the lower extremities	severe osteopenia	progressive kyphosis, scoliosis, arachnodactyly, positive wrist and thumb signs, degeneration of the discus ulnaris orthopedically treated	kyphoscoliosis, arachnodactyly, positive wrist and thumb signs, mild pectus excavatum
Other	hypotonia, delayed motor development, multiple papules (diffuse PoC-like dermatitis, alopecia, patellar instability surgically treated, rotator cuff disease surgically treated, epitrochleitis, subacromial shoulder impingement, hypotrophy of the scapular girdle, gonarthrosis, chronic fatigue, spheroids, piezogenic papules, myopia	delays in walking and acquisition of fine motor skills, impaired temperature sensation, keratoconjunctivitis sicca, piezogenic papules	elbow bursitis, piezogenic papules, sacral dimple, hypertriglyceridemia	hypotonia, diabetes mellitus, cellulitis	NA	alopecia, skin striae	strabismus surgically treated, myopia, astigmatism

*Patients reported by Alazami et al., 2016 [15], Blackburn et al., 2018 [16], and Hebebrand et al., 2018 [17]. P1: A-II:1;P2: B-II:1; P3:C-IV:6; P4: C-IV:4 according to [16]; P5: D-II:1; P6: D-II:2 according to [17]. Abbreviations: + present, - absent, NA not available, MVP mitral valve prolapse

Consistent with the multisystemic presentation of EDS in general, there are also variable features including congenital hip dislocation, hypotonia, delayed motor development, acrogeria, prominent superficial veins in the chest region, hernias, dental anomalies, gastrointestinal (bowel rupture) and vascular complications (mitral valve prolapse, aortic root dilation needing surgery), early-onset varicose veins, and several skeletal anomalies (Table 2). In particular, bone involvement seems a common feature of classical-like EDS type 2 with osteopenia/osteoporosis affecting hips, femurs, and spine that are present, at variable degree, in all of the patients reported so far, with the exception of the two siblings, reported by Hebebrand and coworkers [17], who were not tested for osteopenia. In addition, degenerative arthritis, (kypho)scoliosis, arachnodactyly, positive wrist and thumb signs, mild pectus excavatum, T10 vertebral deformity (our patient), narrowing of the interpedicular distance of the lumbar spine, shortened and squared iliac bones, and remodeling of long bones of the lower extremities are also encountered (Table 2). In addition, all subjects have severe foot deformities including bilateral pes planus, hammertoes, hallux valgus, hindfoot deformity, and sandal gap, which are observed in more than a few other EDS subtypes as well [1]. Although in cEDS patients a variable degree of low bone mineral density and a high prevalence of radiological vertebral fractures were reported [7,36], *AEBP1*-related EDS seems to display a more severe bone involvement that could potentially facilitate the differential with cEDS. Nevertheless, considering the limited number of individuals with *AEBP1* defect known so far, a larger cohort of patients is needed to confirm this preliminary observation.

The adipocyte enhancer binding protein 1 gene (*AEBP1*) encodes a 1158-amino acid secreted aortic carboxypeptidase-like protein (ACLP) composed of an N-terminal signal sequence, a charged lysine, proline, and glutamic acid-rich domain, a collagen-binding discoidin domain and a metallocarboxypeptidase (MCP)-like domain [37,38]. This latter domain is inactive toward standard MCP substrates, as it lacks several critical active sites and substrate-binding residues that are necessary for activity [37,38]. Indeed, ACLP acts as an extracellular matrix (ECM)-binding protein rather than as active MCPs that shows similar embryonic expression pattern as other ECM proteins and is found at high levels particularly in collagen-rich tissues comprising the dermal layer of the skin, the medial layer of blood vessels, the basement membrane of the lung, and the periosteum. Consistently, ACLP plays fundamental roles in both embryonic development and adult tissue homeostasis, particularly in repair processes [38–43]. Indeed, *AEBP1* knock-out mice show ventral wall defects, develop spontaneous skin ulcerations, and have significantly delayed healing of dermal punch wounds [38]. This cutaneous phenotype is consistent with the defective wound healing and abnormal scar formation observed in individuals with *AEBP1* defects and suggest that ACLP has a crucial role in damage sensing and ECM remodeling following injury by regulating fibroblast proliferation and mesenchymal stem cell differentiation into collagen-producing cells [42,43]. Blackburne and coworkers demonstrated that ACLP also binds collagens type I, III, and V and is able to promote the polymerization of collagen type I in vitro [16]. In line with these findings, the ultrastructural study performed by the same authors on a patient's skin biopsy revealed reduced dermal collagen and irregular disrupted collagen fibers, as well as our HF-USG and RCM investigations that disclosed abnormal collagen fibers deposition together with a reduced dermal thickness. Moreover, we recognized an increase of the epidermal thickness likely correlating with the diffuse PoC-like dermatitis, which is probably not related to classical-like EDS type 2. The use of these non-invasive diagnostic techniques may be promising for the investigation of the qualitative and quantitative cutaneous alterations, but further studies including electron microscopy on skin as golden standard of reference on large cohorts of patients are warranted. In our case, we did not perform skin biopsy because of the patient's will due to psychological reluctance for her important skin fragility with delayed wound healing.

The *AEBP1* variants discovered before our patient's characterization were all loss-of-function (LOF) mutations (Table 2) and included compound heterozygous variants [(c.1470del; p.Asn490_Met495delins40) and (c.1743C>A; p.Cys581*) in the first individual (P1); a homozygous variant (c.1320_1326del; p.Arg440Serfs*3) in the second individual (P2); a homozygous splice variant leading to skipping of the last 22 bp of exon 13 (c.1630+1G>A) in the two siblings from the third family (P3, P4), and a homozygous nonsense variant (c.917dup; p.Tyr306*) in the two siblings from the fourth family (P5, P6). Hebebrand and coworkers performed the analysis of all *AEBP1* LOF variants reported in multiple databases showing that these are distributed throughout the protein and by using conservative criteria for pathogenic LOF variants (nonsense, frameshift, canonical splice sites, or initiation-codon) these authors estimated a carrier frequency of 1/829 for the gnomAD database. The analysis of CADD scores for all possible missense variants showed a higher predicted deleteriousness for positions close to the discoidin and the MCP-like domains, whereas the unstructured N-/C-terminal parts showed lower scores. The high deleteriousness scores observed for missense variants within these domains cite evidence in support of additional mutational mechanisms leading to aberrant function and the authors thus argued that the relatively low estimated carrier frequencies could be significantly higher if missense variants contribute to a comparable fraction of disease variants [17]. The present c.1925T>C; p.(Leu642Pro) homozygous variant disclosed within the MCP-like domain of the protein corroborates this hypothesis, since it represents the first likely pathogenic *AEBP1* missense substitution (ACMG class 4) associated with classical-like EDS type 2. The variant is predicted *in silico* to affect the tertiary structure of the protein by disrupting an α-helix located in a highly conserved domain, thus likely interfering with its function in terms of impaired partner binding capability. Nevertheless, a definite proof of variant's causality is lacking, since the effective functional consequences on the ECM organization, particularly of collagens, and on the other not yet well-defined roles of the ACLP protein were not studied, because the patient refused skin biopsy.

5. Conclusions

Our findings expand the knowledge of the clinical phenotype of this recently defined autosomal-recessive EDS subtype, provide the first evidence that missense variants contribute to the allelic repertoire of *AEBP1*, and suggest that in the diagnostic process of a cEDS patient this gene should be investigated when a recessive inheritance is compatible and no causal variant is identified in the other cEDS genes. Further reports are needed to better characterize the *AEBP1*-related phenotype, define specific clinical criteria that might facilitate the differential with the other EDS forms, delineate genotype-phenotype correlations, and collect natural history data for prognostication. Finally, ACLP function needs to be explored more in-depth to provide insights into molecular mechanisms involved in the pathophysiology of *AEBP1*-related EDS that may represent a starting point for identifying potential therapeutic options.

Supplementary Materials: The following are available online at http://www.mdpi.com/2073-4425/10/2/135/s1, Table S1. Primers.

Author Contributions: M.C. and M.R. conceived the study. M.V. and M.C. performed the clinical diagnosis of the patient, genetic counselling and follow-up; M.V. performed skin evaluations; L.P. and A.M.F. investigated bone health parameters; M.R. and V.C. carried out the molecular analyses; M.R. and N.C. researched the literature; M.R., A.M.F., and M.V. prepared the manuscript; M.C. edited and coordinated the manuscript. All authors discussed, read, and approved the manuscript.

Funding: No funding was active on this project.

Acknowledgments: The authors want to thank the patient for her cooperation during the diagnostic process and the Fazzo Cusan family for its generous support.

Conflicts of Interest: All authors declare that there is no conflict of interest concerning this work.

References

1. Malfait, F.; Francomano, C.; Byers, P.; Belmont, J.; Berglund, B.; Black, J.; Bloom, L.; Bowen, J.M.; Brady, A.F.; Burrows, N.P.; et al. The 2017 international classification of the Ehlers-Danlos syndromes. *Am. J. Med. Genet. C* **2017**, *175*, 8–26. [CrossRef] [PubMed]
2. Bowen, J.M.; Sobey, G.J.; Burrows, N.P.; Colombi, M.; Lavallee, M.E.; Malfait, F.; Francomano, C.A. Ehlers-Danlos syndrome, classical type. *Am. J. Med. Genet. Part C* **2017**, *75*, 27–39. [CrossRef] [PubMed]
3. Beighton, P.; De Paepe, A.; Steinmann, B.; Tsipouras, P.; Wenstrup, R.J. Ehlers-Danlos syndromes: Revised nosology, Villefranche, 1997. Ehlers-Danlos National Foundation (USA) and Ehlers-Danlos Support Group (UK). *Am. J. Med. Genet.* **1998**, *77*, 31–37. [CrossRef]
4. Ritelli, M.; Dordoni, C.; Venturini, M.; Chiarelli, N.; Quinzani, S.; Traversa, M.; Zoppi, N.; Vascellaro, A.; Wischmeijer, A.; Manfredini, E.; et al. Clinical and molecular characterization of 40 patients with classic Ehlers-Danlos syndrome: Identification of 18 COL5A1 and 2 COL5A2 novel mutations. *Orphanet J. Rare Dis.* **2013**, *12*, 8–58. [CrossRef] [PubMed]
5. Colombi, M.; Dordoni, C.; Venturini, M.; Ciaccio, C.; Morlino, S.; Chiarelli, N.; Zanca, A.; Calzavara-Pinton, P.; Zoppi, N.; Castori, M.; et al. Spectrum of mucocutaneous, ocular and facial features and delineation of novel presentations in 62 classical Ehlers-Danlos syndrome patients. *Clin. Genet.* **2017**, *92*, 624–631. [CrossRef] [PubMed]
6. Colombi, M.; Dordoni, C.; Cinquina, V.; Venturini, M.; Ritelli, M. A classical Ehlers-Danlos syndrome family with incomplete presentation diagnosed by molecular testing. *Eur. J. Med. Genet.* **2018**, *61*, 17–20. [CrossRef] [PubMed]
7. Mazziotti, G.; Dordoni, C.; Doga, M.; Galderisi, F.; Venturini, M.; Calzavara-Pinton, P.; Maroldi, R.; Giustina, A.; Colombi, M. High prevalence of radiological vertebral fractures in adult patients with Ehlers-Danlos syndrome. *Bone* **2016**, *84*, 88–92. [CrossRef] [PubMed]
8. Borck, G.; Beighton, P.; Wilhelm, C.; Kohlhase, J.; Kubisch, C. Arterial rupture in classic Ehlers-Danlos syndrome with COL5A1 mutation. *Am. J. Med. Genet. Part A* **2010**, *152*, 2090–2093. [CrossRef]
9. Symoens, S.; Syx, D.; Malfait, F.; Callewaert, B.; De Backer, J.; Vanakker, O.; Coucke, P.; De Paepe, A. Comprehensive molecular analysis demonstrates type V collagen mutations in over 90% of patients with classic EDS and allows to refine diagnostic criteria. *Hum. Mutat.* **2012**, *33*, 1485–1493. [CrossRef] [PubMed]
10. Colombi, M.; Dordoni, C.; Venturini, M.; Zanca, A.; Calzavara-Pinton, P.; Ritelli, M. Delineation of Ehlers-Danlos syndrome phenotype due to the c.934C>T, p.(Arg312Cys) mutation in *COL1A1*: Report on a three-generation family without cardiovascular events, and literature review. *Am. J. Med. Genet. Part A* **2017**, *173*, 524–530. [CrossRef] [PubMed]
11. Brady, A.F.; Demirdas, S.; Fournel-Gigleux, S.; Ghali, N.; Giunta, C.; Kapferer-Seebacher, I.; Kosho, T.; Mendoza-Londono, R.; Pope, M.F.; Rohrbach, M.; et al. The Ehlers-Danlos syndromes, rare types. *Am. J. Med. Genet. C* **2017**, *175*, 70–115. [CrossRef] [PubMed]
12. Colombi, M.; Dordoni, C.; Chiarelli, N.; Ritelli, M. Differential diagnosis and diagnostic flow chart of joint hypermobility syndrome/Ehlers-Danlos syndrome hypermobility type compared to other heritable connective tissue disorders. *Am. J. Med. Genet. Part C* **2015**, *169*, 6–22. [CrossRef] [PubMed]
13. Tinkle, B.; Castori, M.; Berglund, B.; Cohen, H.; Grahame, R.; Kazkaz, H.; Levy, H. Hypermobile Ehlers-Danlos syndrome (a.k.a. Ehlers-Danlos syndrome Type III and Ehlers-Danlos syndrome hypermobility type): Clinical description and natural history. *Am. J. Med. Genet. Part C* **2017**, *175*, 48–69. [CrossRef] [PubMed]
14. Castori, M.; Dordoni, C.; Morlino, S.; Sperduti, I.; Ritelli, M.; Valiante, M.; Chiarelli, N.; Zanca, A.; Celletti, C.; Venturini, M.; et al. Spectrum of mucocutaneous manifestations in 277 patients with joint hypermobility syndrome/Ehlers-Danlos syndrome, hypermobility type. *Am. J. Med. Genet. Part C* **2015**, *169*, 43–53. [CrossRef] [PubMed]
15. Alazami, A.M.; Al-Qattan, S.M.; Faqeih, E.; Alhashem, A.; Alshammari, M.; Alzahrani, F.; Al-Dosari, M.S.; Patel, N.; Alsagheir, A.; Binabbas, B.; et al. Expanding the clinical and genetic heterogeneity disorders of connective tissue. *Hum. Genet.* **2016**, *135*, 525–540. [CrossRef] [PubMed]

16. Blackburn, P.R.; Xu, Z.; Tumelty, K.E.; Zhao, R.W.; Monis, W.J.; Harris, K.G.; Gass, J.M.; Cousin, M.A.; Boczek, N.J.; Mitkov, M.V.; et al. Bi-allelic alterations in AEBP1 lead to defective collagen assembly and connective tissue structure resulting in a variant of Ehlers-Danlos syndrome. *Am. J. Hum. Genet.* **2018**, *102*, 696–705. [CrossRef] [PubMed]
17. Hebebrand, M.; Vasileiou, G.; Krumbiegel, M.; Kraus, C.; Uebe, S.; Ekici, A.B.; Thiel, C.T.; Reis, A.; Popp, B. A biallelic truncating AEBP1 variant causes connective tissue disorder in two siblings. *Am. J. Med. Genet. A* **2018**. [CrossRef]
18. GnomAD Database. Available online: http://gnomad.broadinstitute.org/ (accessed on 7 January 2019).
19. Dalgleish, R. The human collagen mutation database 1998. *Nucleic Acids Res.* **1998**, *26*, 253–255. [CrossRef]
20. Reva, B.; Antipin, Y.; Sander, C. Determinants of protein function revealed by combinatorial entropy optimization. *Genome Biol.* **2007**, *8*, R232. [CrossRef]
21. PhD-SNP Web Server. Available online: http://snps.biofold.org/phd-snp/phd-snp.html (accessed on 21 August 2018).
22. Tavtigian, S.V.; Deffenbaugh, A.M.; Yin, L.; Judkins, T.; Scholl, T.; Samollow, P.B.; de Silva, D.; Zharkikh, A.; Thomas, A. Comprehensive statistical study of 452 BRCA1 missense substitutions with classification of eight recurrent substitutions as neutral. *J. Med. Genet.* **2006**, *43*, 295–305. [CrossRef]
23. Sim, N.L.; Kumar, P.; Hu, J.; Henikoff, S.; Schneider, G.; Ng, P.C. SIFT web server: Predicting effects of amino acid substitutions on proteins. *Nucleic Acids Res.* **2012**, *40*, W452–W457. [CrossRef] [PubMed]
24. Schwarz, J.M.; Cooper, D.N.; Schuelke, M.; Seelow, D. MutationTaster2: Mutation prediction for the deep-sequencing age. *Nat. Methods* **2014**, *11*, 361–362. [CrossRef] [PubMed]
25. Adzhubei, I.A.; Schmidt, S.; Peshkin, L.; Ramensky, V.E.; Gerasimova, A.; Bork, P.; Kondrashov, A.S.; Sunyaev, S.R. A method and server for predicting damaging missense mutations. *Nat. Methods.* **2010**, *7*, 248–249. [CrossRef] [PubMed]
26. Choi, Y.; Chan, A.P. PROVEAN web server: A tool to predict the functional effect of amino acid substitutions and indels. *Bioinformatics* **2015**, *31*, 2745–2747. [CrossRef] [PubMed]
27. Pejaver, V.; Urresti, J.; Lugo-Martinez, J.; Pagel, K.A.; Ning Lin, G.; Nam, H.J.; Mort, M.; Cooper, D.N.; Sebat, J.; Iakoucheva, L.M.; et al. MutPred2: Inferring the molecular and phenotypic impact of amino acid variants. *bioRxiv* **2017**, 134981. [CrossRef]
28. Jagadeesh, K.A.; Wenger, A.M.; Berger, M.J.; Guturu, H.; Stenson, P.D.; Cooper, D.N.; Bernstein, J.A.; Bejerano, G. M-CAP eliminates a majority of variants of uncertain significance in clinical exomes at high sensitivity. *Nat. Genet.* **2016**, *48*, 1581–1586. [CrossRef] [PubMed]
29. Rentzsch, P.; Witten, D.; Cooper, G.M.; Shendure, J.; Kircher, M. CADD: Predicting the deleteriousness of variants throughout the human genome. *Nucleic Acids Res.* **2018**, *47*, D886–D894. [CrossRef] [PubMed]
30. Quang, D.; Chen, Y.; Xie, X. DANN: A deep learning approach for annotating the pathogenicity of genetic variants. *Bioinformatics* **2015**, *31*, 761–763. [CrossRef] [PubMed]
31. Shihab, H.A.; Rogers, M.F.; Gough, J.; Mort, M.; Cooper, D.N.; Day, I.N.; Gaunt, T.R.; Campbell, C. An integrative approach to predicting the functional effects of non-coding and coding sequence variation. *Bioinformatics* **2015**, *31*, 1536–1543. [CrossRef]
32. Carter, H.; Douville, C.; Stenson, P.D.; Cooper, D.N.; Karchin, R. Identifying mendelian disease genes with the variant effect scoring tool. *BMC Genom.* **2013**, *14* (Suppl. 3), S3. [CrossRef]
33. Polańska, A.; Dańczak-Pazdrowska, A.; Jałowska, M.; Żaba, R.; Adamski, Z. Current applications of high-frequency ultrasonography in dermatology. *Postepy Dermatol. Alergol.* **2017**, *34*, 535–542. [CrossRef] [PubMed]
34. Calzavara-Pinton, P.; Longo, C.; Venturini, M.; Sala, R.; Pellacani, G. Reflectance confocal microscopy for in vivo skin imaging. *Photochem. Photobiol.* **2008**, *84*, 1421–1430. [CrossRef] [PubMed]
35. Rajadhyaksha, M.; González, S.; Zavislan, J.M.; Anderson, R.R.; Webb, R.H. In vivo confocal scanning laser microscopy of human skin II: Advances in instrumentation and comparison with histology. *J. Invest. Dermatol.* **1999**, *113*, 293–303. [CrossRef] [PubMed]
36. Theodorou, S.J.; Theodorou, D.J.; Kakitsubata, Y.; Adams, J.E. Low bone mass in Ehlers–Danlos syndrome. *Intern Med.* **2012**, *51*, 3225–3226. [CrossRef] [PubMed]
37. Reznik, S.E.; Fricker, L.D. Carboxypeptidases from A to Z: Implications in embryonic development and Wnt binding. *Cell Mol. Life Sci.* **2001**, *58*, 1790–1804. [CrossRef] [PubMed]

38. Layne, M.D.; Yet, S.F.; Maemura, K.; Hsieh, C.M.; Bernfield, M.; Perrella, M.A.; Lee, M.E. Impaired abdominal wall development and deficient wound healing in mice lacking aortic carboxypeptidase-like protein. *Mol. Cell Biol.* **2001**, *21*, 5256–5261. [CrossRef] [PubMed]
39. Ith, B.; Wei, J.; Yet, S.F.; Perrella, M.A.; Layne, M.D. Aortic carboxypeptidase-like protein is expressed in collagen-rich tissues during mouse embryonic development. *Gene Expr. Patterns* **2005**, *5*, 533–537. [CrossRef] [PubMed]
40. Layne, M.D.; Endege, W.O.; Jain, M.K.; Yet, S.F.; Hsieh, C.M.; Chin, M.T.; Perrella, M.A.; Blanar, M.A.; Haber, E.; Lee, M.E. Aortic carboxypeptidase-like protein, a novel protein with discoidin and carboxypeptidase-like domains, is up-regulated during vascular smooth muscle cell differentiation. *J. Biol. Chem.* **1998**, *273*, 15654–15660. [CrossRef] [PubMed]
41. Layne, M.D.; Yet, S.F.; Maemura, K.; Hsieh, C.M.; Liu, X.; Ith, B.; Lee, M.E.; Perrella, M.A. Characterization of the mouse aortic carboxypeptidase-like protein promoter reveals activity in differentiated and dedifferentiated vascular smooth muscle cells. *Circ. Res.* **2002**, *90*, 728–736. [CrossRef]
42. Schissel, S.L.; Dunsmore, S.E.; Liu, X.; Shine, R.W.; Perrella, M.A.; Layne, M.D. Aortic carboxypeptidase-like protein is expressed in fibrotic human lung and its absence protects against bleomycin-induced lung fibrosis. *Am. J. Pathol.* **2009**, *174*, 818–828. [CrossRef]
43. Tumelty, K.E.; Smith, B.D.; Nugent, M.A.; Layne, M.D. Aortic carboxypeptidase-like protein (ACLP) enhances lung myofibroblast differentiation through transforming growth factor β receptor-dependent and -independent pathways. *J. Biol. Chem.* **2014**, *289*, 2526–2536. [CrossRef] [PubMed]

© 2019 by the authors. Licensee MDPI, Basel, Switzerland. This article is an open access article distributed under the terms and conditions of the Creative Commons Attribution (CC BY) license (http://creativecommons.org/licenses/by/4.0/).

Article

Clinical and Molecular Characterization of Classical-Like Ehlers-Danlos Syndrome Due to a Novel *TNXB* Variant

Daisy Rymen [1,*,†], Marco Ritelli [2,†], Nicoletta Zoppi [2], Valeria Cinquina [2], Cecilia Giunta [1], Marianne Rohrbach [1] and Marina Colombi [2]

1. Connective Tissue Unit, Division of Metabolism and Children's Research Centre, University Children's Hospital, 8032 Zürich, Switzerland; Cecilia.Giunta@kispi.uzh.ch (C.G.); Marianne.Rohrbach@kispi.uzh.ch (M.R.)
2. Division of Biology and Genetics, Department of Molecular and Translational Medicine, University of Brescia, 25123 Brescia, Italy; marco.ritelli@unibs.it (M.R.); nicoletta.zoppi@unibs.it (N.Z.); valeria.cinquina1@unibs.it (V.C.); marina.colombi@unibs.it (M.C.)
* Correspondence: daisy.rymen@uzleuven.be
† These authors are contributed equally.

Received: 25 September 2019; Accepted: 23 October 2019; Published: 25 October 2019

Abstract: The Ehlers-Danlos syndromes (EDS) constitute a clinically and genetically heterogeneous group of connective tissue disorders. Tenascin X (TNX) deficiency is a rare type of EDS, defined as classical-like EDS (clEDS), since it phenotypically resembles the classical form of EDS, though lacking atrophic scarring. Although most patients display a well-defined phenotype, the diagnosis of TNX-deficiency is often delayed or overlooked. Here, we described an additional patient with clEDS due to a homozygous *null*-mutation in the *TNXB* gene. A review of the literature was performed, summarizing the most important and distinctive clinical signs of this disorder. Characterization of the cellular phenotype demonstrated a distinct organization of the extracellular matrix (ECM), whereby clEDS distinguishes itself from most other EDS subtypes by normal deposition of fibronectin in the ECM and a normal organization of the α5β1 integrin.

Keywords: Tenascin X; TNXB; Ehlers-Danlos syndrome; EDS; connective tissue; collagen

---

1. Introduction

The Ehlers-Danlos syndromes (EDS) constitute a clinically and genetically heterogeneous group of connective tissue disorders. Patients present with joint hypermobility, skin hyperextensibility and tissue fragility, giving rise to easy bruising and atrophic scarring. Although most EDS are caused by mutations in genes coding for the fibrillar collagens or collagen-modifying enzymes, over the last two decades, several disorders due to defects in other components of the extracellular matrix (ECM) have been delineated [1].

Tenascins comprise a family of glycoproteins, which modulate the adhesion of cells to their ECM. Tenascin X (TNX) is ubiquitously expressed, but the highest levels are found in muscle and loose connective tissue [2]. TNX is thought to regulate fibril spacing by direct binding to the distinct collagen fibrils in the ECM or by indirect binding via decorin [3]. Additional roles for TNX in elastic fiber remodeling and regulating the expression of certain ECM components, e.g., collagen VI, proteoglycans and matrix metalloproteases have been suggested [4–7].

The clinical relevance of TNX-deficiency was first proposed in 1997 by the identification of a patient presenting both congenital adrenal hyperplasia and an EDS-like phenotype. Genetic analysis revealed a contiguous deletion of the partially overlapping genes *CYP21B* and *TNXB* on chromosome

6. Although a TNX *null*-phenotype was verified both on the protein and on the mRNA level, a mutation on the second allele could not be identified [8]. In 2001, Schalkwijk et al. demonstrated that isolated TNX-deficiency resulted in an autosomal recessive form of EDS, resembling the classical type, however lacking atrophic scarring [9]. In 2017, TNX-deficiency was officially classified as "classical-like EDS" (clEDS), with generalized joint hypermobility, hyperextensible, soft and/or velvety skin without atrophic scarring and easy bruising being the typical clinical hallmarks of the disorder [1].

In 2002, Mao et al. developed a TNX-deficient mouse model, mimicking the EDS phenotype. Indeed, as observed in humans, $Tnxb^{-/-}$ mice were morphologically normal at birth, but displayed progressive hyperextensibility of the skin over time. The group showed that the phenotype did not relate to aberrant collagen fibrillogenesis, but was rather due to altered deposition, and therefore, reduced density of collagen in the ECM [10].

Although most TNX-deficient patients display a well-defined clinical phenotype, the diagnosis is often delayed or overlooked. The former is attributed to the molecular analysis of the *TNXB* gene being complicated by the presence of a highly homologous pseudogene and to the fact that the measurement of TNX in serum is not widely available [11]. The latter is mainly caused by poor clinical awareness, which unfortunately applies to many rare disorders.

Here, we reported on an additional patient with clEDS and a novel homozygous disease-causing variant in *TNXB* to further elaborate the clinical phenotype. Furthermore, we reviewed the clinical features of the clEDS patients described to date, in order to create a well-defined description of the phenotype and increase clinical awareness.

2. Materials and Methods

2.1. Ethical Compliance

This study is in accordance with the Helsinki declaration and its following modifications. Ethics approval has been granted (KEK Ref.-Nr. 2014-0300 and Nr. 2019-00811) in the presence of a signed informed consent of the patient for genetic testing, skin biopsy and the publication of clinical pictures. The patient was evaluated at the University Children's Hospital of Zürich. Targeted next-generation sequencing (NGS) panel for 101 connective tissue disorders (Supplementary Table S1) was performed at the Institute of Medical Genetics of the University of Zürich. *TNXB* mutational screening by Sanger sequencing and Multiplex Ligation-dependent Probe Amplification (MLPA) was achieved at the Division of Biology and Genetics, Department of Molecular and Translational Medicine, of the University of Brescia.

2.2. Cell Culture

As part of the diagnostic workup of EDS, a punch biopsy of the patient's skin for establishing fibroblast cultures for collagen biochemical analysis was previously obtained. The biological material was stored in the Biobank of the Division of Metabolism at the Children's Hospital Zürich. Fibroblasts from the patient and from sex- and age-matched healthy individuals were routinely maintained at 37 °C in a 5% CO_2 atmosphere in Earle's Modified Eagle Medium (MEM) supplemented with 2 mM L-glutamine, 10% FBS, 100 µg/ml penicillin and streptomycin (Life Technologies, Carlsbad, CA, USA). Fibroblasts were expanded until full confluency and then harvested by 0.25% trypsin/0.02% EDTA treatment at the same passage number.

2.3. Molecular Analysis

Mutational screening was performed on genomic DNA purified from peripheral blood leukocytes using standard procedures. In particular, all exons and their intron-flanking regions of the *TNXB* gene (NM_0019105.7, NP_061978.6) were PCR amplified with the GoTaq Ready Mix 2X (Promega, Madison, WI, USA) by using optimized genomic primers that were analyzed for the absence of known variants using the GnomAD database (https://gnomad.broadinstitute.org/). For the pseudogene-homolog region (exons 32–44), Sanger sequencing was performed by nested PCR, using a *TNXB*-specific long-range PCR product encompassing the 3′-end of *TNXB* as a template (for details on primer sequences and PCR conditions see Supplementary Table S2). PCR products were purified with ExoSAP-IT (USB Corporation, Cleveland, OH, USA), followed by bidirectional sequencing with the BigDye Terminator v1.1 Cycle Sequencing kit on an ABI3130XL Genetic Analyzer (Thermo Fisher Scientific, South San Francisco, CA, USA). The sequences were analyzed with the Sequencher 5.0 software (Gene Codes Corporation, Ann Arbor, MI, USA) and variants were annotated according to the Human Genome Variation Society (HGVS) nomenclature with the Alamut Visual software version 2.11 (Interactive Biosoftware, Rouen, France). Deletion/duplication analysis of *TNXB* was performed using the MLPA assay P155, according to manufactures' instructions (MRC-Holland, Amsterdam, the Netherlands).

2.4. RNA Extraction and Quantitative Real-Time PCR

Total RNA was purified from skin fibroblasts of the patient and 3 healthy individuals using the Qiagen RNeasy kit, according to the manufacturer's instructions (Qiagen, Hilden, Germany). RNA quality control was performed on an Agilent 2100 Bioanalyzer (Agilent Technologies, Santa Clara, CA, USA). Relative expression levels of the *TNXB* transcript were analyzed by quantitative real-time PCR (qPCR). Of the total RNA, 3 µg was reverse-transcribed with random primers by a standard procedure. qPCR reactions were performed in triplicate with the SYBR Green qPCR Master Mix (Thermo Fisher Scientific, South San Francisco, CA, USA), 10 ng of cDNA, and with 10 µM of each primer set by using the ABI PRISM 7500 Real-Time PCR System with standard thermal cycling conditions. The *HPRT*, *GAPDH* and *ATP5B* reference genes were also amplified for normalization of cDNA loading. Relative mRNA expression levels were normalized to the geometric mean of these reference genes and analyzed using the $2^{-(\Delta\Delta Ct)}$ equation. Amplification plots, dissociation curves and threshold cycle values were generated by ABI Sequence detection system software version 1.3.1. Statistical analyses were performed with the GraphPad Prism software (San Diego, CA, USA). The results were expressed as the mean values of relative quantification ± Standard Error of the Mean (SEM). Statistical significance between groups was determined using one-way ANOVA. P-values were corrected for multiple testing using the Tukey's method.

2.5. Collagen Biochemical, Ultrastructural and Immunofluorescence Microscopy Studies

Collagen steady state analysis in patient's cultured fibroblasts was conducted prior to genetic analysis to assess possible anomalies in collagen biosynthesis and secretion, as previously described [12]. Briefly, radioactively labeled and pepsinized procollagens from the patient and one healthy individual were separated on a 5% Sodium Dodecyl Sulphate Polyacrylamide Gel Electrophoresis (SDS-PAGE) and visualized by autoradiography.

To analyze ECM organization of collagen type I (COLLI), collagen type III (COLLIII), collagen type V (COLLV) and fibronectin (FN), skin fibroblasts derived either from the patient or an unrelated control individual were grown for 72 hours in the presence of 50 µM ascorbic acid (for COLLI only), refed after 24 hours and immunoreacted as previously reported [13–15]. In brief, cold methanol fixed fibroblasts were immunoreacted with 1:100 anti-COLLI, anti-COLLIII, anti-COLLV (Millipore-Chemicon Int., USA), and anti-FN (Sigma Chemicals, St. Louis, MO, USA) antibodies (Ab). The ECM of TNX was investigated on methanol-fixed cells immunoreacting with 2 µg/ml anti-TNX Ab (Santa Cruz Biotec. Inc., USA). For the analyses of the $\alpha 2\beta 1$, $\alpha 5\beta 1$ and $\alpha v\beta 3$ integrin receptors, cells were fixed in 3% paraformaldehyde (PFA)/60 mM sucrose and permeabilized with 0.5% Triton X-100 as previously reported [13,14,16]. In particular, fibroblasts were reacted for 1 hour at room temperature with 4 µg/ml anti-$\alpha 5\beta 1$ (clone JBS5), anti-$\alpha 2\beta 1$ (clone BHA.2) and anti-$\alpha v\beta 3$ (clone LM609) integrin monoclonal Abs (Millipore-Chemicon Int., USA). The cells were then incubated for 1 h with Alexa Fluor 488 anti-rabbit and Alexa Fluor 555 anti-mouse Ab (Thermo Fisher Scientific, South San Francisco, CA, USA), or with rhodamine-conjugated anti-goat IgG (Calbiochem-Merck, Germany). Immunofluorescence (IF) signals were acquired by a black-and-white CCD TV camera (SensiCam-PCO Computer Optics GmbH,

Germany) mounted on a Zeiss fluorescence Axiovert microscope and digitalized by Image Pro Plus software (Media Cybernetics, Silver Spring, MD, USA).

3. Results

3.1. Case Report

The patient was referred to our center at the age of 41 years because of a suspected connective tissue disorder. She was the first child of consanguineous parents of Swiss origin. The family history was unremarkable. The patient displayed a congenital dislocation of the left hip. Due to recurrent dislocations, an arthrodesis was performed at the age of 2 years. Immediately after surgery, a fracture of the left femur occurred, requiring osteosynthesis. Over the years, multiple joint dislocations ensued, mainly of the shoulders. During childhood, the patient developed a progressive kyphoscoliosis, which was left untreated. At present, the patient displays a right sided gibbus deformity (Figure 1). Upon radiography of the spine, a Cobb angle of 22, 38 and 15 degrees was measured at the cervical, thoracic and lumbar level, respectively.

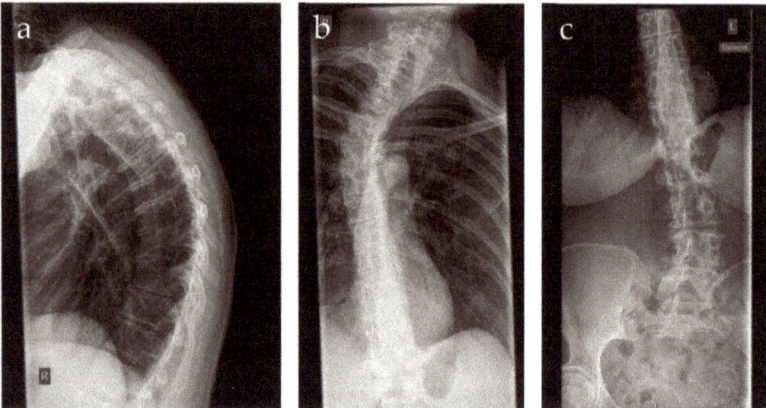

Figure 1. X-ray images of the vertebral column of the patient at the age of 41 years. (**a**) Side view demonstrating the pronounced kyphosis; (**b**) close-up of the pronounced cervical and thoracic scoliosis; (**c**) close-up of the lumbar scoliosis.

In addition, the patient presented a progressive arthrosis of the knee and degenerative changes of the spine. Bone densitometry at the age of 41 years was normal. Generalized joint hypermobility was absent (Beighton score 3/9). The patient reported mild muscular weakness. Computed tomography of the abdomen demonstrated fatty degeneration of the abdominal, gluteal and pelvic floor musculature. The patient was noted to have a soft, dough-like skin texture associated with skin hyperextensibility. In addition, the skin was thin and translucent with multiple varicose veins and edema of the lower extremities (Figure 2c). Redundant and sagging skin was found on both knees (Figure 2a). Striae were already present at young age. Feet and hands were short and broad, with the presence of brachydactyly (Figure 2d,g). Acrogeria of the hands was noticed, with an increased amount of palmar skin creases (Figure 2e). Several subcutaneous nodules were present at the interphalangeal joints of the fingers. Piezogenic papules were present on both feet (Figure 2f). While standing upright, the subcutaneous tissue of the feet was pushed aside, leading to neurogenic pain and the inability to walk barefoot or to go longer distances (Figure 2f). The patient reported large hematomas to occur after minimal trauma. Atrophic scarring was not present (Figure 2b). However, wound healing was clearly delayed with wound dehiscence after surgery. The patient presented with an inguinal and umbilical hernia at the age of 27 and 36 years, respectively, requiring surgical correction. The patient mothered three children, all born through C-section. A urogenital or rectal prolapse did not occur. The patient did, however, suffer from hemorrhoids. Colonoscopy at the age of 40 years demonstrated asymptomatic diverticulosis of the sigmoid colon. The patient presented with one episode of ileitis of unknown origin. In addition, the patient had two episodes of dactylitis and suffers from recurrent aphthous stomatitis. No biochemical indication of an underlying inflammatory disease was found. At the age of 42 years, a spontaneous perforation of the small bowel occurred, necessitating surgery and resection of the involved bowel loop. The intestinal specimen obtained during surgery was reported to be very fragile. An echocardiography at the age of 41 years demonstrated an intact mitral valve and normal dimensions of the aorta. The patient was known with a mild myopia. Hearing was normal to date.

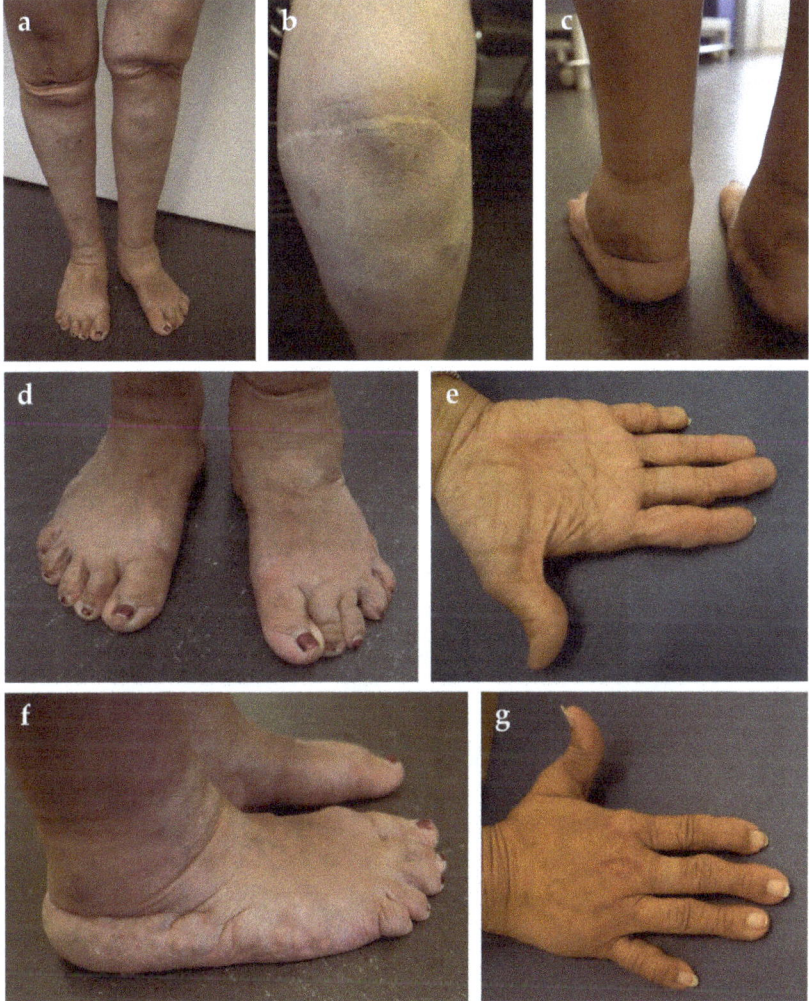

Figure 2. Clinical features of the patient. (**a**) Redundant and sagging skin on both knees; (**b**) widened scar on the left knee after surgery. No atrophic features are present; (**c**) chronic ankle edema in the absence of cardiac failure; (**d**) plane flat feet with broad forefeet and short digits; (**e**) increased palmar skin creases; (**f**) piezogenic papules; (**g**) short and broad acrogeric hands. Note the hyperextensibility of the thumb.

3.2. Molecular and Biochemical Findings

3.2.1. Molecular Analysis

Previous molecular investigations of the patient, all with negative results, included molecular genetic analysis of *COL1A1* and *COL1A2* on cDNA for the suspicion of arthrochalasis EDS, and an NGS panel for 101 connective tissue disorders. Given that the *TNXB* gene was not present in the NGS panel and considering the clinical presentation of the patient that was suggestive for clEDS, we sequenced *TNXB* by traditional Sanger method. Sequencing of all exons and exon-intron boundaries of *TNXB* (NM_0019105.7, NP_061978.6) revealed the homozygous c.5362del, p.(Thr1788Profs*100) variant in exon 15 (Figure 3a), which was absent in all public variant databases. Hence, the novel pathogenic *TNXB* variant (identifier #0000591583) was submitted to the Leiden Open Variation Database (LOVD, https://databases.lovd.nl/shared/genes/TNXB).qPCR analysis on cDNA obtained from patient's dermal fibroblast showed that the c.5362del transcript, which leads to frameshift and formation of a premature termination codon (PTC) p.(Thr1788Profs*100), represents a null-allele undergoing nonsense-mediated mRNA decay (Figure 3b). Consistently, IF analysis on the patient's fibroblasts with a specific Ab against TNX demonstrated a complete absence of the protein in the ECM (Figure 3c).

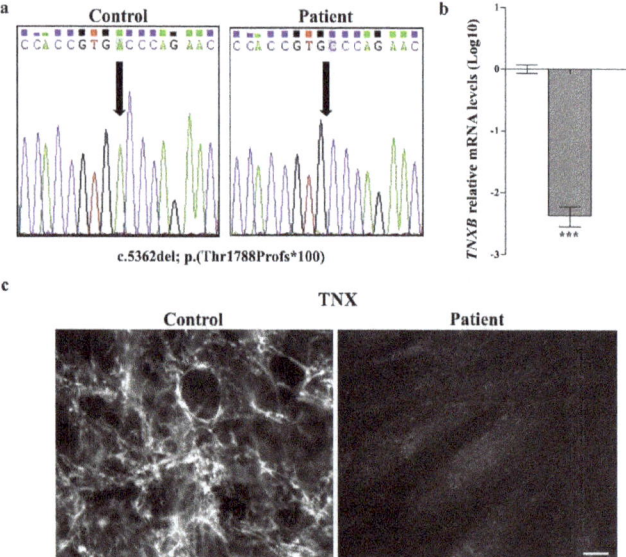

Figure 3. The *TNXB* c.5362del pathogenic variant causes nonsense-mediated mRNA decay and absence of the tenascin X protein in the ECM. (**a**) Sequence chromatograms showing the position of the novel c.5362del; p.(Thr1788Profs*100) variant (arrow) identified in homozygosity in exon 15 of the *TNXB* gene by Sanger sequencing; (**b**) qPCR analysis on cDNA obtained from patient's dermal fibroblasts showed that the transcript with the p.(Thr1788Profs*100) frameshift variant undergoes nonsense-mediated mRNA decay. The relative quantification (RQ) of the *TNXB* transcript levels was determined with the $2^{-(\Delta\Delta Ct)}$ method normalized with the geometric mean of the *HPRT*, *GAPDH* and *ATP5B* reference genes. Bars represent the mean ratio of target gene expression in patients' fibroblasts compared to three unrelated healthy individuals. qPCR was performed in triplicate, and the results are expressed as means ± SEM. The relative mRNA level of *TNXB* in the patient versus controls (about 142-fold decrease) is expressed as Log10 transformed value. Statistical significance (*** $P < 0.001$) was calculated with one-way ANOVA and the Tukey post hoc test. (**c**) IF analysis on patient's skin fibroblasts with a specific antibody against TNX showing the absence of the protein in the ECM compared to control cells. Scale bar: 10 μm.

3.2.2. Collagen Biochemical and Immunofluorescent Analysis

To assess the effects on collagen biosynthesis and secretion, we performed collagen steady-state and pulse-chase analyses in dermal fibroblasts from the patient. Normal relative proportions of COLLI, COLLIII and COLLV were present in the cell lysate and in the medium (Supplementary Figure S1). There were no indications for altered modification of the different collagens.

To investigate the effect of TNX-deficiency on COLLI-, COLLIII-, COLLV- and FN-ECM organization and on the expression of their integrin receptors, dermal fibroblasts from the patient and one healthy individual were investigated by IF (Figure 4). Control fibroblasts organized a reticular ECM of COLLIII and COLLV, rare fibrils of COLLI, an abundant FN-ECM and expression of the canonical $\alpha 2\beta 1$ and $\alpha 5\beta 1$ integrins, whereas patient's fibroblasts showed a lack of organized COLLI, COLLIII, and COLLV in the ECM, though proteins were present at different levels inside the cells. The disorganization of the COLLs-ECM is associated with a strong reduction of their canonical $\alpha 2\beta 1$ integrin receptor (Figure 4a). Contrarily, the FN-ECM organization and the expression of its canonical integrin receptor $\alpha 5\beta 1$ were not affected in TNX-deficient cells. Therefore, the alternative FN-receptor $\alpha v\beta 3$ was almost undetectable both in control and patient cells (Figure 4b).

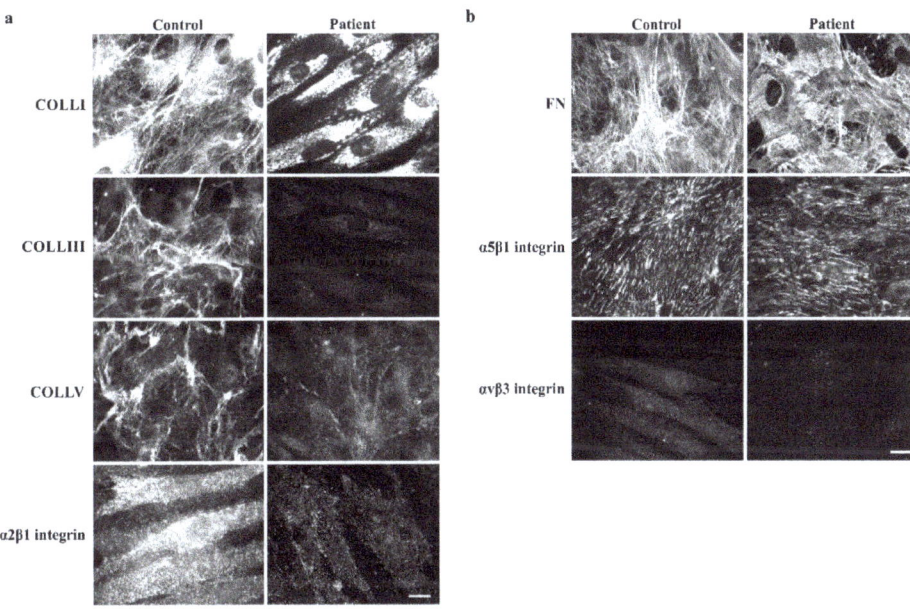

Figure 4. TNX-deficiency leads to the disarray of collagens type I, type III and type V in the ECM together with the reduction of the $\alpha 2\beta 1$ integrin. (**a**) IF analysis with specific Ab demonstrates that TNX-deficient cells are characterized by the lack of organized COLLI, COLLIII and COLLV in the ECM, though proteins are present at different levels inside the cells. The disorganization of the COLLs-ECM is associated with a strong reduction of their canonical $\alpha 2\beta 1$ integrin receptor. (**b**) TNX-deficient cells deposit FN in the ECM and express its canonical integrin receptor $\alpha 5\beta 1$ similarly to control fibroblasts. The alternative $\alpha v\beta 3$ integrin is almost undetectable both in control and patients' cells. Scale bars: 10 μm.

4. Discussion

Most known EDS subtypes are autosomal dominant inherited disorders due to defects in genes coding for the fibrillar collagens or collagen-modifying enzymes. TNX-deficiency differs from the more frequent subtypes not only in its autosomal recessive inheritance pattern but also in the fact that it is caused by a defect in an ECM component other than collagen. The presumed role of TNX in the stabilization and maturation of the collagen- and elastin-ECM is reflected in the cellular and clinical phenotype of the patients described to date [2–7]. Similar to the data in $Tnxb^{-/-}$ mice, we could demonstrate that collagen biosynthesis and secretion were unaffected in our patient's fibroblasts, whereas a disarray of the COLLI-, COLLIII- and COLLV-ECM and a strong reduction of their canonical α2β1 integrin receptor was observed in vitro (Figure 4a) [10]. Unlike the typical EDS cellular phenotype, our patient's fibroblasts were characterized by a normal organization of the FN-ECM and therefore a normal expression of its canonical integrin receptor α5β1 (Figure 4b) [10,17].

Thus far, about 30 patients with complete TNX-deficiency have been described in literature [8,9,18–26]. Most patients displayed either a complete lack of TNX in serum or biallelic mutations in *TNXB* leading to nonsense-mediated mRNA decay (Supplementary Table S3). As expected, those presenting missense mutations had milder or late-onset clinical manifestations [17,24].

As its name implies, TNX-deficiency or classical-like EDS (clEDS) phenotypically resembles the classical form of EDS (cEDS), with the triad of soft/velvety hyperextensible skin (20/20), generalized joint hypermobility (15/19) and a varying degree of tissue fragility (20/20) as its main clinical features (Table 1) [9]. Unlike cEDS, atrophic scarring is absent in patients with clEDS. However, about 50% of TNX-deficient patients present delayed wound healing with wound dehiscence upon suture removal and widened scars (Figure 2b). Data in $Tnxb^{-/-}$ mice suggest that, contrary to cEDS, abnormal wound healing in clEDS is not due to altered matrix deposition in the early phases of wound healing, but is rather caused by decreased ECM stabilization and maturation during the later stages. Indeed, contrary to the high levels of *TNXB* in normal skin, its expression during the early phase of wound healing is non-existing and only increases over time [27].

Whereas joint hypermobility seems to decrease with age, recurrent dislocations of the large joints (18/19) remain a problem over time and can be considered as the most frequent debilitating finding in the TNX-deficient patients described to date (Table 1).

Interestingly, several patients display deformities of the hands and feet, most of which can be ascribed to the underlying connective tissue weakness, such as pes planus, hallux valgus, piezogenic papules and acrogeric hands. Conversely, the presence of brachydactyly and broad hands and feet point to a role of TNX in development [28]. In addition, our patient presented chronic venous insufficiency, varicose veins and non-pitting ankle edema, which has thus far been observed in almost 30% of the TNX-deficient population (Table 1). The venous insufficiency and its consequences might be explained by the aberrant elastic fiber remodeling observed in TNX-deficient dermal fibroblasts and possible changes in endothelial cell proliferation [5,29,30]. Indeed, the interaction between TNX and vascular endothelial growth factor B (VEGF-B) is known to stimulate endothelial cell proliferation [30,31].

Bone is considered to be a target organ of osteogenesis imperfecta rather than EDS. However, premature osteopenia or osteoporosis have been published in various types of EDS and have been mainly attributed to abnormal COLLI fibrillogenesis and ECM deposition [32]. Although bone loss due to an increased number of multinucleated osteoclasts has been found in $Tnxb^{-/-}$ mice, thus far, indications for skeletal fragility in clEDS have not been reported [33]. Moreover, in our patient, bone densitometry at the age of 41 years was normal.

Over time, neuromuscular symptoms have been increasingly recognized in various types of EDS, ranging from muscle weakness to myalgia and easy fatigability [21]. Indeed, a normal composition of the ECM and intact innervation are important for adequate functioning of the muscle [34]. Considering the cellular phenotype of clEDS, neuromuscular involvement is not surprising. Collagen VI, which is deficient in two myopathies, i.e., Ullrich congenital muscular dystrophy and Bethlem myopathy, is downregulated, although with conserved organization of the ECM, in as well TNX-*null* fibroblasts and $Tnxb^{-/-}$ mice [7,34,35]. In addition, TNX is widely distributed in peripheral nerve and its deficiency causes axonal polyneuropathy in almost 40% of the patients (Table 1) [21]. Our patient had a low muscular mass on clinical examination and reported distal muscle weakness. Computed tomography of the abdomen demonstrated fatty degeneration of the muscles as an incidental finding, a well-known phenomenon in myopathies of various origins. Interestingly, the neuromuscular phenotype seems to be mild or absent in children and progresses with age.

Table 1. Summary of clinical features of patients with biallelic *TNXB* variants.

	Present Patient	clEDS*	Total (%)
Major criteria			
Skin hyperextensibility (with velvety skin and absence of atrophic scarring)	+	19/19	20/20 (100)
GJH (Beighton Score > 5/9) (+/− recurrent dislocation)	−	15/18	15/19 (79)
Easy bruising	+	19/19	20/20 (100)
Minor Criteria			
Articular (sub)luxation	+	17/18	18/19 (95)
Feet deformities (broad/short feet, brachydactyly, pes planus, hallux valgus)	+	14/18	15/19 (79)
Hand deformities (acrogeric hands, mallet fingers, clinodactyly and brachydactyly)	+	4/17	5/18 (28)
Mild proximal and distal muscle weakness	+	9/19	10/20 (50)
Polyneuropathy	NA	7/18	7/18 (39)
Edema in the legs in absence of cardiac failure	+	4/17	5/18 (28)
Other features			
Congenital joint dislocation	+	1/17	2/18 (11)
Delayed wound healing	+	7/18	8/19 (42)
Vaginal/uterus/rectal prolapse	−	5/17	5/18 (28)
Inguinal/umbilical/wound herniation	+	4/19	5/20 (25)
Varicose veins	+	3/17	4/18 (22)
Piezogenic papulae	+	13/18	14/19 (74)
High arched/narrow palate +/− dental crowding	+	4/17	5/18 (28)
Refractive error	+	8/17	9/18 (50)
Mitral valve abnormalities	−	4/19	4/20 (20)
Diverticulosis/diverticulitis	+	4/18	5/19 (26)
Spontaneous bowel perforation	+	1/18	2/19 (11)

Note: 19 out of 30 patients reported in literature were included. Patient 3 reported in Schalkwijk et al. [9], the index patient reported in Burch et al. [8], the three patients published in Chen et al. [25] were excluded because of concomitant congenital adrenal hyperplasia. The patient described in Mackenroth et al. [26] was excluded because of concomitant *COL1A1* mutation. Patient 4 and 5 reported in Schalkwijk et al. [9] were excluded because a lack of clinical data. The patient described in O'Connell et al. [24] and the two patients described by Lindor et al. [18] were excluded because no mutation analysis was performed.

Extra-articular symptoms in clEDS, such as gastrointestinal and cardiovascular complications, do not seem to be associated with certain mutations, but are rather related to age and have been reported after the 3rd–4th decade [18,22,23,25]. Gastrointestinal manifestations, although rare, can lead to life-threatening situations. Spontaneous perforation of hollow organs is a complication mainly seen in patients with vascular EDS (vEDS) due to mutations in COLLIII and has only sporadically been described in other EDS subtypes. Compared to other non-vEDS, a higher prevalence seems to exist in clEDS, ranging from tracheal or esophageal rupture to spontaneous perforation of colon, diverticula or small intestine [18,20,22,23,25,36]. Therefore, patients and treating physicians should be aware of this risk in order to minimize the occurrence of iatrogenic perforation and to not delay the diagnosis and treatment of spontaneous events. Contrary to patients with vEDS, aortic root dilatation or aneurysms are not common in clEDS, and have only been described in one non-published case [23]. Peeters et al. suggested that the co-expression of tenascin-C (TNC) in large blood vessels might compensate for the TNX-deficiency, giving rise to a normal arterial vessel wall [37,38]. Conversely, mitral valve involvement (i.e., mitral valve prolapse, thickening and/or insufficiency) has thus far been reported in 20% of clEDS patients, highlighting the need for cardiovascular follow-up (Table 1).

5. Conclusions

Although TNX-deficiency phenotypically resembles cEDS, absent atrophic scarring, the presence of short broad feet, brachydactyly, edema of the lower extremities, acrogeria or the occurrence of hollow organ perforation should initiate targeted diagnostics for clEDS, either by measuring the TNX concentration in serum or by mutation analysis of the *TNXB* gene.

Supplementary Materials: The following are available online at http://www.mdpi.com/2073-4425/10/11/843/s1, Figure S1: Steady-state analysis of collagen in the medium and the cell layer, Table S1: 101 genes analyzed by means of an NGS panel for connective tissue disorders, Table S2: Primer sequences and PCR conditions for *TNXB* sequencing, Table S3: Published *TNXB* pathogenic variants.

Author Contributions: M.C., M.R. (Marianne Rohrbach) and C.G. conceived the study and edited and coordinated the manuscript. The manuscript was drafted by D.R. and M.R. (Marco Ritelli), who organized data contents, reviewed the literature and completed figures and tables. D.R. and M.R. (Marianne Rohrbach) performed the clinical evaluation of the patient. M.R. (Marco Ritelli), N.Z., V.C. and C.G. provided the experimental data. All authors discussed, read and approved the manuscript.

Funding: This project was supported by radiz (rare disease initiative Zurich, Switzerland) to D.R. and has been supported by tshe Swiss National Science Foundation Grant No. 31003A-173183 to C.G. and M.R. (Marianne Rohrbach).

Acknowledgments: The authors want to thank the patient for her cooperation during the diagnostic process and the Fazzo Cusan family for its generous support.

Conflicts of Interest: All authors declare that there is no conflict of interest concerning this work.

References

1. Brady, A.F.; Demirdas, S.; Fournel-Gigleux, S.; Ghali, N.; Giunta, C.; Kapferer-Seebacher, I.; Kosho, T.; Mendoza-Londono, R.; Pope, M.F.; van Damme, T.; et al. The Ehlers–Danlos syndromes, rare types. *Am. J. Med. Genet.* **2017**, *175*, 70–115. [CrossRef] [PubMed]
2. Chiquet-Ehrismann, R.; Tucker, R.P. Tenascins and the importance of adhesion modulation. *Cold Spring Harb. Perspect. Biol.* **2011**, *3*, a004960. [CrossRef] [PubMed]
3. Elefteriou, F.; Exposito, J.Y.; Garrone, R.; Lethias, C. Binding of tenascin-X to decorin. *FEBS Lett.* **2001**, *495*, 44–47. [CrossRef]
4. Zweers, M.C.; van Vlijmen-Willems, I.M.; van Kuppevelt, T.H.; Mecham, R.P.; Steijlen, P.M.; Bristow, J.; Schalkwijk, J. Deficiency of tenascin-X causes abnormalities in dermal elastic fiber morphology. *J. Investig. Dermatol.* **2004**, *122*, 885–891. [CrossRef] [PubMed]
5. Bristow, J.; Carey, W.; Egging, D.; Schalkwijk, J. Tenascin-X, collagen, elastin, and the Ehlers–Danlos syndrome. *Am. J. Med. Genet.* **2005**, *139*, 24–30. [CrossRef] [PubMed]
6. Matsumoto, K.I.; Takayama, N.; Ohnishi, J.; Ohnishi, E.; Shirayoshi, Y.; Nakatsuji, N.; Ariga, H. Tumour invasion and metastasis are promoted in mice deficient in tenascin-X. *Genes Cells* **2001**, *6*, 1101–1111. [CrossRef]
7. Minamitani, T.; Ariga, H.; Matsumoto, K.I. Deficiency of tenascin-X causes a decrease in the level of expression of type VI collagen. *Exp. Cell Res.* **2004**, *297*, 49–60. [CrossRef]

8. Burch, G.H.; Gong, Y.; Liu, W.; Dettman, R.W.; Curry, C.J.; Smith, L.; Miller, W.L.; Bristow, J. Tenascin-X deficiency is associated with Ehlers-Danlos syndrome. *Nature* **1997**, *17*, 104–108. [CrossRef]
9. Schalkwijk, J.; Zweers, M.C.; Steijlen, P.M.; Dean, W.B.; Taylor, G.; van Vlijmen, I.M.; van Haren, B.; Miller, W.L.; Bristow, J. A recessive form of Ehlers-Danlos syndrome caused by tenascin-X deficiency. *N. Engl. J. Med.* **2001**, *345*, 1167–1175. [CrossRef]
10. Mao, J.R.; Taylor, G.; Dean, W.B.; Wagner, D.R.; Afzal, V.; Lotz, J.C.; Rubin, E.M.; Bristow, J. Tenascin-X deficiency mimics Ehlers-Danlos syndrome in mice through alteration of collagen deposition. *Nat. Genet.* **2002**, *30*, 421–425. [CrossRef]
11. Yung, Y.C. Molecular genetics of the human MHC complement gene cluster. *Exp. Clin. Immunogenet.* **1998**, *15*, 213–230.
12. Lindert, U.; Gnoli, M.; Maioli, M.; Bedeschi, M.F.; Sangiorgi, L.; Rohrbach, M.; Giunta, C. Insight into the pathology of a *COL1A1* signal peptide heterozygous mutation leading to severe osteogenesis imperfecta. *Calcif. Tissue Int.* **2018**, *102*, 373–379. [CrossRef] [PubMed]
13. Zoppi, N.; Gardella, R.; de Paepe, A.; Barlati, S.; Colombi, M. Human fibroblasts with mutations in *COL5A1* and *COL3A1* genes do not organize collagens and fibronectin in the extracellular matrix, down-regulate α2β1 integrin, and recruit α vβ3 instead of α5β1 integrin. *J. Biol. Chem.* **2004**, *279*, 18157–18168. [CrossRef] [PubMed]
14. Zoppi, N.; Barlati, S.; Colombi, M. FAK-independent αvβ3 integrin-EGFR complexes rescue from anoikis matrix-defective fibroblasts. *Biochim. Biophys. Acta Mol. Cell Res.* **2008**, *1783*, 1177–1188. [CrossRef] [PubMed]
15. Chiarelli, N.; Carini, G.; Zoppi, N.; Dordoni, C.; Ritelli, M.; Venturini, M.; Castori, M.; Colombi, M. Transcriptome-wide expression profiling in skin fibroblasts of patients with joint hypermobility syndrome/ehlers-danlos syndrome hypermobility type. *PLoS ONE* **2016**, *11*, e0161347. [CrossRef]
16. Zoppi, N.; Chiarelli, N.; Binetti, S.; Ritelli, M.; Colombi, M. Dermal fibroblast-to-myofibroblast transition sustained by α vβ3integrin-ILK-Snail1/Slug signaling is a common feature for hypermobile Ehlers-Danlos syndrome and hypermobility spectrum disorders. *Biochim. Biophys. Acta Mol. Basis Dis.* **2018**, *1864*, 1010–1023. [CrossRef]
17. Zoppi, N.; Chiarelli, N.; Ritelli, M., Colombi, M. Multifaced roles of the αvβ3 integrin in Ehlers–Danlos and arterial tortuosity syndromes' dermal fibroblasts. *Int. J. Mol. Sci.* **2018**, *19*, 982. [CrossRef]
18. Lindor, N.M.; Bristow, J. Tenascin-X deficiency in autosomal recessive Ehlers-Danlos syndrome. *Am. J. Med. Genet.* **2005**, *135*, 75–80. [CrossRef]
19. Pénisson-Besnier, I.; Allamand, V.; Beurrier, P.; Martin, L.; Schalkwijk, J.; van Vlijmen-Willems, I.; Gartioux, C.; Malfait, F.; Syx, D.; Marcorelles, P.; et al. Compound heterozygous mutations of the *TNXB* gene cause primary myopathy. *Neuromuscul. Disord.* **2013**, *23*, 664–669. [CrossRef]
20. Hendriks, A.G.; Voermans, N.C.; Schalkwijk, J.; Hamel, B.C.; van Rossum, M.M. Well-defined clinical presentation of Ehlers-Danlos syndrome in patients with tenascin-X deficiency: A report of four cases. *Clin. Dysmorphol.* **2012**, *21*, 15–18. [CrossRef]
21. Voermans, N.C.; van Alfen, N.; Pillen, S.; Lammens, M.; Schalkwijk, J.; Zwarts, M.J.; van Rooij, I.A.; Hamel, B.C.J.; van Engelen, B.G. Neuromuscular involvement in various types of Ehlers-Danlos syndrome. *Ann. Neurol.* **2009**, *65*, 687–697. [CrossRef] [PubMed]
22. Sakiyama, T.; Kubo, A.; Sasaki, T.; Yamada, T.; Yabe, N.; Matsumoto, K.I.; Futei, Y. Recurrent gastrointestinal perforation in a patient with Ehlers-Danlos syndrome due to tenascin-X deficiency. *J. Dermatol.* **2015**, *42*, 511–514. [CrossRef] [PubMed]
23. Demirdas, S.; Dulfer, E.; Robert, L.; Kempers, M.; van Beek, D.; Micha, D.; van Engelen, B.G.; Hamel, B.; Schalkwijk, J.; Maugeri, A.; et al. Recognizing the tenascin-X deficient type of Ehlers–Danlos syndrome: A cross-sectional study in 17 patients. *Clin. Genet.* **2017**, *91*, 411–425. [CrossRef]
24. O'Connell, M.; Burrows, N.P.; van Vlijmen-Willems, M.J.J.; Clark, S.M.; Schalkwijk, J. Tenascin-X deficiency and Ehlers-Danlos syndrome: A case report and review of the literature. *Br. J. Dermatol.* **2010**, *163*, 1340–1345. [CrossRef] [PubMed]
25. Chen, W.; Perritt, A.F.; Morissette, R.; Dreiling, J.L.; Bohn, M.F.; Mallappa, A.; Xu, Z.; Quezado, M.; Merke, D.P. Ehlers-Danlos syndrome caused by biallelic *TNXB* variants in patients with congenital adrenal hyperplasia. *Hum. Mutat.* **2016**, *37*, 893–897. [CrossRef] [PubMed]

26. Mackenroth, L.; Fischer-Zirnsak, B.; Egerer, J.; Hecht, J.; Kallinich, T.; Stenzel, W.; Spors, B.; von Moers, A.; Mundlos, S.; Gerhold, K.; et al. An overlapping phenotype of osteogenesis imperfecta and Ehlers-Danlos syndrome due to a heterozygous mutation in COL1A1 and biallelic missense variants in TNXB identified by whole exome sequencing. *Am. J. Med. Genet.* **2016**, *170*, 1080–1085. [CrossRef] [PubMed]
27. Egging, D.; van Vlijmen-Willems, I.; van Tongeren, T.; Schalkwijk, J.; Peeters, A. Wound healing in tenascin-X deficient mice suggests that tenascin-X is involved in matrix maturation rather than matric deposition. *Connect. Tissue Res.* **2007**, *48*, 93–98. [CrossRef]
28. Burch, G.H.; Bedolli, M.A.; McDonough, S.; Rosenthal, S.M.; Bristow, J. Embryonic expression of tenascin-X suggests a role in limb, muscle and heart development. *Dev. Dyn.* **1995**, *203*, 491–504. [CrossRef]
29. Naoum, J.J.; Hunter, G.C.; Woodside, K.J.; Chen, C. Current advances in the pathogenesis of varicose veins. *J. Surg. Res.* **2007**, *141*, 311–316. [CrossRef]
30. Ikuta, T.; Ariga, H.; Matsumoto, K. Extracellular matrix tenascin-X in combination with vascular endothelial growth factor B enhances endothelial cell proliferation. *Genes Cells* **2000**, *5*, 913–927. [CrossRef]
31. Ishitsuka, T.; Ikuta, T.; Ariga, H.; Matsumoto, K. Serum tenascin-X binds to vascular endothelial growth factor. *Biol. Pharm. Bull.* **2009**, *32*, 1004–1011. [CrossRef] [PubMed]
32. Formenti, A.M.; Doga, M.; Frara, S.; Ritelli, M.; Colombi, M.; Banfi, G.; Giustina, A. Skeletal fragility: An emerging complication of Ehlers-Danlos syndrome. *Endocrine* **2019**, *63*, 225–230. [CrossRef] [PubMed]
33. Kajitani, N.; Yamada, T.; Kawakami, K.; Matsumoto, K.I. TNX deficiency results in bone loss due to an increase in multinucleated osteoclasts. *Biochem. Biophys. Res. Commun.* **2019**, *512*, 659–664. [CrossRef] [PubMed]
34. Voermans, N.C.; Verrijp, K.; Eshuis, L.; Balemans, M.C.M.; Egging, D.; Sterrenburg, E.; van Rooij, I.A.L.M.; van der Laak, J.A.W.M.; Schalkwijk, J.; Lammens, M.; et al. Mild muscular features in tenascin-X knockout mice, a model of Ehlers-Danlos syndrome. *Connect. Tissue Res.* **2011**, *52*, 422–432. [CrossRef]
35. Kirschner, J.; Hausser, I.; Zou, Y.; Schreiber, G.; Christen, H.J.; Brown, S.C.; Anton-Lamprecht, I.; Muntoni, F.; Hanefeld, F.; Bönnemann, C.G. Ullrich congenital muscular dystrophy: Connective tissue abnormalities in the skin support overlap with Ehlers–Danlos syndromes. *Am. J. Hum. Genet.* **2004**, *132*, 296–301. [CrossRef]
36. Besselink-Lobanova, A.; Maandag, N.J.; Voermans, N.C.; van der Heijden, H.F.; van der Hoeven, J.G.; Heunks, L.M. Trachea rupture in tenascin-X deficient type Ehlers-Danlos syndrome. *Anesthesiology* **2010**, *113*, 746–749. [CrossRef]
37. Peeters, A.C.T.; Kucharekova, M.; Timmermans, J.; van den Berkmortel, F.W.P.J.; Boers, G.H.J.; Novakova, I.R.O.; Egging, D.F.; den Heijer, M.; Schalkwijk, J. A clinical and cardiovascular survey of Ehlers-Danlos syndrome patients with complete deficiency of tenascin-X. *Neth. J. Med.* **2004**, *62*, 160–162.
38. Imanaka-Yoshida, K.; Matsumoto, K. Multiple roles of tenascins in homeostasis and pathophysiology of aorta. *Ann. Vasc. Dis.* **2018**, *11*, 169–180. [CrossRef]

© 2019 by the authors. Licensee MDPI, Basel, Switzerland. This article is an open access article distributed under the terms and conditions of the Creative Commons Attribution (CC BY) license (http://creativecommons.org/licenses/by/4.0/).

Article

Novel *TNXB* Variants in Two Italian Patients with Classical-Like Ehlers-Danlos Syndrome

Lucia Micale [1,†,*], Vito Guarnieri [1,†], Bartolomeo Augello [1], Orazio Palumbo [1], Emanuele Agolini [2], Valentina Maria Sofia [2], Tommaso Mazza [3], Antonio Novelli [2], Massimo Carella [1] and Marco Castori [1]

1. Division of Medical Genetics, Fondazione IRCCS-Casa Sollievo della Sofferenza, 71043 San Giovanni Rotondo (Foggia), Italy; v.guarnieri@operapadrepio.it (V.G.); b.augello@operapadrepio.it (B.A.); o.palumbo@operapadrepio.it (O.P.); m.carella@operapadrepio.it (M.C.); m.castori@operapadrepio.it (M.C.)
2. Laboratory of Medical Genetics, IRCCS-Bambino Gesù Children's Hospital, 00146 Rome, Italy; emanuele.agolini@opbg.net (E.A.); vale17l@gmail.com (V.M.S.); antonio.novelli@opbg.net (A.N.)
3. Unit of Bioinformatics, Fondazione IRCCS-Casa Sollievo della Sofferenza, 71043 San Giovanni Rotondo (Foggia), Italy; t.mazza@operapadrepio.it
* Correspondence: l.micale@operapadrepio.it; Tel.: +39-0882416350; Fax: +39-088241161
† These authors equally contributed to the work.

Received: 26 October 2019; Accepted: 21 November 2019; Published: 25 November 2019

Abstract: *TNXB*-related classical-like Ehlers-Danlos syndrome (*TNXB*-clEDS) is an ultrarare type of Ehlers-Danlos syndrome due to biallelic *null* variants in *TNXB*, encoding tenascin-X. Less than 30 individuals have been reported to date, mostly of Dutch origin and showing a phenotype resembling classical Ehlers-Danlos syndrome without atrophic scarring. *TNXB*-clEDS is likely underdiagnosed due to the complex structure of the *TNXB* locus, a fact that complicates diagnostic molecular testing. Here, we report two unrelated Italian women with *TNXB*-clEDS due to compound heterozygosity for *null* alleles in *TNXB*. Both presented soft and hyperextensible skin, generalized joint hypermobility and related musculoskeletal complications, and chronic constipation. In addition, individual 1 showed progressive finger contractures and shortened metatarsals, while individual 2 manifested recurrent subconjunctival hemorrhages and an event of spontaneous rupture of the brachial vein. Molecular testing found the two previously unreported c.8278C > T p.(Gln2760*) and the c.(2358 + 1_2359 − 1)_(2779 + 1_2780 − 1)del variants in Individual 1, and the novel c.1150dupG p.(Glu384Glyfs*57) and the recurrent c.11435_11524+30del variants in Individual 2. mRNA analysis confirmed that the c.(2358 + 1_2359 − 1)_(2779 + 1_2780 − 1)del variant causes a frameshift leading to a predicted truncated protein [p.(Thr787Glyfs*40)]. This study refines the phenotype recently delineated in association with biallelic *null* alleles in *TNXB*, and adds three novel variants to its mutational repertoire. Unusual digital anomalies seem confirmed as possibly peculiar of *TNXB*-clEDS, while vascular fragility could be more than a chance association also in this Ehlers-Danlos syndrome type.

Keywords: Classical-like; Ehlers-Danlos syndrome; haploinsufficiency; tenascin-X; *TNXB*

1. Introduction

Ehlers-Danlos syndromes (EDS) are a clinically variable and genetically heterogeneous group of hereditary connective tissue disorders mainly featured by abnormal skin texture and repair, tissue and vascular fragility, and joint hypermobility. The 2017 international classification identifies 13 EDS types due to deleterious variants in 19 different genes [1]. More recently, a 14th type of EDS with features overlapping classical type and due to recessive variants in the *AEBP1* gene was added to this nosology [2,3]. Among them, classical, vascular, and hypermobile EDS are the most common, while the others are rarer and their frequency in the general population remains mostly unknown.

Classical-like EDS due to biallelic variants in *TNXB*, encoding tenascin-X (*TNXB*-clEDS), is a second EDS type resembling classical EDS but distinguished from the latter by recessive pattern of inheritance and lack of atrophic scarring [1]. Tenascin-X is a large extracellular matrix–forming glycoprotein, whose deficiency was firstly involved in the etiology of EDS in 1997, by the description of a novel contiguous gene syndrome combining congenital adrenal hyperplasia (CAH) and EDS, and due to the deletion of the *CYP21A2* (OMIM 201910) and *TNXB* neighbouring genes [4]. Subsequently, it was clear that this rare type of EDS is a recessive disorder caused by homozygous or compound heterozygous *null* alleles in *TNXB* [5]. Only 24 individuals with *TNXB*-clEDS have been described to date and most of them are of Dutch origin [6]. A recent cross-sectional study on 17 individuals suggests possible phenotypic hallmarks of *TNXB*-clEDS and hypothesizes that it is underdiagnosed with EDS-like symptoms outside The Netherlands due to the complex molecular structure of the *TNXB* locus [7].

TNXB maps on chromosome region 6p23.1 within the human leukocyte antigen histocompatibility complex in a module characterized by highly homologous sequences between functional genes, *CYP21A2* and *TNXB*, and their corresponding pseudogenes *CYP21A1P* and *TNXA*. This genomic structure is prone to non-homologous recombinations. Misalignment events during meiosis result in pathogenic *CYP21A2/CYP21A1P* and *TNXA/TNXB* chimeric genes. To date, three major types of *TNXA/TNXB* chimera have been identified [4,8]. In particular, CAH-X chimera 1 (CH-1) and CAH-X chimera 2 (CH-2) have *TNXB* exons 35–44 and 40–44, respectively, replaced with *TNXA* [4,9,10]. CH-1 is characterized by a 120-bp deletion (c.11435_11524 + 30del) due to the substitution of *TNXB* exon 35 by *TNXA* that is causative of tenascin-X haploinsufficiency in CAH-X CH-1; this region is the only well-documented discrepancy between *TNXB* and *TNXA* homologous portion. CH-2 is characterized by the variant c.12174C > G p.(Cys4058Trp) derived from the substitution of *TNXB* exon 40 by *TNXA* with a likely dominant negative effect [10]. The third chimera, termed CAH-X chimera 3 (CH-3), has *TNXB* exons 41–44 substituted by *TNXA* and is characterized by a cluster of three pseudogene variants: c.12218G > A p.(Arg4073His) in exon 41, and c.12514G > A p.(Asp4172Asn), and c.12524G > A p.(Ser4175Asn) in exon 43. This chimera has been reported in one patient, and its clinical significance is still preliminary [8].

Due to such a complex molecular architecture of the genomic region surrounding *TNXB*, Demirdas et al. [7] proposed a multistep molecular diagnostics workflow including: (i) a mixed approach of next-generation sequencing (NGS) for the non-homologous *TNXB* sequence and Sanger sequencing for the *TNXA/TNXB* homology region to exclude *TNXB* point variants and rare *TNXA/TNXB* chimera, (ii) followed by *TNXB* deletion/duplication analysis aimed to investigate the presence of the recurrent c.11435_11524 + 30del resulting from the common *TNXA/TNXB* chimeric fusion and other rarer rearrangements. Following this approach, this group was able to identify 12 different *TNXB* variants associated with *TNXB*-clEDS [7].

Here, we describe the first two Italian individuals affected by *TNXB*-clEDS. Molecular testing investigated the full range of possible molecular mechanisms leading to *TNXB* null alleles and found three novel variants.

2. Materials and Methods

Patients were enrolled for this study after obtaining written informed consent for publishing pictures (individual 2) and clinical data (both individuals). This study was approved by the local ethical committee, and is in accordance with the 1984 Helsinki declaration and subsequent versions. Part of the results of molecular investigations presented in this work was obtained from the routine clinical diagnostic activities of the involved institutions.

2.1. Sample Preparation and Next Generation Sequencing Analysis

Genomic DNA was extracted from patients' and unaffected relatives' peripheral blood leucocytes by using Bio Robot EZ1 (Qiagen, Hilden, Germany) according to the manufacturer's instructions.

The DNA was quantified with Qubit spectrophotometer (Thermo Fisher Scientific, Waltham, MA, USA). Probands' DNA first underwent sequencing with a HaloPlex gene panel (Agilent Technologies, Santa Clara, CA, USA) designed to selectively capture known genes associated with the various forms of EDS (*ADAMTS2, AEBP1, B3GALT6, B4GALT7, CHST14, COL1A1, COL1A2, COL3A1, COL5A1, COL5A2, COL12A1, C1R, C1S, DSE, FKBP14, PLOD1, FLNA, PRDM5, SLC39A13, TNXB* and *ZNF469*) according to the current nosology. For *TNXB*, NGS sequencing is effective for exons 1 to 31 only, which correspond to the non-homologous *TNXB* sequence. Libraries were prepared using the Haloplex target enrichment kit (Agilent Technologies, Santa Clara, CA, USA) following manufacturer's instructions. Targeted fragments were then sequenced on a MiSeq sequencer (Illumina, San Diego, CA, USA) using a MiSeq Reagent kit V3 300 cycles flow cell. Reads were aligned to the GRCh37/hg19 reference genome by BWA (v.0.7.17). BAM files were sorted by SAMtools (v.1.7) and purged from duplicates using Mark Duplicates from the Picard suite (v.2.9.0). Mapped reads were locally realigned using GATK 3.8. Reads with mapping quality scores lower than 20 or with more than one-half nucleotides with quality scores less than 30 were filtered out. The GATK's Haplotype Caller tool was used to identify variants. Variants were annotated by ANNOVAR, with information about allelic frequency (1000 Genomes, dbSNP 151, GO-ESP 6500, ExAC, TOPMED, GnomAD, NCI60, COSMIC), reported or computationally estimated pathogenicity (ClinVar, HGMD, LOVD, or SIFT, Polyphen2, LRT, MutationTaster, MutationAssessor, FATHMM, PROVEAN, VEST3, MetaSVM, MetaLR, M-CAP, CADD, DANN, fathmm-MKL, Eigen, GenoCanyon), and amino acids conservation (fitCons, GERP++, phyloP100way, phyloP20way, phastCons100way vertebrate, phastCons20way mammalian, SiPhy 29way). Selected variants were interpreted according to the American College of Medical Genetics and Genomics/Association for Molecular Pathology (ACMGG/AMP) [11]. Specifically, variants without clinical significance at the time of reporting (i.e., benign and likely benign) were excluded by the presence of one or more criteria for benignity. Variants which passed this preliminary selection were selected for further investigation and classified as pathogenic, likely pathogenic or uncertain significance by using the following criteria: (i) null (nonsense, frameshift) variant in a gene previously described as disease-causing by haploinsufficiency or loss-of-function; (ii) missense variant located in a critical and well-established functional domain; (iii) variant affecting canonical splicing sites (i.e., ±1 or ±2 positions); (iv) variant absent in allele frequency population databases; (v) variant reported in allele frequency population databases but with a minor allele frequency significantly lower than the known disease frequency in the general population; (vi) variant predicted as pathogenic/deleterious in ClinVar and/or LOVD; (vii) missense variant predicted as pathogenic/deleterious in most (≥75%) of the selected *in silico* predictors; (viii) variant co-segregating in two or more affected relatives; ix) the predicted pathogenic effect has been confirmed by an appropriate functional study/studies.

2.2. Sanger Sequencing.

The variants identified by NGS were confirmed by Sanger sequencing. Primer sequences are reported in Table 1. The amplified products were subsequently purified by using ExoSAP-IT PCR Product Cleanup Reagent (Thermo Fisher Scientific, Waltham, MA, USA) and sequenced by using BigDye Terminator v1.1 sequencing kit (Thermo Fisher Scientific, Waltham, MA, USA). The fragments obtained were purified using DyeEx plates (Qiagen, Hilden, Germany) and resolved on ABI Prism 3130 Genetic Analyzer (Thermo Fisher Scientific, Waltham, MA, USA). Sequences were analyzed using the Sequencer software (Gene Codes, Ann Arbor, MI, USA). The identified variant was resequenced in independent experiments.

Table 1. Primers used in this study.

Primers	Sequence	Tm (°C)	Size (bp)
TNXB-LongPCR-F	GTCTCTGCCCTGGGAATGA	60	4900
TNXB-LongPCR-R	TGTAAACACAGTGCTGCGA		
TNXB frag1 For	GGCCAAGCCTGGAAGATAAA	60	662
TNXB frag1 Rev	GATTGGAGACAGAAGCACAC		
TNXB frag2 For	CCAGGGAGAGAGGATGGAT	60	671
TNXB frag2 Rev	GTCCCCAGGAATGGAAGT		
TNXB frag3 For	GACCTAGTGCCTCAGCCA	60	733
TNXB frag3 Rev	GGCTCTCTCTACTCCGTG		
TNXB frag4 For	ATGGGTGGGAGTTGAGAG	60	727
TNXB frag4 Rev	TGGAAGCTGAGCAGGTAG		
TNXB frag5 For	TCTCCTCTTCCTGCTTTCCC	60	643
TNXB frag5 Rev	CCCCATCAGTCTCCATGTC		
TNXB frag6 For	CAGGACCAGCACCATCTT	60	741
TNXB frag6 Rev	TTGAGGTTGGCGTAGTGG		
TNXB frag7 For	GCTGTCTCCTACCGAGGG	60	621
TNXB frag7 Rev	GCAGAGAAGGCTTCCTCC		
TNXB_EX3Fw	GGTTGCAGTCAGAGGGGGCG	58	300
TNXB_EX3Rv	GCCCGCGACCTCTACAGTCG		
TNXB_EX35Fw	GGGAGCCTCAGAGTGTGC	58	480
TNXB_EX35Rv	TCAGCCCCTGGAGTTTCTG		

2.3. Analysis of the TNXA/TNXB Homology Region

For the analysis of the *TNXA/TNXB* homology region, the whole genomic sequence of *TNXB*, encompassing exons 32–44 was amplified by employing a Long-PCR reaction using the primers reported in Table 1 and the protocol previously reported [7]. The PCR reaction was performed in a total 50 uL volume reaction containing 5 uL Buffer 1 (10 ×), dNTP (0.625 mM final concentration), primers (10 pmol each), 3U Expand Long Template Enzyme mix (Roche, San Francisco, CA, USA). Cycling conditions were as follows: initial denaturation at 95 °C, 3 min, followed by 30 cycles of 95 °C, 30 s, 62 °C, 30 s, 72 °C, 11 min, final elongation at 72 °C, 7 min. The 4987 bp PCR product was checked on ethidium bromide (EtBr) stained 1% agarose gel and then used as template for seven nested PCR amplifications with the primers listed in Table 1. Reactions were performed in a 25 uL volume reaction containing 2.5 uL Buffer (10 ×), dNTP (0,25 mM final concentration), primers (10 pmol each), 1.25 U AmpliTaq Gold DNA Polymerase (Thermo Fisher Scientific, Waltham, MA, USA). Cycling conditions were the following: initial denaturation at 95 °C, 10 min, followed by 30 cycles of 95 °C, 30 s, 60 °C, 30 s, 72 °C, 30 s, final elongation at 72 °C, 7 min. The overlapping Nested PCR products were then checked on EtBr stained 2% agarose gel, purified with ExoSap-IT kit (Thermo Fisher Scientific, Waltham, MA, USA) and sequenced with a ready reaction kit (BigDye Terminator v1.1 Cycle kit, Thermo Fisher Scientific, Waltham, MA, USA).

2.4. Multiplex Ligation-Dependent Probe Amplification (MLPA) and Quantitative PCR (qPCR) Analysis

MLPA was carried out using the commercial kit (SALSA MLPA KIT P155-D2 Ehlers-Danlos syndrome III & IV, MRC Holland, Amsterdam, The Netherlands). The kit includes 17 probes for *TNXB* with amplification products between 130 and 490 nucleotides. This kit also comprises two probes mapping within the region upstream of *TNXB*, located in the *ATF6B* and *BAK1* genes, and two additional probes mapping within *CYP21A2*. Complete probe sequences and the identity of the genes detected by the reference probes is available online. Hybridization, ligation, and PCR amplification were performed according to the manufacturer's instructions. DNAs from three healthy individuals were included as controls. Coffalyser. Net software (MRC Holland, Amsterdam, The Netherlands) was used for data analysis. Detected deletion was confirmed by qPCR. Primers designed for the amplification of DNA fragments by qPCR, including *TNXB* exon 4 to 8 probes, were listed in Table 1.

The qPCR was performed using Power SYBR Green PCR Master Mix (Thermo Fisher Scientific, Waltham, MA, USA) on ABI 7900HT real time PCR system (Thermo Fisher Scientific, Waltham, MA, USA). Control DNA fragments were located on different chromosomes. Samples were run in triplicate using standard conditions.

2.5. Chromosome Microarrays Analysis (CMA)

CMA of the individual 1 was performed using the CytoScan™ XON array (Thermo Fisher Scientific, Waltham, MA, USA). This array contains 6.85 million empirically selected probes for whole-genome coverage including 6.5 million copy number probes and 300,000 SNP probes for LOH analysis, duo-trio assessment. The sensitivity of the platform is 95% for the detection of exon-level copy number variations (CNVs). The CytoScan™ XON assay was performed according to the manufacturer's protocol, starting with 100 ng of patient DNA. Data analysis was performed using the Chromosome Analysis Suite software version 4.2 (Thermo Fisher Scientific, Waltham, MA, USA). A CNV was validated if at least 25 contiguous probes showed an abnormal log2 ratio.

2.6. Conservation of the Variant.

Evolutionary conservation of p.Gln2760 residue of tenascin-X was investigated with protein sequence alignment generated by Clustal Omega and compared with data provided by UC Santa Cruz Genome Browser

2.7. Total RNA Analysis.

Total RNA was extracted using RNeasy Mini Kit (Qiagen, Hilden, Germany), treated with DNase-RNase free (Qiagen, Hilden, Germany), quantified by Nanodrop (Thermo Fisher Scientific, Waltham, MA, USA) and reverse-transcribed using QuantiTect Reverse Transcription Kit (Qiagen, Hilden, Germany) according to the manufacturer's instructions. PCR amplification and Sanger sequencing were carried out with primers listed in Table 1.

2.8. Variant Designation.

Nucleotide variant nomenclature follows the format indicated in the Human Genome Variation Society (HGVS) recommendations. DNA variant numbering system refers to cDNA. GenBank Accession Number NM_019105.6 was used as reference sequences for *TNXB*. Nucleotide numbering uses +1 as the A of the ATG translation initiation codon in the reference sequence, with the initiation codon as codon 1.

3. Results

3.1. Case Report: Individual 1

This is a 25-year-old woman, first child of healthy and unrelated parents. Her younger brother was described as unaffected. She was born at term, from an uneventful pregnancy and uncomplicated vaginal delivery. Birth parameters were normal. Psychomotor development and education were within normal limits. She recalled congenital contortionism (i.e., positive five-point questionnaire). At 3 years of age, the mother noted trigger fingers of multiple digits in both hands, and this was treated surgically at 16 years of age with complete resolution. She also requested ortodontic therapy for dental crowding and high-arched palate. No visual nor auditory troubles were registered. The patient suffered from constipation since infancy. At 14 years of age, she received the diagnosis of rectal prolapse. She always suffered from easy bruising and tendency to arterial hypotension. More recently, the patient requested rheumatological consultation for proneness to soft-tissue injuries and easy fatigability, although she did never complain of dislocations and fractures. According to the patient's past clinical history, the rheumatologist requested medical genetics consultation for a suspicion of hereditary connective tissue disorder. At examination, she displayed normal upper limb span/height ratio, bilaterally positive wrist sign, bilaterally negative thumb sign, Beighton score (7/9) with genua recurvata, marked hypermobility of the fingers and toes, flatfeet, brachydactyly (due to shortened metatarsals—brachydactyly type E), of the 2nd and 3rd toes bilaterally, brachydactyly type D of the hands (i.e. shortened and broad distal phalanx of the thumbs), soft, doughy and significantly hyperextensible skin, absence of distrophic scars, palmoplantar hyperlinearity, mild palpebral ptosis and bilateral euryblepharon. Of relevance, no other close family members had brachydactyly. Heart ultrasound showed minimal insufficiency of the mitral valve. Bone densitometry showed reduced bone mass at the femoral head (T-score-1.7). Hand X-rays showed "swan neck" deformity of the interphalangeal distal joint of 1st, 2nd and 3rd left fingers.

3.2. Case Report: Individual 2

This is a unique child of unaffected and unrelated parents. Pregnancy and delivery were uneventful, and psychomotor development and education were normal. She came at our medical genetics outpatient service at 26 years of age due to a long history of polyarticular intense pain (visual analogic scale: 7/10) and (sub)luxations of the shoulders, fingers and temporomandibular joints. Additional musculoskeletal and neurological symptoms included moderate pain (visual analogic scale 4-5/10) of the spine, hands and feet, as well as recurrent myalgias, occipital headache, occasional migraine with aura, post-exertional malaise and chronic fatigue. She also reported easy bruising, an episode of spontaneous rupture of the right brachial vein, multiple episodes of ruptures of cystic lesions of the elbows with leakage of a yellowish and filamenotous substance (presumably, molluscoid pseudotumors), eye dryness, recurrent subconjunctival hemorrhages, gastroesophageal reflux with cardia incontinence, esophageal erosions and severe constipation with coprostasis. At examination, at 26 years of age, the patient showed normal anthropometry, Beighton score 7/9, marked hypermobility of the digits, temporomandibular joints and shoulders, bilateral hallux valgus, genua valga, lumbar hyperlordosis, mild thoracic scoliosis, reduced muscle tone, painful movements, soft, doughy and hyperextensible skin (Figure 1a), normal scar formation, residual elbow scars from recurrent ruptures of (likely) mulluscoid pseudotumors (not visible at the time of examination; Figure 1b), a small subcutaneous spheroid of the pretibial region on the left, multiple cutaneous hematomas, piezogenic papules (Figure 1c), palmoplantar hyperlinearity, bluish sclerae, absence of the lingual frenulum and acrocianosis. Heart ultrasound was normal. Bone densitometry showed mildly reduced bone mass at the femoral neck. We were also able to carry out physical examination of individual 2's parents and both had positive Beighton score according to age and sex (i.e., generalized joint hypermobility). The father also showed mildly soft and hyperexensible skin, although both did not report any significantly related complaints.

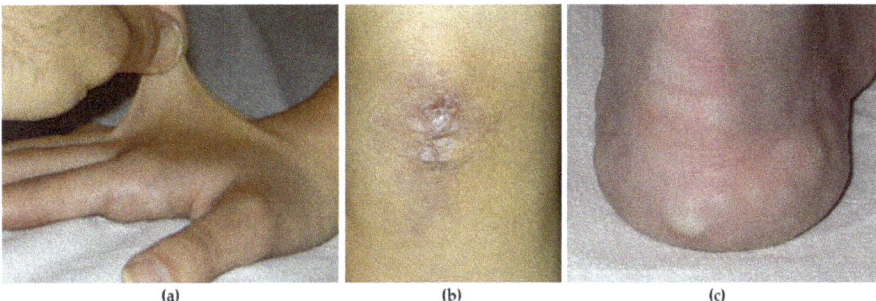

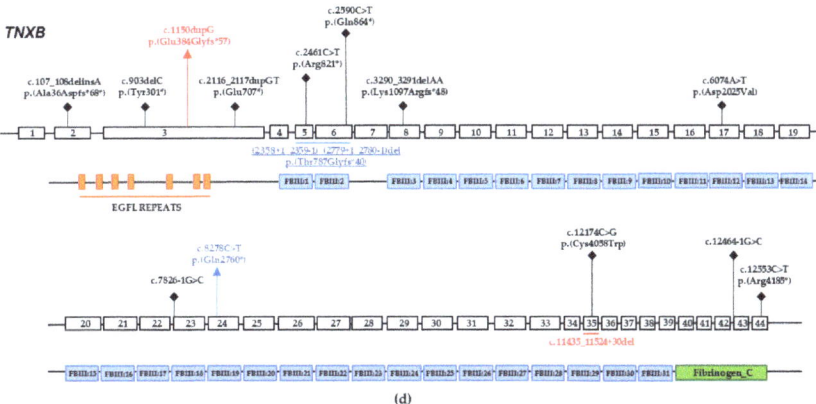

Figure 1. Selected clinical features of individual 2, and schematic diagram showing the genomic structure of *TNXB* and the secondary structure of tenascin-X. (**a**) Markedly hyperextensible skin of the dorsum of the hands. (**b**) Residual scar from recurrent molluscoid pseudotumors of the elbow. (**c**) Piezogenic papules of the heel. (**d**) Coding regions are highlighted with white boxes and introns with black horizontal lines. Tenascin-X is characterized structurally from N-terminus to C-terminus by: (i) an N-terminus with a series of repeats that resemble epidermal growth factor (EGFL repeats); (ii) a stretch of fibronectin type III repeats (FBIII1-31); and (iii) a large C-terminal domain structurally related to fibrinogen (Fibrinogen C). Previously identified variants associated with *TNXB*-clEDS are represented in black. Variants identified in the individuals 1 and 2 are represented in blue and red, respectively.

3.3. Molecular Findings: Individual 1

NGS analysis performed on DNA from individual 1 identified the novel heterozygous c.8278C> T variant located in the exon 24 of *TNXB* (Figure 1d), which is predicted to incorporate a premature stop codon [p.(Gln2760 *)]. No other candidate variants were found in the remaining genes. The c.8278C > T p.(Gln2760 *) variant was not reported in major databases, including dbSNP, ExAC, 1000 Genomes, and GnomAD. This suggests that the variant represents a rare event and was interpreted as likely pathogenic according to the ACMGG/AMP criteria (i.e., a variant absent in population databases and predicted to generate a null allele in a gene previously known as disease-causing with this molecular mechanism). The result was confirmed by direct Sanger sequencing of proband's DNA (Figure 2a). Protein sequences alignment of the homologous regions including the Gln2760 residue of human *TNXB* was generated by using the Clustal Omega tool and showed that the affected residue was evolutionarily conserved (Figure 2b). The Gln2760 residue is located in the 19th fibronectin domain of tenascin-X (Figure 1d) and, thus, the truncated protein presumably loses the last multiple 19–31 fibronectin domains as well as the fibrinogen C motif.

As *TNXB*-clEDS is caused by a complete lack of tenascin-X due to biallelic inactivating variants in *TNXB*, in order to detect the potential presence of a second variant in *TNXB*, we simultaneously performed the long PCR/Sanger sequencing analysis of the *TNXA/TNXB* homologous region and MLPA analysis. The long PCR/Sanger sequencing analysis did not reveal any variant in exons 32–44. On the contrary, MLPA analysis detected a *TNXB* intragenic deletion which includes entirely the exon 5 (Figure 2c). To narrow the proximal deletion breakpoints within the region encompassing the exon 5, qPCR analysis was employed on DNA extracted from patient's, unaffected parents, and control individuals blood. This approach detected a *TNXB* deletion which include both the whole exons 5 and 6 (Figure 2d).

To better molecularly refine the extension of the deletion, we performed a chromosome microarrays analysis using the CytoScan™ XON array. CMA confirmed an interstitial heterozygous microdeletion at chromosome 6p21.33, covered by 62 array probes and spanning 5 Kb, which encompasses the exons 5 and 6 and flanking intronic regions of *TNXB* (Figure 2e). The molecular karyotype of the patient, accordingly with the International System for Human Cytogenomic Nomenclature 2016 was arr [GRCh37] 6p21.33 (32056115_32061375) x1.

Next, we characterized the deletion at the transcriptional level by direct DNA sequencing of in vitro amplified cDNA product generated from total RNA extracted from patients' peripheral blood leucocytes (Figure 2f). We showed that the variant c.(2358 + 1_2359 − 1)_(2779 + 1_2780 − 1)del generates a frameshift with the insertion of a premature stop codon in exon 7 [p.(Thr787Glyfs*40)] (Figure 1d). In light of its absence in population databases, the predicted generation of a null allele and the subsequent demonstration by mRNA study, this variant was interpreted as pathogenic according to the ACMGG/AMP rules. Segregation analysis in both unaffected parents was performed by Sanger sequencing and MLPA/qPCR analysis. We detected the c.8278C > T and (2358 + 1_2359 − 1)_(2779 + 1_2780 − 1)del variants in the proband's father and mother, respectively (Figure 2a,d,f). Both *TNXB* c.8278C > T p.(Gln2760 *) and (2358 + 1_2359 − 1)_(2779 + 1_2780 − 1)del p.(Thr787Glyfs*40) variants have been submitted to the LOVD (Leiden Open Variation Database, https://databases.lovd.nl/shared/variants/0000598484, individual ID # 00266303 https://databases.lovd.nl/shared/variants/0000598485, individual ID # 00266303, respectively).

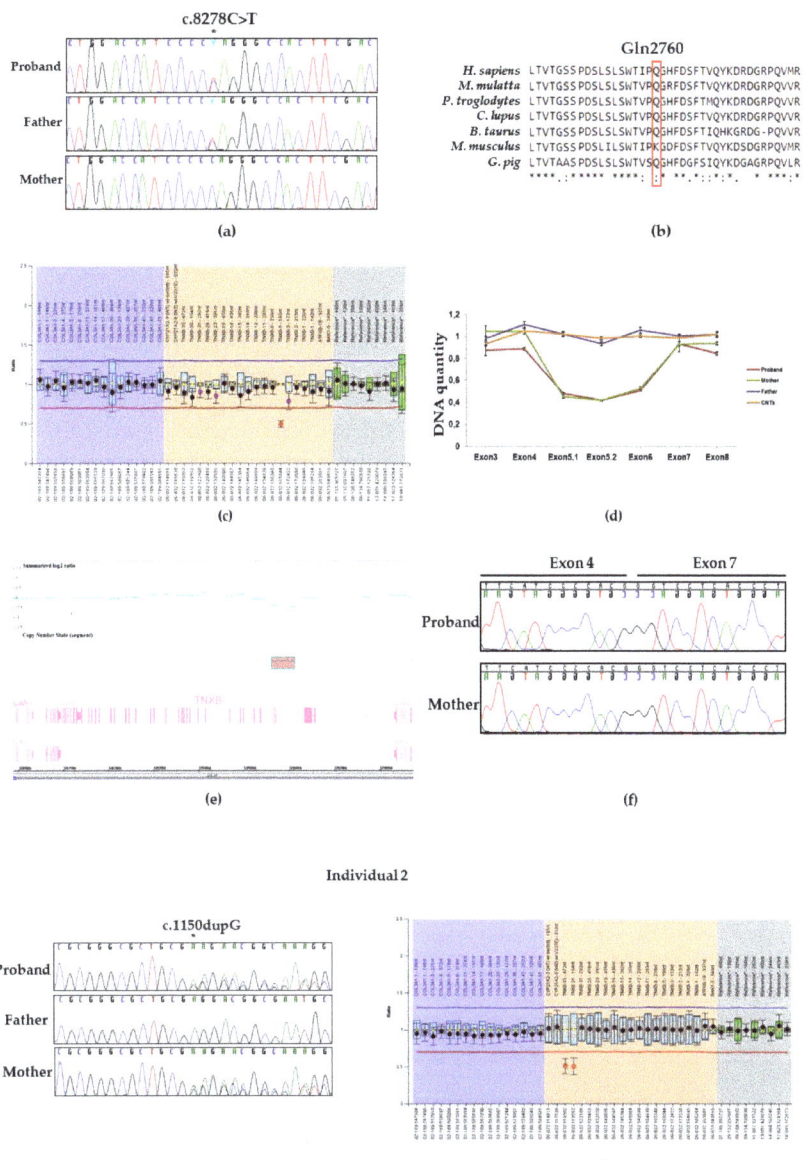

Figure 2. Molecular findings of individuals 1 and 2. (**a**) Electropherograms showing DNA sequencing analysis of PCR product amplified with primers targeting exon 8 of *TNXB*. Nucleotide sequences are provided. The position of the identified variant is indicated with an asterisk. (**b**) Protein sequence alignment of *TNXB* generated by Clustal Omega showed that the affected Gln 2760 residue of tenascin-X is evolutionary conserved. (**c**) Bar chart generated by Coffalyser.net- MLPA analysis software. MLPA was performed on DNA from the individual 1, her unaffected parents and controls (data not showed). A probe ratio of 1 indicates a normal DNA copy number; a probe ratio of 0.5 indicates a heterozygous deletion. MLPA analysis reveals a deletion of exon 5 of *TNXB* in individual 1. (**d**) Profiles of qPCR assay performed to map the deletion breakpoints within the region encompassing the exons 3 to 8 of *TNXB*. Relative DNA quantity of each exon was determined for the patient (red), her asymptomatic

mother and father, (green and purple, respectively), and a pool of DNA controls (CNTs, orange). (e) Results of chromosomal microarray analysis in the Individual 1. Intensity data (Summarized log 2 ratio value) of each probe is drawn along chromosome 6 from 32,000 to 32,080 kb (USCS Genome Browser build February 2009, hg19). The red box indicates the interstitial microdeletion (62 probes with decreased signal) identified, encompassing the exons 5 and 6 of the TNXB gene (lower panel). (f) Electropherograms showing cDNA Sanger sequencing of a transcript region of *TNXB* amplified with primers targeting exon 3 to 8 of Individual 1 and her mother. (g) Sanger sequence of a PCR product amplified with primers targeting exon 3 of *TNXB* of individual 2 and her unaffected parents. The position of the identified variant is indicated with an asterisk. (h) Bar chart generated by Coffalyser.ne-MLPA analysis performed on DNA from the individual 2, her unaffected father and controls (data not showed). MLPA analysis reveals the common partial deletion of exon 35 of *TNXB* in individual 2.

3.4. Molecular Findings: Individual 2

NGS platform targeted for EDS genes revealed that individual 2 carries out a single base deletion c.1150dupG located in exon 3 of *TNXB* (Figure 1d). This heterozygous variant was predicted to generate a premature stop codon at residue 441 [p.(Glu384Glyfs*57)]. No other candidate variants were found in the remaining genes. The c.1150dupG variant was not reported in major databases. Therefore, the variant was interpreted as likely pathogenic according to the ACMGG/AMP guidelines. This result was confirmed by direct Sanger sequencing of proband's DNA. The novel variant was also detected in the proband's mother while it was absent in the father (Figure 2g). MLPA analysis (Figure 2h) detected the recurrent pseudogene-derived 120 bps deletion including the exon 35, previously described by Schalkwijk et al. as the likely result of a common *TNXA/TNXB* fusion gene (CAHX-CH1). This variant was inherited from the healthy carrier father. Both TNXB c.1150dupG p.(Glu384Glyfs*57) and c.11435_11524 + 30del variants have been submitted to the LOVD (https://databases.lovd.nl/shared/variants/0000598486, individual ID # 00266304, https://databases.lovd.nl/shared/variants/0000598487, individual ID #00266304, respectively).

4. Discussion

In this work, we described the first two Italian patients with *TNXB*-clEDS, confirming a wider distribution of this rare EDS type in Europe, and the efficacy and reproducibility of the diagnostic approach published by Demirdas et al. [7]. We also identified three novel disease-alleles in *TNXB*, which expand the known mutational spectrum of *TNXB* associated with clEDS (Figure 1d).

These two adults manifest the previously defined phenotypic spectrum of *TNXB*-clEDS. Scarring was apparently normal in our patients, which is in line with the lack of atrophic/dystrophic scarring as a distinguishing feature, together with recessive inheritance, of *TNXB*-clEDS from classical EDS. Intriguingly, individual 2 reported a history of recurrent ruptures of elbow cystic lesions resembling molluscoid pseudotumors, which are additional cutaneous features considered highly suggestive for classical EDS. This expands the cutaneous similarities between *TNXB*-clEDS and classical EDS; a fact that complicates the differential diagnosis of these disorders on clinical groups and reinforces the opportunity to consider *TNXB* molecular testing in all individuals with a clinical diagnosis of classical EDS resulted negative to *COL5A1*, *COL5A2*, and *COL1A1* (recurrent variants) analysis. Individual 2 also testifies for a possible vascular involvement in *TNXB*-clEDS. In fact, this patient reported recurrent subconjunctival hemorrhages, a feature previously annotated in multiples subjects by Demirdas et al. [7], as well as spontaneous rupture of the brachial vein. The latter is an apparently novel feature of *TNXB*-clEDS and could indicate, if confirmed by other observations, a more severe vascular involvement in this condition.

Demirdas et al. [7] pointed out a peculiar appendicular phenotype of *TNXB*-clEDS featured by foot brachydactyly and small joint (apparently acquired) contractures. Our individual 1 supports this hypothesis, as she showed shortened metatarsals and a history of multiple acquired finger contractures with residual swan neck deformities of the left fingers. Furthermore, constipation and evacuation troubles represented major complaints in both individuals. While these features are not rare within the EDS community of syndromes and are highly represented in adults with hypermobile EDS and hypermobility spectrum disorders [12], this observation in *TNXB*-clEDS confirms the opportunity to better investigate the long-range manifestations of these disorders in order to improve quality of life of individuals with EDS.

To date, a total of 15 different *TNXB* deleterious variants, including the three novel reported in this paper were described. These variants are frameshift (7/15), stop codon (4/15), or splicing (2/15) and lead to the insertion of a premature stop codon with a presumed loss of expression of the protein. Two out of these 15 variants (2/15) are missense which have detrimental effects on the proper protein folding and stability [10,13]. Among the 15 variants, 11 are identified in single patients/families. A 2 bp deletion (c.3290_3291del), a 30Kb deletion generating a *TNXB/TNXA* fusion gene, and a pseudogene-derived missense variant (c.12174C > G) were found in more than one patient (Figure 1d). Nevertheless, the current nosology of EDS and related disorders clearly states that only "*null* alleles" in *TNXB* can be considered causative of *TNXB*-clEDS [1,6]. Therefore, missense *TNXB* variants should be considered supportive of the diagnosis in a clinical setting only if they appear convincing for haploinsufficiency.

In this study, individual 1 carries two novel *TNXB* variants, c.8278C>T located in the exon 24 and (2358 + 1_2359 − 1)_(2779 + 1_2780 − 1)del which results in the non in frame deletion of whole exons 5 and 6. Both variants are predicted to generate a premature stop codon. Individual 2 is a compound heterozygote for c.1150dupG and c.11435_11524 + 30del variants. The c.1150dupG is a novel variant located in exon 3 of *TNXB* and is predicted to create a premature stop codon. The c.11435_11524 + 30del variant, which abolishes part of exon 35 and intron 35 of *TNXB*, has been previously described and *TNXA/TNXB* chimeric recombination type 1 [5,7,10]. This deleted region represents the only large *TNXB*-specific sequence in the *TNXA*-homolog region of *TNXB*. The *TNXA*-derived variation is a molecular event which often takes place between a functional gene and a pseudogene. Although this *TNXA/TNXB* fusion gene has been previously characterized, its molecular effect is not yet known. However, tenascin-X serum measurement in affected individuals by previous studies indicate that this variant likely results in a *null* allele [5,9]. In this work, we were not able to carry out a serum dosage of tenascin-X in our patients. Nevertheless, we are confident that the molecular features of the identified variants are convincing for the generation of a not functional allele.

Due to the complex nature of the genomic region spanning around *TNXB*, the underdiagnosis of *TNXB*-clEDS in the routine diagnostic activities of most laboratories is a likely scenario. In fact, molecular testing of *TNXB* is challenging due to the presence of the pseudogene *TNXA*, which is more than 97% identical to *TNXB* at its 3' end (exons 32–44). With the only exception of exon 35, which partially shows a *TNXB* specific sequence (see above), exon and intron sequences in this region are (nearly) identical in both *TNXB* and *TNXA*. Our experience confirms the need of a multistep and multi-technique approach (comprising NGS for the non-homologous region, Sanger sequencing with a long-PCR and nested-PCR system for the *TNXA/TNXB* homologous segment, and quantitative analysis for intragenic and intergenic rearrangement) for an efficient analysis of the entire *TNXB* coding region, with only slight modifications from the methodology proposed by Demirdas et al. [7] Tenascin-X serum concentration measurement in patients with suspected *TNXB*-clEDS is an alternative in the absence of effective molecular diagnostic facilities.

In summary, we reported two additional individuals with *TNXB*-clEDS. Our findings support the previously defined phenotype, which shows similarities with classical EDS but also include some possible distinguishing features and potentially underreported, clinically relevant manifestations. We also expanded the mutational spectrum of *TNXB* and highlighted the need of a high level of specialty for an efficacious *TNXB* molecular screening in a clinical setting.

Author Contributions: M.C. (Marco Castori), L.M., and V.G. designed the study and wrote the manuscript. L.M., V.G., E.A., and A.N. performed NGS analysis and studies on cDNA. B.A., E.A., V.M.S., and A.N. performed MLPA and qPCR analysis. O.P. and M.C. (Massimo Carella) performed CMA analysis. M.C. (Marco Castori), V.G., and L.M. interpreted molecular data. T.M. carried out the bioinformatics analysis. M.C. (Marco Castori) provided clinical evaluation of the patients. All authors contributed to the writing and reviewing the manuscript and approved its final version.

Funding: This work was supported by the Ricerca Corrente 2018–2019 Program from the Italian Ministry of Health.

Acknowledgments: The authors thank the family for their kind availability in sharing the findings within the scientific community.

Conflicts of Interest: The authors declare no conflict of interest. The funders had no role in study design, data collection and analysis, decision to publish, or preparation of the manuscript.

References

1. Malfait, F.; Francomano, C.; Byers, P.; Belmont, J.; Berglund, B.; Black, J.; Bloom, L.; Bowen, J.M.; Brady, A.F.; Burrows, N.P.; et al. The 2017 international classification of the Ehlers-Danlos syndromes. *Am. J. Med. Genet. C Semin. Med. Genet.* **2017**, *175*, 8–26. [CrossRef] [PubMed]
2. Ritelli, M.; Cinquina, V.; Venturini, M.; Pezzaioli, L.; Formenti, A.M.; Chiarelli, N.; Colombi, M. Expanding the Clinical and Mutational Spectrum of Recessive AEBP1-Related Classical-Like Ehlers-Danlos Syndrome. *Genes (Basel)* **2019**, *10*, 135. [CrossRef] [PubMed]
3. Syx, D.; De Wandele, I.; Symoens, S.; De Rycke, R.; Hougrand, O.; Voermans, N.; De Paepe, A.; Malfait, F. Bi-allelic AEBP1 mutations in two patients with Ehlers-Danlos syndrome. *Hum. Mol. Genet.* **2019**, *28*, 1853–1864. [CrossRef] [PubMed]
4. Burch, G.H.; Gong, Y.; Liu, W.; Dettman, R.W.; Curry, C.J.; Smith, L.; Miller, W.L.; Bristow, J. Tenascin-X deficiency is associated with Ehlers-Danlos syndrome. *Nat. Genet.* **1997**, *17*, 104–108. [CrossRef] [PubMed]
5. Schalkwijk, J.; Zweers, M.C.; Steijlen, P.M.; Dean, W.B.; Taylor, G.; van Vlijmen, I.M.; van Haren, B.; Miller, W.L.; Bristow, J. A recessive form of the Ehlers-Danlos syndrome caused by tenascin-X deficiency. *N. Engl. J. Med.* **2001**, *345*, 1167–1175. [CrossRef] [PubMed]
6. Brady, A.F.; Demirdas, S.; Fournel-Gigleux, S.; Ghali, N.; Giunta, C.; Kapferer-Seebacher, I.; Kosho, T.; Mendoza-Londono, R.; Pope, M.F.; Rohrbach, M.; et al. The Ehlers-Danlos syndromes, rare types. *Am. J. Med. Genet. C Semin. Med. Genet.* **2017**, *175*, 70–115. [CrossRef] [PubMed]
7. Demirdas, S.; Dulfer, E.; Robert, L.; Kempers, M.; van Beek, D.; Micha, D.; van Engelen, B.G.; Hamel, B.; Schalkwijk, J.; Loeys, B.; et al. Recognizing the tenascin-X deficient type of Ehlers-Danlos syndrome: A cross-sectional study in 17 patients. *Clin. Genet.* **2017**, *91*, 411–425. [CrossRef] [PubMed]
8. Chen, W.; Perritt, A.F.; Morissette, R.; Dreiling, J.L.; Bohn, M.F.; Mallappa, A.; Xu, Z.; Quezado, M.; Merke, D.P. Ehlers-Danlos Syndrome Caused by Biallelic TNXB Variants in Patients with Congenital Adrenal Hyperplasia. *Hum. Mutat.* **2016**, *37*, 893–897. [CrossRef] [PubMed]
9. Merke, D.P.; Chen, W.; Morissette, R.; Xu, Z.; Van Ryzin, C.; Sachdev, V.; Hannoush, H.; Shanbhag, S.M.; Acevedo, A.T.; Nishitani, M.; et al. Tenascin-X haploinsufficiency associated with Ehlers-Danlos syndrome in patients with congenital adrenal hyperplasia. *J. Clin. Endocrinol. Metab.* **2013**, *98*, E379–E387. [CrossRef] [PubMed]
10. Morissette, R.; Chen, W.; Perritt, A.F.; Dreiling, J.L.; Arai, A.E.; Sachdev, V.; Hannoush, H.; Mallappa, A.; Xu, Z.; McDonnell, N.B.; et al. Broadening the Spectrum of Ehlers Danlos Syndrome in Patients With Congenital Adrenal Hyperplasia. *J. Clin. Endocrinol. Metab.* **2015**, *100*, E1143–E1152. [CrossRef] [PubMed]
11. Richards, S.; Aziz, N.; Bale, S.; Bick, D.; Das, S.; Gastier-Foster, J.; Grody, W.W.; Hegde, M.; Lyon, E.; Spector, E.; et al. Standards and guidelines for the interpretation of sequence variants: A joint consensus recommendation of the American College of Medical Genetics and Genomics and the Association for Molecular Pathology. *Genet. Med.* **2015**, *17*, 405–424. [CrossRef] [PubMed]
12. Copetti, M.; Morlino, S.; Colombi, M.; Grammatico, P.; Fontana, A.; Castori, M. Severity classes in adults with hypermobile Ehlers-Danlos syndrome/hypermobility spectrum disorders: A pilot study of 105 Italian patients. *Rheumatology (Oxford)* **2019**, *58*, 1722–1730. [CrossRef] [PubMed]
13. Kaufman, C.S.; Butler, M.G. Mutation in TNXB gene causes moderate to severe Ehlers-Danlos syndrome. *World J. Med. Genet.* **2016**, *6*, 17–21. [CrossRef] [PubMed]

© 2019 by the authors. Licensee MDPI, Basel, Switzerland. This article is an open access article distributed under the terms and conditions of the Creative Commons Attribution (CC BY) license (http://creativecommons.org/licenses/by/4.0/).

Case Report

Absence of Collagen Flowers on Electron Microscopy and Identification of (Likely) Pathogenic *COL5A1* Variants in Two Patients

Chloe Angwin [1], Angela F. Brady [1], Marina Colombi [2], David J. P. Ferguson [3,4], Rebecca Pollitt [5], F. Michael Pope [6], Marco Ritelli [2], Sofie Symoens [7], Neeti Ghali [1] and Fleur S. van Dijk [1,*]

1. National Complex Ehlers-Danlos Syndrome Service London, North West Health Care NHS Trust, Harrow HA1 3UJ, UK; chloeangwin@nhs.net (C.A.); angela.brady@nhs.net (A.F.B.); neeti.ghali@nhs.net (N.G.)
2. Division of Biology and Genetics, Department of Molecular and Translational Medicine, University of Brescia, 25123 Brescia, Italy; marina.colombi@unibs.it (M.C.); marco.ritelli@unibs.it (M.R.)
3. Nuffield Department of Clinical Laboratory Sciences, University of Oxford, John Radcliffe Hospital, Oxford OX3 9DU, UK; david.ferguson@ndcls.ox.ac.uk
4. Department Biological & Medical Sciences, Oxford Brookes University, Oxford OX3 0BP, UK
5. Connective Tissue Disorders Service, Sheffield Diagnostic Genetics Service, Sheffield S10 2TQ, UK; rebeccapollitt@nhs.net
6. Department of Dermatology, Chelsea and Westminster Hospital, London, SW10 9NH UK; fmandhpope@googlemail.com
7. Center for Medical Genetics, Ghent University Hospital, B-9000 Ghent, Belgium; Sofie.Symoens@UGent.be
* Correspondence: fleur.dijk@nhs.net; Tel.: +44-208-869-3166; Fax: +44-208-869-3106

Received: 25 August 2019; Accepted: 20 September 2019; Published: 27 September 2019

Abstract: Two probands are reported with pathogenic and likely pathogenic *COL5A1* variants (frameshift and splice site) in whom no collagen flowers have been identified with transmission electron microscopy (TEM). One proband fulfils the clinical criteria for classical Ehlers-Danlos syndrome (cEDS) while the other does not and presents with a vascular complication. This case report highlights the significant intrafamilial variability within the cEDS phenotype and demonstrates that patients with pathogenic *COL5A1* variants can have an absence of collagen flowers on TEM skin biopsy analysis. This has not been previously reported in the literature and is important when evaluating the significance of a TEM result in patients with clinically suspected cEDS and underscores the relevance of molecular analysis.

Keywords: classical Ehlers-Danlos Syndrome; electron microscopy; collagen flowers; *COL5A1*

1. Introduction

The Ehlers-Danlos syndromes (EDS) consist of 13 subtypes with overlapping features including joint hypermobility, skin, and vascular fragility and generalised connective tissue friability [1,2]. Current major criteria for classical EDS (cEDS) are (1) skin hyperextensibility and atrophic scars and (2) joint hypermobility. Minor criteria are easy bruising, soft doughy skin, skin fragility, molluscoid pseudotumors, subcutaneous spheroids, hernia(s), epicanthal folds, complications of joint hypermobility, and an affected first degree relative. The minimal criteria for a diagnosis of cEDS are major criterion 1 plus either major criterion 2 or 3 of the 9 minor criteria [3].

In patients who satisfy the main criteria of cEDS according to the Villefranche criteria [4], the variant detection rate in either *COL5A1* or *COL5A2* is over 90% [5,6]. However, intrafamilial variability in classical EDS has been reported [7]. There are 194 reported unique variants reported in the *COL5A1* gene [8]. These genes encode collagen type V, a fibrillar heterotrimer ($[\alpha1(V)]_2$ $\alpha2(V)$) that is present in a wide variety of tissues but is particularly prevalent in bone, skin and tendon [9]. Collagen type V accounts for approximately 5% of total body collagen and has a role in maintaining the deposition and structure of other more abundant collagens, particularly collagen type I [9]. Rarely, a diagnosis of cEDS is due to dominant variants in *COL1A1* or *COL1A2* [3,10,11].

For many years, skin biopsies for electron microscopy (EM) have been recommended as a first line of investigation to confirm or exclude a diagnosis of cEDS [3,12]. This was due to the occurrence of collagen flowers visible by EM. Collagen flowers in individuals with classical EDS were described by Vogel et al. who reported collagen fibrils with an abnormally large diameter and a highly irregular and lobulated contour interspersed with normal appearing fibrils with a mean diameter larger than that of collagen fibrils in normal skin [13]. A longitudinal section showed that the large atypical fibrils were seen to be poorly integrated filamentous aggregates. Variation in frequency of very large highly irregular fibrils differed per patient but in general constituted approximately 5% [13]. These very large highly-irregular fibrils are often described as longitudinally splayed and loosely packed fibrils, which in cross section produce the collagen flower pattern. Although it is known that collagen flowers can be found in other collagen disorders including osteogenesis imperfecta and Ullrich congenital muscular dystrophy, typical collagen flowers were thought to be invariably present in people with cEDS [3,14,15]. Given the high detection rates of pathogenic variants in cEDS, current recommendations are that electron microscopy based on a skin biopsy should no longer the first line of investigation but could be used to clarify inconclusive molecular results, or to guide further testing if initial molecular testing is negative [3]. Interestingly, it was mentioned that "the absence of typical collagen flowers would go against the diagnosis, as there are no known reports of patients with type V collagen abnormalities without collagen flowers on EM" [3]. Here, we report for the first time two patients, one fulfilling the clinical diagnosis of classical EDS and one not, with (likely) pathogenic variants in *COL5A1* and absence of collagen flowers on EM.

2. Materials and Methods

2.1. Skin Biopsy

Skin biopsies in probands 1 and 2 were taken in the form of a punch biopsy from the medial surface of the arm. P1 was 44 years and P2 was 39 years when the skin biopsy was performed. The sample for transmission electron microscopy (TEM) (3 mm) was preserved in 4% glutaraldehyde in 0.1 M phosphate buffer. This publication does not constitute research and does not require formal Research Ethics approval or Research and Development Approval as stipulated by the UK Policy Framework for Health and Social Care Research and the Health Research authority decision tool.

2.2. DNA Analysis

Sanger sequencing of the *COL5A1* and *COL5A2* genes and MLPA of the *COL5A1* gene was performed in proband 1. Sanger sequencing and MLPA of the *COL3A1* gene was performed in proband 2. Proband 2 was included in a large research project aimed at identifying new pathogenic variants with next-generation sequencing in patients with an EDS phenotype [16]. The sequencing data of patient 2 have been deposited to the EDS and OI databases: https://www.le.ac.uk/genetics/collagen.

3. Clinical Report

3.1. Family 1

Proband 1 is a 47-year-old woman, born at 36 weeks gestation, with a childhood history of clumsiness, tissue fragility, and joint hypermobility (Figure 1A–C). She would bruise easily and recalls some episodes of significant swelling after mild injury, occasionally requiring drainage when this involved the knees. There was no history of dislocation or fracture. As an adult she reports myalgia and arthralgia in the ankles, hips, wrists, and shoulders. Other medical history includes endometrial polyps removed without complication at age 32, varicose veins treated surgically at age 41, hiatus hernia, diverticular disease, a urethral cyst, and hypotension associated with fainting episodes. Neurocardiology investigations found a high tendency towards POTS (postural orthostatic tachycardia syndrome) with no evidence of autonomic failure. Cardiac investigations have shown right bundle branch block and trivial aortic regurgitation, tricuspid regurgitation, and mitral regurgitation.

She fulfilled the criteria for cEDS on physical examination. There were no craniofacial dysmorphic features except for epicanthic folds. She had soft, doughy, and hyperextensible skin—particularly at elbows, neck, and knees—with redundancy at the knees and Achilles tendons. She had scarring over the forehead, wrist, and lower legs. Lower leg scars were widened and atrophic with erythema and hemosiderin deposition. There were thread veins and varicose veins in the lower limbs. Her feet had bilateral hallux valgus deformities with piezogenic papules. Hands had bilateral hitchhikers' thumb and increased palmar markings. She had generalized joint hypermobility (gJHM) with a (Beighton score 5/9).

Family history: The proband's parents, four older brothers and their children are healthy with no reported features of cEDS. One brother was tested for the COL5A1 variant as well. He did not meet the criteria for cEDS and was found not to have the COL5A1 variant.

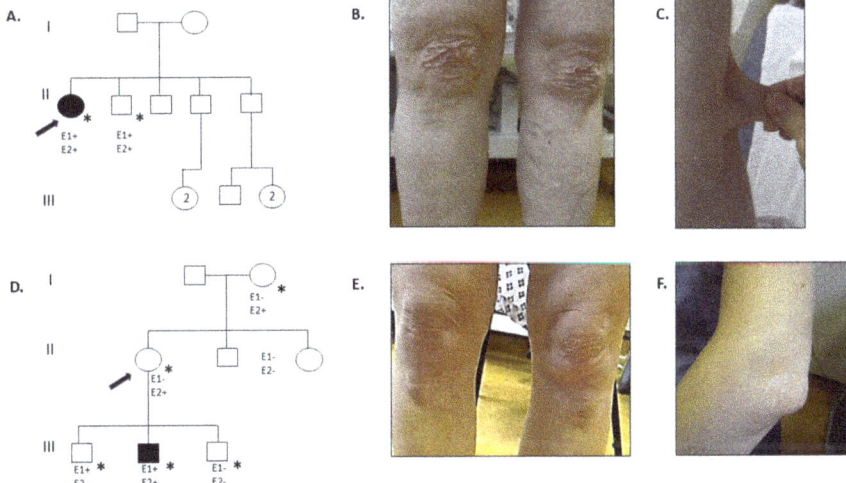

Figure 1. Pedigrees and clinical photographs of probands 1 and 2. (**A**) Family pedigree of proband 1. E1 denotes clinical diagnosis of cEDS, E2 denotes presence of (likely) pathogenic COL5A1 variant, * denotes documented evaluation. Shading denotes having a clinical diagnosis of cEDS and presence of (likely) pathogenic COL5A1 variant. (**B**) Knee area of proband 1 demonstrating redundancy of skin, atrophic and widened scarring and varicose veins. (**C**) Hyperextensible skin at the elbow in proband 1. (**D**) Family pedigree of proband 2. (**E**) Knees area of proband 2 show no atrophic scarring or haemosiderin deposition but minor redundancy of skin. (**F**) Widened, atrophic scar at the elbow.

3.2. Family 2

Proband 2 is a woman who was diagnosed with an acute, spontaneous, left carotid artery dissection presenting with left sided headache, and Horner's syndrome at 37 years old (Figure 2D–F). This was successfully treated conservatively. She had no past medical history and was a non-smoker with normal cholesterol levels. Her past surgical history included bilateral bunion surgery at age 13 and three Caesarean sections, the first for fetal distress and the others as routine following without complication.

On physical examination she did not have craniofacial dysmorphic features [17]. Her skin was mildly hyperextensible over face, neck, and elbows with delayed recoil and slight redundancy of the skin around the knees. The skin was not thin and no bruising was visible. A widened, atrophic scar at a site of trauma over the elbow and thin, well healed post-operative scars from bunion surgery were observed. There was bilateral hallux valgus She did not have gJHM. (Beighton score 3/9) and did not fulfil the current clinical criteria for cEDS. She had a normal echocardiogram.

Family history: The mother of the proband has the *COL5A1* variant and had a healthy childhood but developed hypertension later in life. Findings on clinical examination at the age of 81 were that of skin hyperextensibility with significantly delayed recoiling probably due to her advanced age, abdominal striae, a mild spinal scoliosis, and bilateral hallux valgus. The father of the proband had a history of joint hypermobility and hallux valgus and passed away at the age of 81. He had a healthy childhood and later in life developed hypertension, bowel cancer, skin cancer, and cardiovascular disease, requiring a double coronary artery bypass graft post myocardial infarction.

The proband has two siblings, a sister and a brother. There is no clinical information available regarding her brother. The proband's sister has a cardiovascular history with bicuspid aortic valve requiring replacement, and aortic root dilatation. She does not have the *COL5A1* variant found in her mother and sister.

The proband has three sons, all born by caesarean section. The oldest son, 9 years old, experienced foetal distress and oxygen deprivation in the womb, requiring emergency caesarean section and developed cerebral palsy with significant spasticity. On examination, there was minimal bruising, atrophic scarring over the forehead and occiput, mild hyperextensibility of the skin, and a Beighton score of 2/9 and therefore did not meet the current clinical criteria for cEDS. He had a normal echocardiogram and was found to have the *COL5A1* variant. The middle son, 7 years old, had had a healthy childhood. On examination he had hyperextensible skin, atrophic scarring on the forehead and lower legs, bruising over the shins and a Beighton score of 7/9 and therefore met the current clinical criteria for cEDS. He had a normal echocardiogram and he was found to have the *COL5A1* variant. The youngest son, 4 years old, had had a healthy childhood and did not have any clinical features of cEDS. He therefore did not meet the current clinical criteria for cEDS and was found not to have the *COL5A1* variant.

4. Results

4.1. Skin Biopsy TEM Results

Proband 1: TEM was concluded to be relatively normal but showed some minor deformation of the outline of certain collagen fibrils (Figure 2a). No large collagen flowers were observed. Elastic fibres and fibroblasts appeared normal.

Proband 2: TEM showed collagen clumps within the reticular dermis. Within these, the collagen fibrils themselves were of relatively even diameter and had a symmetrical circular outline (Figure 2b), although longitudinally sectioned fibres showed occasional kinking. No abnormal collagen flowers were seen within the biopsy (Figure 2b). The other structures including elastic fibres and fibroblasts were normal in appearance. The conclusion was one of a normal skin biopsy. After the molecular result, re-examination identified rare, slightly deformed fibrils (Figure 2b) with slight variation in collagen fibril diameter but no further abnormalities and no typical collagen flowers as described previously (Figure 2c). It should be noted that collagen flowers are concentrated within the reticular dermis and rarely observed in the papillary dermis. Therefore, care is required to ensure the correct area of the skin biopsy was examined.

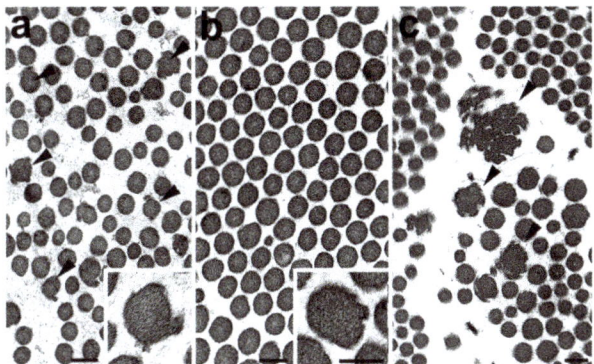

Figure 2. Electron micrographs of the reticular dermis of skin biopsy from proband 1 (**a**), proband 2 (**b**) and a case of classical EDS (**c**). Bars represent 100nm. (**a**) Cross section of collagen fibres from proband 1 showing a number of fibres presenting an irregular outline (arrowheads). Insert: detail of an irregular fibre. (**b**) Cross section through a clump if collagen fibres of proband 2 showing a relatively normal circular outline although very rare slightly irregular fibres were observed (insert). (**c**) Cross section of collagen fibres from a case of classical EDS showing a number of irregular fibres and large collagen flowers (arrowheads).

4.2. DNA Analysis

Proband 1: DNA analysis identified the heterozygous *COL5A1* variant c.4414del, p.(Leu1472Serfs*16). This variant results in a pathogenic frameshift in exon 57 of *COL5A1* (NM_000093.4; NP_000084.3), predicted to result in a premature termination codon, nonsense-mediated RNA decay, and consequent type V collagen haploinsufficiency.

Proband 2: DNA analysis of *COL3A1* in which variants cause vascular EDS, did not identify pathogenic variants. The patient was referred for further sequencing as part of a large research project looking at Mendelian inheritance [16] and the likely pathogenic variant c.4068G>A; p.(Ala1356=) in the *COL5A1* gene was identified. This silent variant has been previously published is predicted to result in abnormal splicing of exon 51 of the *COL5A1* gene [7]. Splice site prediction software (Alamut visual software) indicated that the variant would destroy the splice donor site leading to disruption of normal splicing with skipping of exon 51, leading to the production of a shortened protein [7]. The same variant was identified in the mother of the proband as well as the proband's two older sons, associating with the clinical diagnosis of cEDS. The variant was not identified in the youngest son (Figure 1B).

5. Discussion

We report two patients with a (likely) pathogenic *COL5A1* variant who did not have collagen flowers on electron microscopy although minor deformation of the outline of certain collagen fibrils were observed in proband 1, which are similar to the initial stage of collagen flower formation (cf Figure 2a,c). Large collagen flowers on electron microscopy have previously been used to confirm or exclude a clinical diagnosis of cEDS. One other report describes a 12-year-old boy with marked clinical features of cEDS, without collagen flowers on electron microscopy. Interestingly, investigation showed disorganisation, variation of fibril diameter, and irregular fibril outlines. However, this patient did not have molecular investigations, and the underlying molecular basis is unknown [18].

5.1. Phenotype

Proband 1 clearly fulfilled the criteria for cEDS. Proband 2 did not fulfil the clinical criteria for cEDS. Moreover, she presented with a vascular complication (carotid artery dissection) which is not a common feature in cEDS [19]. In a recent systemic review, 12/110 (11%) patients with *COL5A1/2* variants had vascular complications of which six had arterial dissection, three had arterial aneurysms, and one had intracerebral haemorrhage. It is not certain whether the vascular complication in this proband is linked to the *COL5A1* variant. It has been hypothesized that glycine substitutions near or at the C-terminal end of collagen type V, may predispose to vascular events [20], but at this point there is not enough evidence to support this hypothesis.

There is clear intrafamilial variability across the generations of the family of proband 2, as the mother of the proband was 81 years old with skin hyperextensibility as the only major feature whereas one grandson with the same variant had hyperextensible skin and atrophic scarring, clearly fulfilling the clinical criteria for cEDS. Wide variability between family members with the identical *COL5A1* variant has been previously reported [6,17].

5.2. Transmission Electron Microscopy

In proband 1, who fulfilled the clinical criteria for cEDS, minor deformation of the outline of certain collagen fibrils were observed but no collagen flowers were present. The appearances were similar to early changes seen in fibrils in cEDS (Figure 2a,c). These changes were sufficient for the abnormality to be highlighted in the EM report but not significant enough to conclude that it concerned an abnormal biopsy. In proband 2, no collagen flowers or other abnormalities were observed. Unfortunately, the family members of proband 2, who had pathogenic *COL5A1* variant (mother and sons of proband), did not consent for a skin biopsy to be performed and as such no TEM results are available. While the presence or absence of collagen flowers is reported, it should also be noted that incidence and size/complexity of the collagen flowers can vary markedly between patients but the significance of this to the clinical features and the underlying genetic cause is unknown (Ferguson unpublished observations).

5.3. COL5A1 Variants

The variant identified in proband 1, who fulfils clinical criteria for cEDS, is a pathogenic frameshift variant in exon 57 predicted to result in a premature termination codon leading to haploinsufficiency. The likely pathogenic variant in *COL5A1* c.4068G>A identified in proband 2, who does not fulfil clinical criteria for cEDS, is a splice site variant and expected to result in skipping of exon 51 which is in-frame and as such would lead to a shortened protein and exert a dominant-negative effect. One variant was described by Symoens et al., *COL5A1* c.4068G>T which demonstrated skipping of exon 51 on mRNA analysis; p.(Gly1339_Ala1356del). The proband had skin hyperextensibility, atrophic scarring, joint hypermobility, and no history of vascular events. TEM was not performed [5]. Colombi et al, identified an identical variant. The proband in this case did not fulfil the clinical criteria for cEDS, and reported joint instability, gastrointestinal symptoms, and soft skin with a few small atrophic scars over the knees. She had historical generalised joint hypermobility. Predictive software (Alamut visual software) projected a similar functional outcome as the previous *COL5A1* c.4068G>T variant [7], both reduce the donor site strength to similar extents so a similar protein effect would be expected. According to the ACMG guidelines [21] this variant is classified as likely pathogenic fulfilling *PP3*: Multiple lines of computational evidence support a deleterious effect on the gene or gene product (conservation, evolutionary, splicing impact, etc.); *PM2*: Absent from controls (or at extremely low frequency if recessive) in Exome Sequencing Project, 1000 Genomes Project, or Exome Aggregation Consortium and *PS1*: Same amino acid change as a previously established pathogenic variant regardless of nucleotide change.

Hypotheses for this incomplete cEDS phenotype included protective lifestyle factors, potential for this variant to result in only partial activation of abnormal splicing of exon 51 and the existence of protective variants in other genes, which may counteract the loss of collagen V function during the deposition of structural collagens [7].

Of the 194 unique variants reported in the *COL5A1* gene, 31 have been reported as splice site variants [8]. Most pathogenic *COL5A1* variants (including splice site variants that introduce a premature stop codon) lead to haploinsufficiency for COL5A1 mRNA. This is expected to be the case for the variant reported in P1. Structural variants exerting a dominant-negative effect are a minority and most commonly involve splice site variants resulting in exon skipping which is the case for the variant reported in P2 and variants that result in the substitution for glycine in the triple-helical region [22]. Collagen type V is thought to perform a regulatory function in collagen fibrillogenesis. It has been hypothesized that the final common pathway for all COL5A1 variants is reduced availability of collagen type V, and that clinical phenotypes result from disrupted fibrillogenesis [5,8,23,24].

6. Conclusions

In conclusion, we present two probands with (likely) pathogenic COL5A1 variants (frameshift and splice site) in whom no collagen flowers were identified, although minor deformation of the outline of certain collagen fibrils was observed in proband 1. Proband 1 fulfils the clinical criteria for cEDS but proband 2 does not and presents with a vascular complication. The mother and two sons of proband 2 also have the COL5A1 variant, one of whom fulfils clinical criteria of cEDS. This case report highlights the significant intrafamilial variability within the cEDS phenotype. We demonstrate that patients with (likely) pathogenic COL5A1 variants can have an absence of collagen flowers on biopsy. It is currently unclear whether the absence of collagen flowers can be linked to the (severity of) clinical features and/or the specific genetic cause. It is also uncertain whether the vascular complication in P2 is caused by the COL5A1 variant. Absence of collagen flowers in patients with (likely) pathogenic COL5A1 variants has not been previously reported in the literature but is important when evaluating the significance of a TEM result in patients with suspected cEDS and underscores the relevance of molecular analysis.

Author Contributions: Conceptualization, F.S.v.D., N.G.; Methodology, D.J.P.F.; Formal Analysis, C.A., D.J.P.F., R.P., S.S., F.S.v.D.; Investigation, D.J.P.F., F.M.P., N.G., and F.S.v.D.; Writing—Original Draft Preparation, C.A. and F.S.v.D.; Writing—review and editing, A.F.B., M.C., D.J.P.F., F.P., M.R., and N.G.; Visualization, C.A., D.J.P.F., and F.S.v.D.; Supervision, F.S.v.D.

Funding: This research received no external funding.

Acknowledgments: We would like to thank the families for their kind cooperation.

Conflicts of Interest: All authors state that there is no conflict of interest.

References

1. Malfait, F.; Francomano, C.; Byers, P.; Belmont, J.; Berglund, B.; Black, J.; Bloom, L.; Bowen, J.M.; Brady, A.F.; Burrows, N.P.; et al. The 2017 international classification of the Ehlers–Danlos syndromes. *Am. J. Med. Genet. Part C Semin. Med. Genet.* **2017**, *175*, 8–26. [CrossRef] [PubMed]
2. Malfait, F. Vascular aspects of the Ehlers-Danlos Syndromes. *Matrix Biol.* **2018**. [CrossRef] [PubMed]
3. Bowen, J.M.; Sobey, G.J.; Burrows, N.P.; Colombi, M.; Lavallee, M.E.; Malfait, F.; Francomano, C.A. Ehlers–Danlos syndrome, classical type. *Am. J. Med. Genet. Part C Semin. Med. Genet.* **2017**, *175*, 27–39. [CrossRef] [PubMed]
4. Beighton, P.; De Paepe, A.; Steinmann, B.; Tsipouras, P.; Wenstrup, R.J. Ehlers-danlos syndromes: Revised nosology, Villefranche, 1997. *Am. J. Med. Genet.* **1998**, *77*, 31–37. [CrossRef]
5. Symoens, S.; Syx, D.; Malfait, F.; Callewaert, B.; De Backer, J.; Vanakker, O.; Coucke, P.; De Paepe, A. Comprehensive molecular analysis demonstrates type V collagen mutations in over 90% of patients with classic EDS and allows to refine diagnostic criteria. *Hum. Mutat.* **2012**, *33*, 1485–1493. [CrossRef] [PubMed]
6. Ritelli, M.; Dordoni, C.; Venturini, M.; Chiarelli, N.; Quinzani, S.; Traversa, M.; Zoppi, N.; Vascellaro, A.; Wischmeijer, A.; Manfredini, E.; et al. Clinical and molecular characterization of 40 patients with classic Ehlers-Danlos syndrome: Identification of 18 COL5A1 and 2 COL5A2 novel mutations. *Orphanet J. Rare Dis.* **2013**, *8*, 58. [CrossRef]
7. Colombi, M.; Dordoni, C.; Cinquina, V.; Venturini, M.; Ritelli, M. A classical Ehlers-Danlos syndrome family with incomplete presentation diagnosed by molecular testing. *Eur. J. Med. Genet.* **2018**, *61*, 17–20. [CrossRef]
8. Dalgleish, R. Ehlers-Danlos Syndrome Variant Database: Collagen, Type V, alpha 1 (COL5A1). Available online: https://eds.gene.le.ac.uk/home.php?select_db=COL5A1 (accessed on 23 August 2019).

9. Sun, M.; Chen, S.; Adams, S.M.; Florer, J.B.; Liu, H.; Kao, W.W.Y.; Wenstrup, R.J.; Birk, D.E. Collagen V is a dominant regulator of collagen fibrillogenesis: Dysfunctional regulation of structure and function in a corneal-stroma-specific Col5a1-null mouse model. *J. Cell Sci.* **2011**, *124*, 4096–4105. [CrossRef]
10. Colombi, M.; Dordoni, C.; Venturini, M.; Zanca, A.; Calzavara-Pinton, P.; Ritelli, M. Delineation of Ehlers–Danlos syndrome phenotype due to the c.934C>T, p.(Arg312Cys) mutation in COL1A1: Report on a three-generation family without cardiovascular events, and literature review. *Am. J. Med. Genet. Part A* **2017**, *173*, 524–530. [CrossRef]
11. Malfait, F.; Symoens, S.; Goemans, N.; Gyftodimou, Y.; Holmberg, E.; López-González, V.; Mortier, G.; Nampoothiri, S.; Petersen, M.B.; De Paepe, A. Helical mutations in type i collagen that affect the processing of the amino-propeptide result in an Osteogenesis Imperfecta/Ehlers-Danlos Syndrome overlap syndrome. *Orphanet J. Rare Dis.* **2013**, *8*, 78. [CrossRef]
12. Hausser, I.; Anton-Lamprecht, I. Differential ultrastructural aberrations of collagen fibrils in Ehlers-Danlos syndrome types I-IV as a means of diagnostics and classification. *Hum. Genet.* **1994**, *93*, 394–407. [CrossRef] [PubMed]
13. Vogel, A.; Holbrook, K.A.; Steinmann, B.; Gitzelmann, R.; Byres, P.H. Abnormal collagen fibrilstructure in the Gravis Form (Type I) of Ehlers-Danlos Syndrome. *Lab. Investig.* **1979**, *40*, 201–206. [PubMed]
14. De Almeida, H.L.; Bicca, E.; Rocha, N.M.; de Castro, L.A.S. Light and Electron Microscopy of Classical Ehlers-Danlos Syndrome. *Am. J. Dermatopathol.* **2013**, *24*, 45–54. [CrossRef] [PubMed]
15. Balasubramanian, M.; Wagner, B.E.; Peres, L.C.; Sobey, G.J.; Parker, M.J.; Dalton, A.; Arundel, P.; Bishop, N.J. Ultrastructural and histological findings on examination of skin in osteogenesis imperfecta: A novel study. *Clin. Dysmorphol.* **2015**, *24*, 45–54. [CrossRef] [PubMed]
16. Weerakkody, R.A.; Vandrovcova, J.; Kanonidou, C.; Mueller, M.; Gampawar, P.; Ibrahim, Y.; Black, H.A. Targeted next-generation sequencing makes new molecular diagnoses and expands genotype-phenotype relationship in Ehlers-Danlos syndrome. *Genet. Med.* **2016**, *18*, 119. [CrossRef] [PubMed]
17. Colombi, M.; Dordoni, C.; Venturini, M.; Ciaccio, C.; Morlino, S.; Chiarelli, N.; Zanca, A.; Calzavara-Pinton, P.; Zoppi, N.; Castori, M.; et al. Spectrum of mucocutaneous, ocular and facial features and delineation of novel presentations in 62 classical Ehlers-Danlos syndrome patients. *Clin. Genet.* **2017**, *92*, 624–631. [CrossRef] [PubMed]
18. Bicca, E.D.B.C.; De Almeida, F.B.; Pinto, G.M.; De Castro, L.A.S.; De Almeida, H.L., Jr. Classical Ehlers-Danlos syndrome: Clinical, Histological and ultrastructural aspects. *An. Bras. Dermatol.* **2011**, *86*, 164–167. [CrossRef] [PubMed]
19. D'Hondt, S.; Van Damme, T.; Malfait, F. Vascular phenotypes in nonvascular subtypes of the Ehlers-Danlos syndrome: A systematic review. *Genet. Med.* **2018**, *20*, 562. [CrossRef] [PubMed]
20. Monroe, G.R.; Harakalova, M.; van der Crabben, S.N.; Majoor-Krakauer, D.; Bertoli-Avella, A.M.; Moll, F.L.; Oranen, B.I.; Dooijes, D.; Vink, A.; Knoers, N.V.; et al. Familial Ehlers-Danlos syndrome with lethal arterial events caused by a mutation in COL5A1. *Am. J. Med. Genet. Part A* **2015**, *167*, 1196–1203. [CrossRef]
21. Richards, S.; Aziz, N.; Bale, S.; Bick, D.; Das, S.; Gastier-Foster, J.; Grody, W.W.; Hegde, M.; Lyon, E.; Spector, E.; et al. Standards and guidelines for the interpretation of sequence variants: A joint consensus recommendation of the American College of Medical Genetics and Genomics and the Association for Molecular Pathology. *Genet. Med.* **2015**, *17*, 405. [CrossRef]
22. Malfait, F.; Wenstrup, R.; De Paepe, A. Classic Ehlers-Danlos Syndrome. Available online: https://www.ncbi.nlm.nih.gov/books/NBK1244/ (accessed on 10 February 2017).
23. Malfait, F.; De Paepe, A. Molecular genetics in classic Ehlers-Danlos syndrome. *Am. J. Med. Genet. Semin. Med. Genet.* **2005**, *139*, 17–23. [CrossRef] [PubMed]
24. Burrows, N.P.; Nicholls, A.C.; Richards, A.J.; Luccarini, C.; Harrison, J.B.; Yates, J.R.W.; Pope, F.M. A point mutation in an intronic branch site results in aberrant splicing of COL5A1 and in Ehlers-Danlos syndrome type II in two British families. *Am. J. Hum. Genet.* **1998**, *63*, 390–398. [CrossRef] [PubMed]

© 2019 by the authors. Licensee MDPI, Basel, Switzerland. This article is an open access article distributed under the terms and conditions of the Creative Commons Attribution (CC BY) license (http://creativecommons.org/licenses/by/4.0/).

Article

Arterial Elasticity in Ehlers-Danlos Syndromes

Amanda J. Miller [1,*], Jane R. Schubart [2], Timothy Sheehan [3], Rebecca Bascom [4] and Clair A. Francomano [5]

1. Department of Neural and Behavioral Sciences, Penn State College of Medicine, Hershey, PA 17033, USA
2. Department of Surgery, Penn State College of Medicine, Hershey, PA 17033, USA; jschubart@pennstatehealth.psu.edu
3. Department of Neurology, Medical University of South Carolina, Charleston, SC 29425, USA; sheehant@musc.edu
4. Department of Medicine, Penn State College of Medicine, Hershey, PA 17033, USA; rbascom@pennstatehealth.psu.edu
5. Department of Medical and Molecular Genetics, Indiana University School of Medicine, Indianapolis, IN 46202, USA; cfrancom@iu.edu
* Correspondence: aross1@pennstatehealth.psu.edu; Tel.: +1-717-531-7676

Received: 13 November 2019; Accepted: 2 January 2020; Published: 4 January 2020

Abstract: Ehlers-Danlos Syndromes (EDS) are a group of heritable disorders of connective tissue (HDCT) characterized by joint hypermobility, skin hyperextensibility, and tissue fragility. Orthostatic intolerance (OI) is highly prevalent in EDS however mechanisms linking OI to EDS remain poorly understood. We hypothesize that impaired blood pressure (BP) and heart rate control is associated with lower arterial stiffness in people with EDS. Orthostatic vital signs and arterial stiffness were assessed in a cohort of 60 people with EDS (49 female, 36 ± 16 years). Arterial elasticity was assessed by central and peripheral pulse wave velocity (PWV). Central PWV was lower in people with EDS compared to reference values in healthy subjects. In participants with EDS, central PWV was correlated to supine systolic BP ($r = 0.387$, $p = 0.002$), supine diastolic BP ($r = 0.400$, $p = 0.002$), and seated systolic BP ($r = 0.399$, $p = 0.002$). There were no significant correlations between PWV and changes in BP or heart rate with standing ($p > 0.05$). Between EDS types, there were no differences in supine hemodynamics or PWV measures ($p > 0.05$). These data demonstrate that increased arterial elasticity is associated with lower BP in people with EDS which may contribute to orthostatic symptoms and potentially provides a quantitative clinical measure for future genotype-phenotype investigations.

Keywords: Ehlers-Danlos syndromes; pulse wave velocity; blood pressure; orthostatic intolerance

---

1. Introduction

Ehlers-Danlos syndromes (EDS) are a collection of heritable disorders of connective tissue characterized by joint hypermobility, mild skin hyperextensibility, and tissue fragility [1]. Common symptoms of EDS include joint instability, chronic pain, gastrointestinal issues, and sleep disturbances [2]. Many people with EDS have persistent symptoms of orthostatic intolerance (OI) including lightheadedness, fatigue, nausea, and palpitations [3]. Additionally, the prevalence of EDS is higher in patients with orthostatic intolerance compared to the general population [4]. The association between EDS and autonomic cardiovascular dysfunction is most prevalent in people with hEDS [5–7], but there is also evidence of orthostatic intolerance in classical EDS [8]. The high prevalence of OI in EDS demonstrates a need to understand cardiovascular pathophysiology in all EDS types, as the pathophysiology explaining the high rate of OI in EDS is unknown. The leading theory connecting the two disorders is that generalized connective tissue laxity in EDS increases vascular compliance, leading to insufficient vasoconstriction and venous insufficiency when upright resulting in symptoms of OI [3].

Despite its wide acceptance, there is only data to support this theory in small samples of people with vascular EDS and there is no published evidence to support this theory across other types of EDS.

Pulse wave velocity (PWV) has emerged as the gold standard method for measuring stiffness of the arteries because of its reliability and reproducibility [9]. PWV is a non-invasive technique that involves placing pressure transducers on the skin that can sense the velocity of blood traveling in the arteries, which is a function of the stiffness or elasticity of the arteries. Central PWV, the most widely used and accepted measurement for PWV, measures the stiffness or elasticity of the central cardiovascular system from the carotid to femoral arteries. Using this technique, increased PWV (implying increased arterial stiffness) has been shown to predict future hypertension, coronary heart disease, stroke, adverse cardiovascular events, and mortality [10].

While PWV is well accepted as a measure of arterial stiffness, it has been used far less often to measure arterial elasticity, which is the mathematical inverse of stiffness. More distensible arteries will stretch more as pulse waves travel, resulting in lower (slower) pulse wave velocity. Few studies have sought to identify people, including those with EDS, with suspected increased arterial elasticity and hence decreased PWV. One study evaluated PWV in nine people with comorbid hypermobile EDS and postural tachycardia syndrome (POTS) and found PWV measurements were not different compared to healthy controls [11]. Two studies examined PWV in people with vascular EDS. One study found decreased PWV in about 20% of genetically related people with vascular EDS [12]. The other study found that PWV measurements in people with vascular EDS were similar to those of healthy volunteers [13].

Therefore, the current study is the first assessment of PWV measurements in a large heterogeneous sample of people with EDS. We hypothesized that the collagen changes in EDS would confer an increased distensibility of the vasculature in all EDS types, and that this would contribute to orthostatic intolerance. In this study, we investigated central and peripheral arterial stiffness in people with EDS using the non-invasive measurement of pulse-wave velocity (PWV). We hypothesize that impaired blood pressure (BP) and heart rate control is associated with increased arterial elasticity in people with EDS.

2. Materials and Methods

The National Institute on Aging (NIA) study Clinical and Molecular Manifestations of HDCT was designed to investigate the natural history of the most common HDCT. Emphasis was placed on the cardiovascular, musculoskeletal, and neurological complications of HDCT and the natural history of these complications. The original study protocol was designed to collect clinical and family history data, and to use this information to clarify the clinical distinctions between diagnoses. Consenting participants were initially classified based on diagnostic criteria in place at the time of their clinical visit at the NIA (2001–2013). Subjects contributing only biological samples were diagnosed either through a limited onsite evaluation or through review of submitted medical records. The HDCT NIA Dataset v. 2016 consented cohort includes 1009 participants with an average age of 39 ± 18 years (range 2–95, median 40). One hundred and ninety-four participants were 18 years or younger.

The NIA study Clinical and Molecular Manifestations of HDCT began by assembling consented cohorts with a wide range of heritable HDCT, under an umbrella protocol (Protocol 2003-086, later changed to 03-AG-N330). After the study was closed to enrollment, the Institutional Review Board approved the reorganization and migration of the data into a relational database repository and approved re-contacting participants to determine if they would be interested in participating in future research. The HDCT cohort data are provided in SAS datasets, PDF, Excel, MRI DICOM file formats and are now under the umbrella of protocol 11-AG-N079, Sample and Data Repository Protocol for NIA Studies. Participants were recruited from the pool of patients previously seen by the principal investigators and from patient support groups nationally. An authorized guardian provided consent for minor participants, with age-appropriate assent by the minor. In 2016, a signed Data Transfer Agreement between NIA and Penn State University resulted in transfer of a copy of the

HDCT NIA Dataset v.2016 data repository to the Penn State University Clinical Translational Science Institute (PSU-CTSI). Datasets were accompanied by copies of original CRFs and SAS dataset codebook descriptions [14].

Participants were stratified by EDS type including: classical, hypermobile, vascular, or other or unclassified according to the Villefranche nosology [15] as previously described for this cohort [14]. Briefly, classical EDS was determined by joint laxity and skin that is extremely hyperextensible, fragile, bruises easily, and has thin atrophic scars. Hypermobile EDS was classified by history of dislocations, generalized joint laxity, and velvety texture of skin with an absence of extreme skin extensibility and profoundly abnormal scars. Vascular EDS was determined by genetic testing for variation in the COL3A1, the gene encoding type III collagen. The other and unclassified EDS category included patients with the rarer types of Ehlers–Danlos syndromes. A molecular diagnosis was used for the arthrochalasia and kyphoscoliotic types. Some patients had features overlapping with two or more types of EDS, and classification proved to be difficult in those cases, and such patients were diagnosed as "EDS, unclassified". If there was the clinical impression of EDS but they did not meet the diagnostic criteria for any of the known types, we assigned a diagnosis of "EDS, unclassified".

The analytic cohort for the present study was a subset of the EDS cohort from this NIA study of HDCT consisting of 60 participants who had both orthostatic BP recordings and PWV measurements. BP and heart rate were measured by brachial artery oscillometry in triplicate following 5 min in the supine, seated, and standing postures. Central arterial stiffness was measured by carotid to femoral PWV and peripheral stiffness by carotid to radial PWV.

2.1. PWV Measurements

The methods used to assess PWV in this study were the same methods used in the Baltimore Longitudinal Study of Aging [16]. In short, PWV data were collected using a SphygmoCor device (AtCor Medical) that utilizes an EKG and high-fidelity tonometer to acquire waveforms from carotid, femoral, and radial pulses. The software determines the velocity of the pulse wave, i.e., estimated time that it takes the pulse wave to travel between pulse sites divided by the distance between sites. Central PWV is calculated by measuring pulse waves at the carotid and femoral arteries, representing the stiffness of the central vascular tree. Peripheral PWV is calculated by measuring pulse waves at the carotid and radial arteries indicating blood flow to peripheral vascular beds. Reference values of pulse wave velocity in healthy humans were collected using similar methods (pulse wave tonometry divided by distance between sites) [16].

2.2. Orthostatic Vital Sign Measurements

Orthostatic vital signs were measured by a brachial artery BP cuff on both arms. BP was measured supine then seated then during standing. Study participants stayed in each posture (supine, sitting, and standing) for 5 min prior to BP recordings. BP was measured in triplicate in each position with one minute between recordings. If BP varied by 15 mmHg or heart rate by 10 beats/minute in one position, a fourth recording was measured. All BP and heart rate measurements on the left arm were averaged for each participant in each posture.

2.3. Data Analysis

Descriptive statistics include demographic data and EDS type. Comparison of characteristics among types was performed using ANOVA with post-hoc Tukey-Kramer tests when justified. Pearson's correlations were run between BP, heart rate, and PWV measurements for the entire cohort. We performed a stratified analysis of central PWV measurements by age in the EDS participants of all types and compared those values to age-matched reference values from a large cohort of healthy participants ($n = 1455$, Reference Values for Arterial Stiffness, 2010) [17].

3. Results

Overall our data set included 60 (49 female) EDS participants age 13–70 years. There were no differences in age, height, weight, and body mass index between EDS participants of different types (Table 1).

3.1. Pulse Wave Velocity in EDS

Arterial elasticity did not differ by EDS type (Table 1). Grouped together, central PWV is lower in participants with EDS (4.73 ± 0.16 cm/s) compared to reference values in a large sample of healthy participants (Figure 1). PWV increases with age in healthy populations but the increase in arterial stiffness with aging is attenuated in people with EDS.

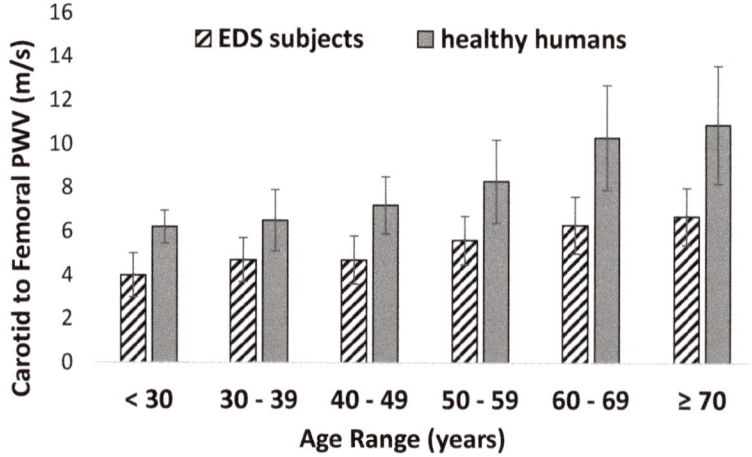

Figure 1. Pulse wave velocity (PWV) by age in participants with Ehlers-Danlos syndromes (EDS) compared to reference values in healthy humans' data from normal subjects in Reference Values for arterial Stiffness Collaboration (RVASC) [17] ($n = 1455$). Data are shown as mean ± standard deviation.

3.2. Orthostatic Blood Pressure in EDS

In the supine posture, BP and heart rate did not vary by EDS type (Table 1). In the standing posture, there was more variability in BP and heart rate measurements within each EDS type as shown by higher standard deviations compared to supine measurements demonstrating a wide range in responses to orthostasis (Table 1). Systolic BP in the standing posture was different between EDS types (ANOVA, $p = 0.003$). Post-hoc analysis showed that standing systolic BP was lower in participants with vascular EDS compared to those with hypermobile EDS ($p = 0.021$) and other/unspecified EDS ($p = 0.002$). Standing diastolic BP and heart rate also trended lower in the vascular EDS group ($p = 0.087$, Table 1.)

3.3. Correlations between Pulse Wave Velocity and Blood Pressure

Correlations of central and peripheral PWV to BP and heart rate are shown in Table 2. Central PWV did not correlate to HR or orthostatic BP changes over a 5 min period. Central PWV correlated significantly with supine ($r = 0.387$) and seated ($r = 0.399$) systolic BPs and supine diastolic BP ($r = 0.400$). Peripheral PWV did not correlate to HR or orthostatic BP. Peripheral PWV was correlated to diastolic BP in the supine ($r = 0.322$), seated ($r = 0.383$), and standing ($r = 0.323$) postures. All significant correlations were positive indicating that lower PWV (more elasticity) is associated with lower BP in our cohort of EDS participants.

Table 1. Analysis of Ehlers–Danlos syndromes by type

	Classical (n = 10)	Hypermobile (n = 13)	Vascular (n = 8)	Other/Unspecified (n = 29)	P Value	All Patients (n = 60)
Age (years)	42 (15–63)	34 (13–51)	38 (13–70)	34 (13–66)	0.557	40 (13–70)
Sex (M/F)	0/9	2/11	2/7	7/22	-	11/49
Height (m)	1.63 ± 0.04	1.67 ± 0.05	1.62 ± 0.12	1.66 ± 0.08	0.336	1.65 ± 0.08
Weight (kg)	74.6 ± 23.5	68.8 ± 12.6	65.3 ± 16.4	68.0 ± 21.3	0.804	70.1 ± 21.8
BMI (kg/m^2)	27.9 ± 8.0	24.5 ± 4.0	24.5 ± 4.1	24.7 ± 7.7	0.593	25.1 ± 6.6
Supine Hemodynamics						
Systolic BP (mmHg)	116 ± 11	121 ± 7	117 ± 11	120 ± 13	0.655	119 ± 11
Diastolic BP (mmHg)	68 ± 8	68 ± 10	62 ± 11	67 ± 7	0.356	66 ± 8
HR (beats/min)	77 ± 10	77 ± 10	66 ± 8	74 ± 14	0.200	74 ± 13
Central PWV (m/s)	4.91 ± 0.56	4.82 ± 0.38	4.89 ± 0.47	4.59 ± 0.20	0.873	4.73 ± 0.16
Peripheral PWV (m/s)	7.45 ± 0.39	7.24 ± 0.33	7.39 ± 0.29	7.12 ± 0.17	0.810	7.23 ± 1.00
Standing Hemodynamics						
Systolic BP (mmHg)	110 ± 19	118 ± 21	93 ± 24	121 ± 14	0.003	115 ± 20
Diastolic BP (mmHg)	75 ± 11	81 ± 12	69 ± 14	73 ± 10	0.087	75 ± 12
HR (beats/min)	91 ± 14	92 ± 12	85 ± 17	89 ± 16	0.719	90 ± 15

Body mass index (BMI), blood pressure (BP), mean arterial pressure (MAP), heart rate (HR), pulse wave velocity (PWV). Data are shown as mean (min-max) or mean ± standard deviation.

Table 2. Correlations of pulse wave velocity to orthostatic hemodynamics in Ehlers-Danlos syndromes

	Carotid to Femoral PWV	Carotid to Radial PWV
SBP (supine)	0.387 *	0.076
SBP (seated)	0.399 *	0.098
SBP (standing)	0.199	0.008
Δ SBP (standing-seated)	−0.077	−0.066
DBP (supine)	0.400 *	0.322 *
DBP (seated)	0.204	0.383 *
DBP (standing)	0.078	0.323 *
Δ DBP (standing-seated)	−0.158	−0.062
HR (supine)	0.015	0.185
HR (seated)	0.044	0.234
HR (standing)	−0.039	0.165
Δ HR (standing-seated)	−0.111	−0.048

Pearson's r-correlations are shown. Change in (Δ), Systolic blood pressure (SBP), diastolic blood pressure (DBP), heart rate (HR). * Significant correlation at $p \leq 0.05$ level.

4. Discussion

4.1. Overall Findings

This study used PWV to evaluate arterial stiffness in a diverse sample of people with different EDS types. This study provides three novel findings. We demonstrated that PWV is lower in people with EDS compared to reference values in the healthy population implying that their arteries are more elastic. We also found that lower PWV (indicating greater elasticity) is associated with lower systolic and diastolic BP in people with EDS. These findings may help explain the connection between EDS and impaired autonomic cardiovascular control. We also found no differences in PWV measurements among EDS types which suggests that the elasticity of the vasculature is similar among the diverse types of EDS.

4.2. Significance of Decreased Pulse Wave Velocity in Ehlers-Danlos Syndrome

The clinical association between EDS and orthostatic intolerance was identified in 1999 by Rowe et al. who first hypothesized that the mechanism connecting these two disorders is an increased enhanced elasticity in the arteries of people with EDS, predisposing them to OI [8]. Two decades later, this theory has become widely accepted despite the lack of empirical data to support it [3].

PWV has become the gold standard for assessing arterial structure because it is reproducible and aligns with more invasive measures. Its ease of use means it is available for testing in larger cohorts [9]. PWV has become a popular and validated method to assess increased stiffness of the central and peripheral vascular system in healthy humans and disease populations ranging from cardiovascular to neurological disease [9,10]. However, this technique is less commonly used to assess populations with increased arterial elasticity.

Three studies have previously measured PWV in people with EDS [11–13]. In a single family of 27 people with vascular EDS, Francois et al. utilized an older method for measuring pulse wave velocity involving piezo crystal microphones over the carotid, femoral, and dorsal arteries, and reported significantly decreased PWV (outside 2 standard deviations of normal values) in 5/27 participants studied [12]. A more recent study used the ultrafast ultrasound technique in 102 healthy participants and 37 vascular EDS participants and found that that central PWV was not significantly different in vascular EDS participants compared to controls [13]. Cheng, et al. employed a similar tonometry technique as was used in the current study to assess PWV in nine people with comorbid hypermobile EDS and POTS and nine age, sex, and BMI matched healthy controls, and found a trend to lower central PWV measurements in the people with EDS/POTS compared to controls [11]. Our study adds to the current literature by measuring both central and peripheral PWV in a larger and more heterogeneous group of people with EDS. In contrast to prior studies, we found that central PWV was significantly decreased in people with EDS compared to reference ranges for healthy subjects. This is likely due to our larger and more diverse sample.

The issue of age-associated changes in vascular function in people with vascular EDS was addressed by Mirault et al. using ultrafast ultrasound imaging (a method used by this group to measure PWV). They reported that the age-associated increase in vascular stiffness was attenuated in the vascular EDS participants [13]. We observed a similar phenomenon, namely that the PWV increased very little with progressive age deciles (Figure 1) which differs from reference values derived from healthy humans. High PWV (implying increased arterial stiffness) is related to adverse cardiovascular events in large epidemiology studies [18–20]. While one may speculate that lower PWV may be cardio-protective, it is unclear whether increased arterial elasticity is beneficial in people with EDS. Whether PWV has prognostic value in EDS deserves further investigation.

4.3. Association between Pulse Wave Velocity and Blood Pressure

Overall, lower PWV is related to lower BP measurements but is not directly indicative of orthostatic tolerance in EDS. These findings are consistent with measurements in healthy subjects and in other patient populations in which PWV tracks similarly to BP [10]. All significant correlations were positive indicating that lower PWV (more elasticity) is associated with lower BP in our cohort of people with EDS. We cannot infer causation from these data.

4.4. Comparisons between EDS Types

We did not see a difference in most orthostatic vital signs between EDS types. Systolic BP was slightly lower in vascular EDS which may reflect a difference in physiology or medications taken. Overall, there was a huge range in BP and heart rate responses to orthostasis which demonstrates inconsistent hemodynamic responses in this population and may reflect the presence of different types of OI. According to Roma, et al., about half of people with EDS have POTS (increase in heart rate of 30 beats/minute while standing) but others have orthostatic hypotension or hypertension [3]. It has been thought that vascular EDS was unique in terms of increased arterial distensibility. Our data are the first to compare arterial elasticity among EDS types in a single study, and demonstrate no difference in PWV among types of EDS. This is an important point, and it provides a possible explanation for the common presence of orthostatic intolerance in all EDS types.

4.5. Strengths

Strengths of this this study are the large sample size with a diverse EDS cohort including several EDS types, and the concomitant measurement of orthostatic vital signs and PWV. We compared central PWV measurements to published reference values in a large cohort. To our knowledge, there are no peripheral PWV reference values from large populations. The methods used in this study were consistent with study protocols used in the Baltimore Longitudinal Study of Aging [16].

4.6. Limitations

This study had several limitations. First, we did not include a contemporaneous control group in this study. However, we used reference values from a large cohort of healthy volunteers for comparison [16]. Second, participants were accessed while on medications which may impact BP, heart rate, and PWV assessments. Orthostatic vital signs were measured following 5 min in the supine posture then after 5 min sitting then 5 min of standing. This limits the ability to diagnose orthostatic intolerance as current diagnostic criteria for orthostatic intolerance involves hemodynamic measurements from the supine to standing posture after at least 10 min [21]. Finally, we acknowledge the heterogeneity of our EDS participants as a potential problem. Since these data were collected prior to the 2017 reclassification of EDS [1,14,15]. It is possible that some participants classified as having hypermobile or unspecified EDS would be categorized as hypermobility spectrum disorders using current criteria.

5. Conclusions

Overall, this is the first report of increased arterial elasticity in all types of EDS. The increased arterial elasticity was associated with lower supine and seated systolic and diastolic blood pressure in all types of EDS. We did not see differences in PWV in different types of EDS but standing systolic and diastolic blood pressure were lower in vascular EDS compared to the hypermobile and unspecified types. Our findings suggest that increased arterial elasticity may be related to impaired blood pressure control in EDS. Further studies are needed to determine whether this pathophysiological finding relates to orthostatic symptoms in people with EDS.

Author Contributions: Conceptualization: A.J.M., J.R.S., R.B., C.A.F.; Formal analysis: A.J.M., J.R.S., T.S.; Funding acquisition: J.R.S., R.B., C.A.F.; Supervision: J.R.S., R.B., C.A.F.; Writing—original draft: A.J.M., R.B., C.A.F.; Writing—review and editing: A.J.M., J.R.S., T.S., R.B., C.A.F. All authors have read and agreed to the published version of the manuscript.

Funding: This research was funded by the National Institute on Aging Intramural Research Program, under Protocol 2003-086 (later changed to 03-AG-N330) and by the National Center for Advancing Translational Sciences, National Institutes of Health, through Grant UL1 TR002014. The content is solely the responsibility of the authors and does not necessarily represent the official views of the National Institutes of Health. This project was also supported by the Patient-Centered Outcomes Research Institute (PCORI) Eugene Washington Engagement Award (Title: The STRETCH Project: To Build Capacity Advancing Patient-centered Research in Ehlers-Danlos Syndrome).

Conflicts of Interest: The authors declare no conflict of interest.

Abbreviations

BP	Blood pressure
EDS	Ehlers–Danlos syndromes
OI	Orthostatic intolerance
POTS	Postural orthostatic tachycardia syndrome
PWV	Pulse wave velocity

References

1. Malfait, F.; Francomano, C.; Byers, P.; Belmont, J.; Berglund, B.; Black, J.; Bloom, L.; Bowen, J.M.; Brady, A.F.; Burrows, N.P.; et al. The 2017 international classification of the Ehlers-Danlos syndromes. *Am. J. Med. Genet. C Semin. Med. Genet.* **2017**, *175*, 8–26. [CrossRef] [PubMed]
2. Murray, B.; Yashar, B.M.; Uhlmann, W.R.; Clauw, D.J.; Petty, E.M. Ehlers-Danlos syndrome, hypermobility type: A characterization of the patients' lived experience. *Am. J. Med. Genet. A* **2013**, *161*, 2981–2988. [CrossRef] [PubMed]
3. Roma, M.; Marden, C.L.; De Wandele, I.; Francomano, C.A.; Rowe, P.C. Postural tachycardia syndrome and other forms of orthostatic intolerance in Ehlers-Danlos syndrome. *Auton. Neurosci.* **2018**, *215*, 89–96. [CrossRef] [PubMed]
4. Wallman, D.; Weinberg, J.; Hohler, A.D. Ehlers-Danlos Syndrome and Postural Tachycardia Syndrome: A relationship study. *J. Neurol. Sci.* **2014**, *340*, 99–102. [CrossRef]
5. Gazit, Y.; Nahir, A.M.; Grahame, R.; Jacob, G. Dysautonomia in the joint hypermobility syndrome. *Am. J. Med.* **2003**, *115*, 33–40. [CrossRef]
6. De Wandele, I.; Rombaut, L.; De Backer, T.; Peersman, W.; Da Silva, H.; De Mits, S.; De Paepe, A.; Calders, P.; Malfait, F. Orthostatic intolerance and fatigue in the hypermobility type of Ehlers-Danlos Syndrome. *Rheumatology* **2016**, *55*, 1412–1420. [CrossRef]
7. Celletti, C.; Camerota, F.; Castori, M.; Censi, F.; Gioffre, L.; Calcagnini, G.; Strano, S. Orthostatic Intolerance and Postural Orthostatic Tachycardia Syndrome in Joint Hypermobility Syndrome/Ehlers-Danlos Syndrome, Hypermobility Type: Neurovegetative Dysregulation or Autonomic Failure? *BioMed Res. Int.* **2017**, *2017*, 9161865. [CrossRef]
8. Rowe, P.C.; Barron, D.F.; Calkins, H.; Maumenee, I.H.; Tong, P.Y.; Geraghty, M.T. Orthostatic intolerance and chronic fatigue syndrome associated with Ehlers-Danlos syndrome. *J. Pediatr.* **1999**, *135*, 494–499. [CrossRef]

9. Laurent, S.; Cockcroft, J.; Van Bortel, L.; Boutouyrie, P.; Giannattasio, C.; Hayoz, D.; Pannier, B.; Vlachopoulos, C.; Wilkinson, I.; Struijker-Boudier, H.; et al. Expert consensus document on arterial stiffness: Methodological issues and clinical applications. *Eur. Heart J.* **2006**, *27*, 2588–2605. [CrossRef]
10. Hirata, K.; Kawakami, M.; O'Rourke, M.F. Pulse wave analysis and pulse wave velocity: A review of blood pressure interpretation 100 years after Korotkov. *Circ. J.* **2006**, *70*, 1231–1239. [CrossRef]
11. Cheng, J.L.; Au, J.S.; Guzman, J.C.; Morillo, C.A.; MacDonald, M.J. Cardiovascular profile in postural orthostatic tachycardia syndrome and Ehlers-Danlos syndrome type III. *Clin. Auton. Res.* **2017**, *27*, 113–116. [CrossRef] [PubMed]
12. Francois, B.; De Paepe, A.; Matton, M.T.; Clement, D. Pulse wave velocity recordings in a family with ecchymotic Ehlers-Danlos syndrome. *Int. Angiol.* **1986**, *5*, 1–5. [PubMed]
13. Mirault, T.; Pernot, M.; Frank, M.; Couade, M.; Niarra, R.; Azizi, M.; Emmerich, J.; Jeunemaitre, X.; Fink, M.; Tanter, M.; et al. Carotid stiffness change over the cardiac cycle by ultrafast ultrasound imaging in healthy volunteers and vascular Ehlers-Danlos syndrome. *J. Hypertens.* **2015**, *33*, 1890–1896. [CrossRef] [PubMed]
14. Bascom, R.; Schubart, J.R.; Mills, S.; Smith, T.; Zukley, L.M.; Francomano, C.A.; McDonnell, N. Heritable disorders of connective tissue: Description of a data repository and initial cohort characterization. *Am. J. Med. Genet. A* **2019**, *179*, 552–560. [CrossRef]
15. Beighton, P.; De Paepe, A.; Steinmann, B.; Tsipouras, P.; Wenstrup, R.J. Ehlers-Danlos syndromes: Revised nosology, Villefranche, 1997. Ehlers-Danlos National Foundation (USA) and Ehlers-Danlos Support Group (UK). *Am. J. Med. Genet.* **1998**, *77*, 31–37. [CrossRef]
16. David, M.; Malti, O.; AlGhatrif, M.; Wright, J.; Canepa, M.; Strait, J.B. Pulse wave velocity testing in the Baltimore longitudinal study of aging. *J. Vis. Exp.* **2014**, e50817. [CrossRef]
17. Reference Values for Arterial Stiffness Collaboration. Determinants of pulse wave velocity in healthy people and in the presence of cardiovascular risk factors: 'establishing normal and reference values'. *Eur. Heart J.* **2010**, *31*, 2338–2350. [CrossRef]
18. Mitchell, G.F.; Hwang, S.J.; Vasan, R.S.; Larson, M.G.; Pencina, M.J.; Hamburg, N.M.; Vita, J.A.; Levy, D.; Benjamin, E.J. Arterial stiffness and cardiovascular events: The Framingham Heart Study. *Circulation* **2010**, *121*, 505–511. [CrossRef]
19. Willum-Hansen, T.; Staessen, J.A.; Torp-Pedersen, C.; Rasmussen, S.; Thijs, L.; Ibsen, H.; Jeppesen, J. Prognostic value of aortic pulse wave velocity as index of arterial stiffness in the general population. *Circulation* **2006**, *113*, 664–670. [CrossRef]
20. Laurent, S.; Boutouyrie, P.; Asmar, R.; Gautier, I.; Laloux, B.; Guize, L.; Ducimetiere, P.; Benetos, A. Aortic stiffness is an independent predictor of all-cause and cardiovascular mortality in hypertensive patients. *Hypertension* **2001**, *37*, 1236–1241. [CrossRef]
21. Sheldon, R.S.; Grubb, B.P.; Olshansky, B.; Shen, W.K.; Calkins, H.; Brignole, M.; Raj, S.R.; Krahn, A.D.; Morillo, C.A.; Stewart, J.M.; et al. 2015 heart rhythm society expert consensus statement on the diagnosis and treatment of postural tachycardia syndrome, inappropriate sinus tachycardia, and vasovagal syncope. *Heart Rhythm* **2015**, *12*, e41–e63. [CrossRef] [PubMed]

© 2020 by the authors. Licensee MDPI, Basel, Switzerland. This article is an open access article distributed under the terms and conditions of the Creative Commons Attribution (CC BY) license (http://creativecommons.org/licenses/by/4.0/).

Review

Cellular and Molecular Mechanisms in the Pathogenesis of Classical, Vascular, and Hypermobile Ehlers-Danlos Syndromes

Nicola Chiarelli, Marco Ritelli, Nicoletta Zoppi and Marina Colombi *

Division of Biology and Genetics, Department of Molecular and Translational Medicine, University of Brescia, 25121 Brescia, Italy
* Correspondence: marina.colombi@unibs.it

Received: 26 June 2019; Accepted: 9 August 2019; Published: 12 August 2019

Abstract: The Ehlers-Danlos syndromes (EDS) constitute a heterogenous group of connective tissue disorders characterized by joint hypermobility, skin abnormalities, and vascular fragility. The latest nosology recognizes 13 types caused by pathogenic variants in genes encoding collagens and other molecules involved in collagen processing and extracellular matrix (ECM) biology. Classical (cEDS), vascular (vEDS), and hypermobile (hEDS) EDS are the most frequent types. cEDS and vEDS are caused respectively by defects in collagen V and collagen III, whereas the molecular basis of hEDS is unknown. For these disorders, the molecular pathology remains poorly studied. Herein, we review, expand, and compare our previous transcriptome and protein studies on dermal fibroblasts from cEDS, vEDS, and hEDS patients, offering insights and perspectives in their molecular mechanisms. These cells, though sharing a pathological ECM remodeling, show differences in the underlying pathomechanisms. In cEDS and vEDS fibroblasts, key processes such as collagen biosynthesis/processing, protein folding quality control, endoplasmic reticulum homeostasis, autophagy, and wound healing are perturbed. In hEDS cells, gene expression changes related to cell-matrix interactions, inflammatory/pain responses, and acquisition of an in vitro pro-inflammatory myofibroblast-like phenotype may contribute to the complex pathogenesis of the disorder. Finally, emerging findings from miRNA profiling of hEDS fibroblasts are discussed to add some novel biological aspects about hEDS etiopathogenesis.

Keywords: autophagy; collagen III; collagen V; Ehlers-Danlos syndrome; endoplasmic reticulum; extracellular matrix; fibroblast-to-myofibroblast transition; miRNA; transcriptome; wound healing

---

1. The Extracellular Matrix: An Overview

Connective tissues have an extracellular matrix (ECM) with a specific composition generated during embryogenesis and maintained in adult life. The ECM is a complex network that provides a structural scaffold to the surrounding cells and is a reservoir of bioactive molecules such as cytokines and growth factors that control cell behavior [1]. The main ECM components include proteoglycans, hyaluronic acid, adhesive glycoproteins such as fibronectin and laminins, and fibrous proteins like collagens and elastin [2]. Matricellular proteins such as thrombospondins, osteopontin, periostin, and tenascins are non-structural ECM proteins, primarily acting as mediators of cell–matrix interactions, which are abundantly expressed during embryonic development, wound healing, and tissues renewal [3].

The human matrisome consists of about 300 macromolecules comprising the "core matrisome", which is composed of many different collagens, proteoglycans (e.g., aggrecan, versican, perlecan, and decorin), and glycoproteins (e.g., laminins, elastin, fibronectin, thrombospondins, and tenascins) [4]. Matrisome also includes many matrisome-associated proteins and ECM-regulators, i.e., ECM-cross-linking (e.g., lysyl oxidases, transglutaminases) and ECM-modifying enzymes (e.g., proteases and their inhibitors) together with secreted factors including transforming growth factor β (TGFβ), wingless integrated (Wnt), and multiple cytokines [5].

Collagens represent the major ECM structural components and play a central role in providing the structural integrity of several connective tissues (e.g., cartilage and bone) and various organ systems including skin, lungs, blood vessels, and cornea. Collagens are also involved in cell adhesion, chemotaxis, and migration [1,2]. The dynamic interplay between cells and collagens regulates tissue remodeling during growth, differentiation, morphogenesis, and wound healing [6,7]. The common molecular feature of collagens is their triple helical structure, which consists of three collagenous α-chains with the typical recurring (Gly-Xaa-Yaa)$_n$ tripeptide sequence. The presence of glycine residues in the collagenous domain is essential for stability and correct assembly of the triple helix. Collagen biosynthesis, assembly, and maturation require a sequence of well-controlled intracellular and extracellular events (for review see [8,9]). Collagen I is the most abundant type expressed in bone, cornea, dermis, and tendon. Collagen III is primarily present in the tunica media of the blood vessels and hollow organs (e.g., uterus, intestine). Collagen V is widely distributed, especially in dermis, tendons, and muscles, playing a central role in collagen I fibrillogenesis [9]. These fibrillar collagens form structures necessary to ensure the strength and structural integrity of the ECM of all connective tissues and organs of the body [10].

Fibronectin is a dimeric and fibrillar glycoprotein ubiquitously organized in the ECM of all tissues and is also present in soluble form in the plasma. Cellular fibronectin self-assembles in fibers and binds collagens, fibrin, proteoglycans, and cell surface receptors, providing cell growth, adhesion, and migration. During wound healing, it forms a provisional matrix with fibrin and enters in the granulation tissue formation in the late phase of re-epithelization [11].

In addition to ensuring physical support and structural integrity, the proper ECM composition and organization are crucial for cell health. ECM undergoes a continuous turnover either under physiological or in pathological circumstances, and its homeostasis is critical for connective tissues architecture and function [12,13]. Integrins are specific cell surface receptors that mediate the complex cell-matrix interactions. These bridging molecules, which are heterodimeric transmembrane receptors containing α and β subunits, connect ECM to cytoskeleton by interacting via their extracellular domain with collagens and other matrix molecules and via their cytoplasmic tails with cytoskeleton components (e.g., actin, vinculin, talin, paxillin), thus mediating cell adhesion and motility [14,15].

2. Pathological ECM Remodeling and Perturbation of Cellular Homeostasis

Cell-matrix interaction via integrins is crucial for cell survival and tissue homeostasis. Prolonged loss of integrin-mediated cell–ECM adhesion leads to anoikis [16]. Under physiological conditions, ECM detachment triggers anti-apoptotic signals as a cell survival mechanism to delay the onset of anoikis. One of such signaling pathways is autophagy, which is a highly conserved cellular catabolic process that promotes homeostasis and mitigates the stress due to ECM detachment [17]. Autophagy is essential for cellular maintenance and homeostasis by promoting the turnover of macromolecules and organelles via the lysosomal degradative pathway [18]. Physiological and pathological changes in the ECM composition play a crucial role in modulating autophagy activity [19,20]. For instance, deficiency of collagen VI, which is associated with a spectrum of different myopathic conditions, perturbs ECM architecture, impairs the autophagic flux, and activates pro-apoptotic signals [21]. Autophagy, in turn, contributes to the maintenance of endoplasmic reticulum (ER) function by mediating its turnover through the autophagic sequestration of ER fragments into autophagosomes, the so-called ER-phagy process [22].

ECM components also modulate immune cell migration into inflamed tissues and their activation and proliferation [23]. The stimulation of the innate immunity results from the recognition of specific mediators, namely pattern recognition receptors, which, in turn, recognize molecules, referred as danger-associated molecular patterns, which are released from damaged tissues [24]. It is well documented that different ECM components or their fragments including the fibronectin 1 extra domain

A, one of the alternative spliced regions of fibronectin encoding gene, tenascin-C, fibrinogen, and several proteoglycans, serve as danger signals and trigger immune responses following tissue damage or in response to pathological ECM remodeling [24,25]. In fibrotic conditions, increased ECM production, accumulation of ECM fragments, augmented secretion of cytokines, fibroblast-to-myofibroblast transition, and activation of immune responses, dependent on toll-like receptors, occur [26]. The regulation of ECM synthesis and remodeling is central for human health, as recognized in different heritable connective tissue disorders [1,27]. Indeed, molecular defects in a large range of ECM-related genes, including those encoding enzymes involved in biosynthesis or processing of ECM proteins, cause a myriad of connective tissue disorders, e.g., Ehlers–Danlos syndromes, Osteogenesis imperfecta, Marfan syndrome, Loeys–Dietz syndromes, arterial tortuosity syndrome, and numerous skeletal dysplasias [1,27]. These disorders are characterized by a multisystem involvement in terms of cardiovascular, skeletal, and cutaneous features [28], highlighting the functional relevance of the ECM in ensuring the integrity and function of several connective tissues.

The pathological consequences of defects in ECM components depend on the balance between extracellular effects, e.g., reduced protein secretion and export of misfolded proteins, and intracellular consequences such as apoptosis activation, ER dysfunction, and autophagy perturbation that impact in different ways on the molecular pathology and disease severity [27,29–31].

3. Ehlers-Danlos Syndromes

Ehlers-Danlos syndromes (EDS) represent a clinically and genetically heterogeneous group of conditions that share a variable combination of skin hyperextensibility, joint hypermobility, and internal organ and vessel fragility [32]. The 2017 international classification of the Ehlers-Danlos syndromes recognizes 13 subtypes, which are caused by pathogenic variants in 19 different genes, mainly encoding fibrillar collagens and collagens-modifying proteins [32]. EDS types are grouped based on the underlying genetic and pathogenetic mechanisms in disorders related to (i) collagens primary structure and processing (*COL1A1*, *COL1A2*, *COL3A1*, *COL5A1*, *COL5A2* and *ADAMTS2*), (ii) collagens folding and cross-linking (*PLOD1* and *FKBP14*), (iii) structure and function of the myomatrix, i.e., the specialized ECM of muscle (*TNXB* and *COL12A1*), (iv) glycosaminoglycans biosynthesis (*B4GALT7*, *B3GALT6*, *CHST14*, and *DSE*), (v) complement pathway (*C1S* and *C1R*), and (vi) intracellular processes (*SLC39A13*, *ZNF469*, and *PRDM5*). The classical (cEDS), vascular (vEDS) and the molecularly unsolved hypermobile (hEDS) EDS forms account for more than 90% of patients. Recently, a new and very rare EDS variant has been identified that is caused by biallelic mutations in the *AEBP1* gene (Table 1) [33–36].

The new nosology proposed for each subtype a set of major, minor, and minimal criteria addressing clinical suspicion for a specific EDS type and confirmatory molecular testing. For a comprehensive overview of all EDS forms see the landmark work by Malfait and colleagues [32].

The decrease in the tensile strength and integrity of skin, joints, and hollow organs is a common disease mechanism shared by the different EDS types [37]. This mechanical weakness is considered the driving factor of connective tissue fragility, even if it is likely that multiple cell-matrix interplays and involvement of distinct intracellular signaling pathways contribute to the molecular pathology of the different EDS phenotypes [38].

In the following chapters, we will review and expand the results derived from our previous transcriptome and in vitro studies on cEDS, vEDS, and hEDS patients' dermal fibroblasts. Taken together, these studies highlighted that the alteration of the ECM structural integrity is a common disease factor contributing to the pathogenesis of all these conditions.

Table 1. EDS types grouped according to the underlying genetic defect and pathomechanisms.

EDS Type	IP	Gene	Protein
Group A: disorders of collagen primary structure and collagen processing			
Classical EDS (cEDS)	AD	Major: COL5A1, COL5A2 Rare: COL1A1	COLLV COLLI
Vascular EDS (vEDS)	AD	COL3A1	COLLIII
Arthrochalasia EDS (aEDS)	AD	COL1A1, COL1A2	COLLI
Dermatosparaxis EDS (dEDS)	AR	ADAMTS2	ADAMTS-2
Cardiac-valvular EDS (cvEDS)	AR	COL1A2	COLLI
Classical-like 2 EDS [A] (cl2EDS)	AR	AEBP1	ACLP
Group B: disorders of collagen folding and collagen cross-linking			
Kyphoscoliotic EDS (kEDS)	AR	PLOD1 FKBP14	LH1 FKBP22
Group C: disorders of structure and function of myomatrix			
Classical-like EDS (clEDS)	AR	TNXB	Tenascin X
Myopathic EDS (mEDS)	AD/AR	COL12A1	COLLXII
Group D: disorders of glycosaminoglycan biosynthesis			
Spondylodysplastic EDS (spEDS)	AR	B4GALT7 B3GALT6	β4GalT7 β3GalT6
Musculocontractural EDS (mcEDS)	AR	CHST14 DSE	D4ST1 DSE
Group E: disorders of complement pathway			
Periodontal EDS (pEDS)	AD	C1R C1S	C1r C1s
Group F: disorders of intracellular processes			
Spondylodysplastic EDS (spEDS)	AR	SLC39A13	ZIP13
Brittle Cornea Syndrome (BCS)	AR	ZNF469 PRDM5	ZNF469 PRDM5
EDS type molecularly unsolved			
Hypermobile EDS (hEDS)	AD	Unknown	Unknown

[A] New EDS variant recently defined in [33–36]. AD: autosomal dominant; AR: autosomal recessive; IP: inheritance pattern.

In cEDS and vEDS fibroblasts, the ECM disarray is a direct consequence of molecular defects in respectively collagen V and collagen III that impair common molecular functions essential to guarantee adequate folding and maturation of proteins and biological processes crucial for cell survival and homeostasis. The ECM disorganization observed in hEDS cells may be a consequence of an excessive pathological turnover, mainly due to ECM-degrading enzymes and other so far unknown factors, which might be primary contributors involved in the transition to a pro-inflammatory myofibroblast-like phenotype. Consistently, the perturbation of distinct transcriptional patterns observed in cEDS, vEDS, and hEDS fibroblasts pointed out different disease mechanisms underlying the pathophysiology of these EDS cell types.

Altogether, these insights represent a starting point for future investigations on the numerous pathobiological aspects underlying these conditions. An overview of the biological findings emerged from transcriptome and in vitro studies on dermal fibroblasts from cEDS, vEDS, and hEDS patients is summarized in Table 2.

Table 2. Overview of the biological processes dysregulated in cEDS, vEDS, and hEDS patients' dermal fibroblasts emerged from transcriptome and in vitro studies.

Insights in the Pathogenesis of cEDS, vEDS, and hEDS	Transcriptome Findings on Patients' Fibroblasts			In Vitro Studies on Patients' Fibroblasts		
Perturbed Biological Processes	cEDS	vEDS	hEDS	cEDS	vEDS	hEDS
ECM disorganization	+	+	+	+	+	+
Altered cell-matrix interactions	−	−	+	+	+	+
Disturbed cell-cell contacts	−	−	+	−	−	+
Fibroblast-to-myofibroblast transition	−	−	+	−	−	+
Altered inflammatory responses	+	−	+	−	na	+
Perturbed cell migration	+	+	−	+	+	+
Defective wound healing	+	+	−	+	na	na
Survival from anoikis	−	−	−	+	+	na
Collagens biosynthesis/processing	+	+	−	na	+	na
ER homeostasis/protein folding	+	+	−	na	+	na

+: detected by transcriptome or in vitro studies, −: not experimentally detected by transcriptome or in vitro studies, na: not ascertained.

4. Classical Ehlers-Danlos Syndrome

Classical EDS (cEDS, OMIM#130000) is characterized by marked skin involvement, generalized joint hypermobility, and abnormal wound healing [32,39]. Most patients harbor point mutations or chromosomal rearrangements in *COL5A1* or *COL5A2* genes encoding the collagen V [40,41]. This collagen is abundantly distributed in a variety of tissues as heterotrimers, which co-assemble with collagen I to form heterotypic fibrils [42].

Collagen V knockout mice synthesize and secrete normal amounts of collagen I, but collagen fibrils are absent, and the animals die at the onset of organogenesis, supporting the crucial role of collagen V for embryonic development [43].

Collagen V haploinsufficiency is the most common molecular defect caused by *COL5A1* null alleles, whereas rare *COL5A1* variants and the majority of *COL5A2* mutations reported so far affect collagen V structural integrity by exerting a dominant negative effect [40,41].

5. Altered ECM Turnover, Wound Healing, and Inflammation in cEDS Fibroblasts

Although the reduced availability of collagen V is crucial in the pathogenesis of cEDS, the molecular aspects contributing to the pathophysiology of the disorder remain poorly characterized. Our in vitro findings demonstrated that cEDS patients' fibroblasts show disassembly of many ECM components, including collagen V and III, fibronectin, and fibrillins, and disorganization of collagen- and fibronectin-specific α2β1 and α5β1 integrin receptors [44–47]. cEDS cells also exhibit a reduced in vitro migration capability, an abnormal wound healing response, and a crosstalk involving the αvβ3 integrin and epidermal growth factor (EGF) receptor that rescues them from anoikis [44–50]. In line with these in vitro findings, *Col5a1* and *Col5a2* deficient mice show a defective wound healing response and reduced cell migration [51,52].

Transcriptome profiling of cEDS fibroblasts added new insights into the complex molecular mechanisms involved in the maintenance of ECM homeostasis and proper wound healing, since patients' cells showed the dysregulated expression of many genes encoding matricellular and soluble proteins with prominent functions in cell proliferation and migration, collagen assembly and ECM remodeling during wound healing, i.e., *SPP1*, *POSTN*, *EDIL3*, *IGFBP2*, and *C3* [47]. Wound healing is a highly controlled multistep process involving several growth factors, cytokines, matrix metalloproteases, and cellular receptors, as well as proper crosstalk of different ECM constituents essential for ensuring tissue regeneration [53]. In addition to ECM glycoproteins and collagens, many other matricellular proteins including osteopontin, periostin, and tenascins are required for the formation of a provisional ECM during wound repair [54,55].

Osteopontin encoded by *SPP1*, which shows a decreased expression in patients' cells, is involved in several physiological processes related to inflammation, biomineralization, cell viability, and wound healing [56,57]. Through its interaction with the $\alpha v \beta 3$ integrin, osteopontin facilitates the adhesion of bone cells during bone tissue formation by stimulating a mineralized collagen ECM [58,59]. Finally, the functional role of osteopontin in the ECM reorganization during wound healing is crucial, since *Opn*-deficient mice show ECM disorganization and disassembly of collagen fibrils in the deep layers of wound sites [60].

In cEDS fibroblasts, the decreased expression of periostin encoded by *POSTN* may also contribute to the generalized ECM disarray and in vitro poor wound healing [44,48]. Indeed, periostin plays an important role in ECM structure and organization and particularly in collagen assembly, by acting as a scaffold protein for the bone morphogenetic protein 1, which facilities the proteolytic activation of lysyl oxidase that, in turn, catalyzes the covalent cross-link formation of collagens [61]. Consistently, *Postn*-deficient mice exhibit marked reduction of collagen cross-linking and increased levels of collagen fragments owing to proteolytic digestion [62,63]. Periostin also acts as a pro-survival protein in many cellular circumstances by interacting with $\alpha v \beta 3$ and $\alpha v \beta 5$ integrin receptors and mediating the activation of several intracellular signaling pathways [64]. In wound sites, it promotes activation of fibroblasts during wound contraction and stimulates collagen assembly and ECM reorganization [65,66]. *POSTN* not only shows a diminished expression in vEDS cells but also in dermal fibroblasts from patients with *FKBP14*-kEDS [67], further emphasizing the crucial role of periostin as a scaffold matricellular protein necessary for collagen assembly and ECM stability.

EDIL3 (EGF-like repeat- and discoidin I-like domain-containing protein 3), the most down-regulated transcript in cEDS fibroblasts, encodes an ECM-associated protein that promotes angiogenesis in vitro through binding to $\alpha v \beta 3$ and $\alpha v \beta 5$ integrins [68]. It stimulates cell migration and proliferation, mediates apoptotic cell phagocytosis, regulates neutrophil recruitment to the inflamed tissue, and prevents chondrocyte anoikis through its interaction with the $\alpha v \beta 3$ integrin [69–71].

IGFBP2 (insulin-like growth factor-binding protein 2), the most up-regulated gene in cEDS cells, enhances cell migration in different cell types through its binding to $\alpha v \beta 3$ and $\alpha 5 \beta 1$ integrins [72,73]. Its high expression in cEDS cells might represent a transcriptional response in the attempt to counteract, at least in vitro, their reduced migration capability [48,49].

Of note is also the marked up-regulation of the complement factor C3 belonging to a complex network of plasma and membrane proteins involved in the innate immunity [74]. Complement can modulate the inflammatory response during wound healing to restore tissue injury; however, its unbalanced or prolonged activation can exacerbate inflammation, delaying the physiological wound healing [75]. Specifically, C3 functions as a negative regulator of tissue healing, since C3-deficient mice exhibit an increased wound healing and angiogenesis [76].

Taken together, these gene expression abnormalities expand the current understanding of altered molecular mechanisms underlying the deficient wound healing response observed in cEDS cells. Additional functional work might help to establish the concrete involvement of these proteins including the $\alpha v \beta 3$ integrin in the impaired wound healing, which likely leads to the cutaneous manifestations of cEDS [39,77].

6. Perturbation of ER Homeostasis and Autophagy in cEDS Fibroblasts

ER is a fundamental cellular organelle involved in the maintenance of numerous aspects of cell health by ensuring folding and exporting of secretory or transmembrane proteins [78]. The biosynthesis, processing, and integrity of collagens and other ECM structural constituents are critical for intracellular proteostasis [79]. To restore intracellular equilibrium, the ER counteracts the accumulation of aggregated or misfolded proteins by means of quality control mechanisms such as unfolded protein response, ER-associated degradation, and autophagy [80].

In cEDS fibroblasts, a possible unbalance of ER homeostasis and autophagy was assumed given the decreased expression of many associated genes such as *DNAJB7*, *ATG10*, *CCPG1*, and *SVIP* [47]. *DNAJB7* encodes a member of the J protein/heat shock protein family acting as ER chaperones in the quality control of aggregate protein [81]. *ATG10* is a member of the autophagy related proteins family that participates to the generation and expansion of autophagosomes [22]. The protein encoded by *CCPG1* acts as an ER-phagy cargo receptor facilitating the attachment to growing autophagosomes of the microtubule-associated LC3 protein that is crucial for autophagosome maturation [22,78,82].

CCPG1 plays a key role in the ER proteostasis, since its deficiency causes accumulated insoluble proteins and consequent ER dilation [83]. A recent study reported the contribution of ER-phagy in the selective degradation of misfolded procollagen I molecules via a calnexin-FAM134B complex [84]. Furthermore, inefficient procollagen folding in the ER may induce autophagy as a cytoprotective mechanism [85]. Based on these findings, it is reasonable to speculate a possible role of CCPG1 in this autophagy-dependent mechanism and that its decreased expression in cEDS fibroblasts might impair the ER quality control. The disturbance of ER homeostasis in cEDS is also suggested by the decreased transcription of the small VCP/p97-interacting protein (SVIP), which is a modulator of the ER-associated degradation pathway [86,87]. Previous and recent data highlighted the contribution of SVIP in the regulation of autophagy, since its overexpression is associated with increased LC3 lipidation and attenuation of hepatic fibrosis by the induction of the autophagic flux [88,89].

The impairment of ECM organization, matrix-cell interactions, and the activation of ECM-dependent intracellular signaling may elicit autophagy [18,19]. Cell detachment from the ECM activates the autophagy pathway that, in turn, protects cells from anoikis [90]. Moreover, depletion of autophagy regulators is associated with induction of pro-apoptotic signals, decrease of collagen degradation via lysosome pathway, and regulation of cell adhesion [90,91]. In line with these observations, the aberrant expression of collagen V and defective remodeling of ECM in cEDS cells might affect ER homeostasis and autophagy, and consequently activate a pro-survival mechanism mediated by a crosstalk between $\alpha v \beta 3$ integrin and EGF receptor [45,50]. Additional studies are needed to better elucidate the contribution of these processes in the molecular pathology of cEDS.

7. Vascular Ehlers-Danlos Syndrome

Among the different EDS forms, vascular EDS (vEDS, OMIM#130050) is the most severe type and is primarily characterized by life-threatening features of tissue fragility leading to arterial dissection or aneurysm, gastrointestinal ruptures, and pregnancy complications at a young age [92,93].

vEDS is caused by mutations in *COL3A1* encoding collagen III that shows a predominant expression in blood vessels and hollow organs [94]. Most disease-causing variants in *COL3A1* are glycine substitutions that destroy the triple helical winding, thus altering the structural integrity of collagen III due to misfolded procollagen III in the ER, and thereby impairing the secretion and deposition into the ECM of functionally mature molecules [95,96].

Col3a1 deficient mice show a reduced lifespan mainly due to arterial ruptures and abnormalities of collagen fibril organization in several collagen-rich organs, i.e., aorta, skin, lung, and bowel [10].

Our previous protein findings on cultured patients' fibroblasts showed that dominant negative mutations in *COL3A1* lead to the reduced secretion of collagen I into the ECM [44], consistent with the known regulatory role of collagen III in synthesis and deposition of heterotypic fibrils that largely contain collagen I [96].

8. Disturbance of ECM Organization, Collagens Processing, and ER Homeostasis in vEDS Fibroblasts

Although it is well known that the disruption of the collagen III triple helical structure leads to abnormal protein folding, different biological aspects of the vEDS pathogenesis are not yet fully studied. In line with cEDS and hEDS transcriptome profiling, vEDS fibroblasts show the differential expression of several genes encoding structural constituents of the ECM, further supporting the notion that abnormal ECM remodeling is a common denominator of these conditions [50,97].

vEDS cells showed a marked decrease in expression of the fibrillin 2 encoding gene (*FBN2*). Fibrillins are essential structural ECM components involved in the organization of blood vessels and dermis, and in combination with elastic fibers they act as scaffold to ensure tissue elasticity [98]. Moreover, fibrillins interact with elastin microfibril interface-located proteins (EMILINs) and facilitate their incorporation into the dermal ECM [99].

Fibrillins also regulate the bioavailability of the TFGβ through the interaction with latent TFGβ binding proteins [99]. Specifically, fibrillin 2 plays a role in bone and soft connective tissue morphology by influencing the collagen cross-linking [100].

Besides its role in elastogenesis and ECM stability, fibrillin 2 also has a role during wound healing [101]. About this, vEDS fibroblasts share with cEDS cells the decreased expression of the related periostin-encoding gene and show reduced migration capability [49]. Consistently, both altered wound healing and reduced total collagen content were reported in a *Col3a1* transgenic mouse model [102].

Our protein findings confirmed the pathological ECM remodeling of vEDS fibroblasts, as a generalized fibrillin disarray in combination with the disassembly of EMILINs and elastin network was revealed, consistent with the extreme vascular fragility observed in vEDS patients [97]. The disorganization of core proteins of the proteoglycans perlecan, versican, and decorin, which are involved in the formation of collagen fibrils, further emphasizes the widespread ECM disarray and altered collagens biosynthesis/secretion of vEDS cells that are consequent to collagen III defect [97].

The biosynthetic pathway of fibrillar collagens is a highly regulated process involving folding enzymes, molecular chaperones, and post-translational modifications essential for proper protein assembly, stability of collagen fibrils, and their transport to the cell surface [8,9]. In vEDS cells, this complex machinery seems to be perturbed given the reduced expression of many ER-resident enzymes involved in different steps of collagen biosynthesis, i.e., *P4HA2*, *P4HA3*, *LOXL3*, and *FKBP14*.

P4HA2 and *P4HA3* encode the α-subunit of the collagen prolyl-4-hydroxylase, which catalyzes the hydroxylation of collagen prolyl residues necessary to provide thermal stability to the collagen triple helix. Lysyl oxidase-like 3 (*LOXL3*) stabilizes the formation of intra- and intermolecular crosslinks during assembly of collagen and elastin fibrils. *FKBP14* encodes a peptidyl-prolyl cis-trans isomerase (FKBP22) that catalyzes in the ER lumen collagen folding and it functions as a molecular chaperone for different collagens including collagen III [8,103]. Dermal fibroblasts of *FKBP14*-deficient patients show a generalized perturbation of protein folding and a consequent enlargement of ER cisternae [104]. The marked decrease of FKBP22 protein levels observed in vEDS cells suggests an ER accumulation of misfolded proteins, consistent with the possible dilation of ER cisternae evinced by immunofluorescence analysis with the ER marker PDI [97].

Structural mutations in different collagen types disturb the assembly into hetero- or homotrimers or lead to abnormal triple helix folding. The consequent accumulation of misfolded collagen molecules into the ER lumen activates the proteasomal degradation system to re-establish ER proteostasis [27]. In vEDS cells, this quality control machinery might not work properly, given the decreased transcription of several genes encoding different catalytic and non-catalytic subunits of the proteasome complex, such as *PSMA6*, *PSMB6*, *PSMC3*, and *PSMD2*. In addition, the reduced transcription of members belonging to the DnaJ heat shock protein family, i.e., *DNAJB7*, *DNAJB11*, *DNAJC3*, *DNAJC10*, and *DNAJC24*, and to the thioredoxin superfamily, i.e., *TXN*, *PDIA4*, *PDIA5*, and *PDIA6*, which all act as intracellular mediators for correct protein folding and intracellular redox balance [105], further corroborates a perturbed ER proteostasis in vEDS fibroblasts.

This imbalance can be overcome by the activation of stress-related pathways, such as unfolded protein response and autophagy to restore basal cellular equilibrium [80]. The alteration of the ER redox state may also trigger pro-death signals through the regulation of members of the Bcl-2 family and activation of caspase-dependent apoptosis [106,107]. Consistently, we previously demonstrated that vEDS fibroblasts are in a pre-apoptotic state, due to downregulation of the Bcl-2 anti-apoptotic protein and increased levels of caspase enzymes, and activate a cell survival mechanism through an $\alpha v \beta 3$-EGFR crosstalk [45,50]. However, in vEDS cells an enhanced expression of unfolded protein response-related genes was not identified, consistent with the recent findings on cultured dermal fibroblasts from *Col3a1* transgenic mice that did not show elevated levels of the unfolded protein response markers *Bip* and *Chop* [102]. In line with this evidence, a recent transcriptome analysis of dermal fibroblasts from *FKBP14*-kEDS patients did not reveal a high expression of genes associated with ER stress and unfolded protein response activation [67], though early data on this EDS cell type suggested an enlargement of ER [104]. Nevertheless, given that different reports highlighted the role of ER stress in the pathogenesis of several collagenopathies [27,108,109], further work is warranted to explore the possible ER perturbation as a disease mechanism of vEDS to identify novel potential therapeutic targets.

9. Hypermobile Ehlers-Danlos Syndrome

Hypermobile EDS (hEDS, OMIM#130020), mainly characterized by generalized joint hypermobility and its complications, minor skin changes, and apparently segregating with an autosomal dominant pattern, is still without a known molecular basis. The phenotypic spectrum of hEDS is wide-ranging and heterogeneous and further complicated by multiple associated symptoms shared with other heritable or acquired (autoimmune) connective tissue disorders and chronic inflammatory systemic diseases [32,110,111].

Despite the significant advances in molecular genetic techniques, attempts to disclose the genetic cause(s) of hEDS have been so far inconclusive. Several studies struggled to define its genetic etiology but without compelling evidence, corroborating the hypothesis of a high genetic heterogeneity of the condition [112–115]. The introduction of more selective clinical criteria for hEDS in the novel classification aimed to minimize heterogeneity allows for the formation of homogeneous cohorts to facilitate scientific research to discover the underlying genetic cause(s) of the condition [32]. Nowadays, hEDS is considered at one end of a continuous spectrum of phenotypes, which originates from isolated non-syndromic joint hypermobility and passing through the recently defined hypermobility spectrum disorders (HSD) [111]. HSD refers to patients who present symptomatic joint hypermobility but do not fulfill the new diagnostic criteria of hEDS. Recently, given the clinical continuity between hEDS and HSD and our data on patients' dermal fibroblasts [49], it was proposed that these disorders might be considered as a single entity, referred to as hEDS/HSD [116], as already occurred for the hypermobility type of EDS and joint hypermobility syndrome [117]. Until now, no validated biological biomarkers have been identified for recognizing hEDS/HSD, which are dominated by extremely variable phenotypes and chronic disability affecting patients' quality of life [111,116].

In this intricate scenario, the integration of various biological knowledge could be an effective strategy to delineate molecular mechanisms contributing to the disease pathophysiology. Transcriptome and proteome profiling can be useful to reveal specific biological signatures, thus providing insights not only for the understanding of the pathomechanisms but also for the identification of reliable tools for therapeutic options [118–122].

Previous findings based on transcriptome and protein studies on a cohort of hEDS/HSD patient-derived dermal fibroblasts represent up to now the main effort to unravel their complex etiopathogenesis [48,123]. Proteome profiling of patients' cells is currently ongoing to corroborate these data, since the gene expression profiling and cellular studies on patients' fibroblasts provided significant clues that are likely relevant for the disease pathogenesis. In the following paragraphs we review our past findings and discuss some novel emerging aspects, offering future perspectives for molecular research in this field.

10. Pathological ECM Remodeling and Defective Cell-Cell Interactions in hEDS/HSD Cells

Although hEDS/HSD etiology remains elusive, patients' skin fibroblasts show a disorganization of the ECM like that observed in cells derived from the other EDS types. In particular, hEDS/HSD, cEDS, and vEDS fibroblasts exhibit a marked disorganization of collagen and fibronectin ECM and their specific α2β1 and α5β1 integrin receptors and showed the preferential expression of the αvβ3 integrin [44,49,50,123].

Transcriptome of hEDS/HSD fibroblasts revealed the dysregulated expression of several genes encoding either ECM glycoproteins such as elastin (*ELN*) and sparc/osteonectin (*SPOCK*), ECM regulators, i.e., metalloproteinases (*MMP16*, *PAPPA2*) and transglutaminase (*TGM2*), or ECM associated secreted factors such as secreted frizzled-related protein 2 (*SFRP2*) and transforming growth factor alpha (*TGFA*). This ECM signature is in common with cEDS and vEDS cells, underlining that the matrix perturbation may act as a key driving factor for the EDS pathogenesis, irrespective of the underlying molecular defects [50].

Cells sense the intrinsic mechanical properties of the ECM and convert these stimuli into intracellular signals [124]. In addition to integrins that primarily mediate this cell response, intracellular signals may be triggered also through the cadherin superfamily, which are calcium-dependent transmembrane proteins forming complex adhesions and connect to the actin cytoskeleton via numerous proteins [125,126]. Interestingly, transcriptome profiling of hEDS/HSD cells revealed a differential expression of many adhesion molecule-encoding genes including members of cadherins and protocadherins, i.e., *CDH2*, *CDH10*, *PCDH9*, *PCDHB16*, *PCDHB18*, claudins (*CLDN11*), and desmosomes (desmoplakin, *DSP*), which are involved in the formation of specialized cell-cell junction complexes essential for maintaining epithelial integrity, morphogenesis, and tissue architecture [127,128]. Since these adhesion proteins can act as signaling modulators of intracellular pathways, such as Wnt, Hippo, NF-kB, JAK-STAT that are crucial for development and organogenesis [128], their altered expression could impact on multiple biological processes essential for embryogenesis and tissue homeostasis.

These transcriptional changes suggested a fibroblast-to-myofibroblast transition of hEDS/HSD cells. This phenomenon induces the formation of cells with muscle-like features that are characterized by increased cell contractility, formation of alpha smooth muscle actin (α-SMA)-stress fibers, together with the reorganization of cell-matrix and cell-cell contacts and cytoskeletal architecture [127,129–131]. Our in vitro studies confirmed the phenotypic conversion of hEDS/HSD fibroblasts into migrating myofibroblast-like cells, since they express the typical markers α-SMA and cadherin-11 and show augmented levels of the protease MMP9 and an altered expression of the inflammation mediators CYR61 and CTGF [49]. This phenotypic switch is elicited by a signal transduction pathway involving the αvβ3 integrin that signals through the integrin linked kinase (ILK) and the transcription factor Snail1 [49,50]. This myofibroblast-like phenotype observed in vitro might reflect a persistent in vivo inflammatory-like condition consistent with the patients' systemic clinical manifestations, comprising gastrointestinal dysfunction, increased susceptibility to osteoarthritis, chronic generalized musculoskeletal pain, inflammatory soft-tissue lesions, and neurological features [110,111].

Activation of myofibroblasts is itself part of physiological wound repair following tissue injury, whereas in chronic injury and inflammatory fibrotic conditions their persistent activation exacerbates the disease progression [132–135]. During fibroblast-to-myofibroblast transition, a complex mechanochemical signaling is activated involving profibrotic secreted factors such as TGFβ and Wnt and ECM-degrading enzymes and intracellular effectors required for stress fiber contractility [127,131]. At molecular level, the cytokine TGFβ is considered the master regulator of profibrotic processes. A growing body of evidence has highlighted the regulation of the Wnt/β-catenin pathway by TGFβ as well as the involvement of their downstream molecular effectors in the fibroblast-to-myofibroblast transition and fibrotic responses [136].

As revealed by transcriptome analysis, several signaling pathways essential for cell growth and proliferation were found to be likely perturbed in hEDS/HSD fibroblasts, i.e., TGFβ, TNF, Jak-STAT, and PI3K-Akt [123]. Transcriptomics data also suggested the differential expression of different Wnt-related genes including the up-regulated frizzled receptor 3 (*FZD3*) and the down-regulated Wnt negative regulators *PRICKLE1* and *SFRP2*. *SFRP2*, the most down-regulated transcript in patients' cells, acts as a critical Wnt modulator, since it directly binds to Wnt proteins and prevents their interactions with FZD receptors [137].

The synergistic crosstalk between TGFβ and Wnt signaling in the myofibroblast activation is documented as well as the inhibitory role of SFRP2 in the TGFβ-dependent myofibroblast formation and post-inflammatory fibrosis [138,139]. In this view, a possible involvement of these signaling pathways in the pathomechanisms of hEDS/HSD can be envisaged. Our findings may offer further clues to address important questions concerning the activation of these pathological mechanisms, and it remains to be clarified which growth factors, i.e., TGFβ, CTGF, and key regulatory pathways sustain the fibroblast-to-myofibroblast transition of hEDS/HSD cells.

11. Differential Expression of Genes Involved in Inflammatory, Immune, and Pain Response in hEDS/HSD Cells

Over the past few years, clinical research described the presence of comorbidities in hEDS/HSD patients, such as functional gastrointestinal and eosinophilic disorders [140,141], increased prevalence of asthma [142], and chronic pain syndromes, i.e., chronic fatigue, fibromyalgia, irritable bowel disease, and inflammatory joints conditions [111,143], though specific underlying causes and mechanisms remain to be explored. In this regard, transcriptome of hEDS/HSD cells revealed the aberrant transcription of a range of genes related to inflammation, pain, and immune responses, i.e., *AQP9*, *CFD*, *SPON2*, *PRLR*, and *NR4A* receptors, which might impair biological functions and molecular pathways with a potential role in the disease's pathogenesis. Among them, patients' cells showed the enhanced expression of *AQP9*, a member of the family of water-selective membrane channels that play a role both in antimicrobial defense and skin barrier permeability [144]. A high expression of this transporter was detected in synovial tissues and fibroblast-like synoviocytes from osteoarthritis and rheumatoid arthritis patients and may have a role in the pathogenesis of inflammatory synovitis [145,146].

In line with this finding, patients' cells also showed increased mRNA levels of complement factor D (*CFD*), a component of the alternative complement pathway [147], which is involved in pathophysiological mechanisms of osteoarthritis and is considered as a potential predictive biomarker of joint pain in patients with hip and knee osteoarthritis [148–150].

In hEDS/HSD no reliable biomarkers have been identified. A previous study identified elevated basal serum tryptase levels due to increased *TPSAB1* copy number associated with hereditary alpha tryptasemia in individuals with multisystem complaints, i.e., joint hypermobility, sleep disruption, irritable bowel syndrome, body pain, headache, arthralgia, and chronic gastroesophageal reflux, partly overlapping with those frequently observed in hEDS/HSD patients [151]. In our hEDS/HSD patients, no elevated basal serum tryptase level was observed, suggesting the absence of the association between their clinical features and copy number variations in the *TPSAB1* gene.

Other inflammation-related genes dysregulated in hEDS/HSD include *SPON2*, up-regulated in patients' cells, which encodes an ECM protein with multifunctional properties in the innate immune system and inflammatory cell recruitment [152–154], and *PRLR*, showing a decreased expression in hEDS/HSD cells, which encode the prolactin receptor implicated in inflammatory responses and immune cells regulation [155,156]. The prolactin-PRLR axis contributes to the activation of pain-related pathways through the sensitization of transient receptor potential channels that promote painful sensations [157,158]. Chronic pain represents a common complaint among hEDS/HSD patients affecting their quality of life [111,116], though specific molecular pathways or mediators of pain are still unknown.

As further evidence of unbalanced inflammatory responses in patients' cells, transcriptome revealed a decreased expression of the *NR4A* nuclear receptors (*NR4A1*, *NR4A2*, *NR4A3*), which act as transcriptional regulators of inflammatory responses mediated by NF-kB signaling [159–161]. These receptors attenuate inflammatory events through inhibition of the NF-kB nuclear translocation and induction of the expression of its inhibitor NFKBIA, which, in turn, blocks the NF-kB nuclear localizing sequence [159]. The concomitant decreased mRNA levels of *NR4A1* and *NFKBIA* in patients' cells may be related to aberrant NF-kB signaling.

Despite additional work being required to support these hypotheses, our findings depict the complex sequence of transcriptional events that should stimulate more investigations to provide new insights into the pathomechanisms underlying the molecular networks related to aberrant inflammatory responses associated with hEDS/HSD.

12. Emerging Aspects of hEDS/HSD Pathophysiology by microRNAs Profiling

Transcriptome analysis may be a valuable strategy also to delineate distinct molecular signatures related to the differential expression of microRNAs (miRNAs). miRNAs are small non-coding RNA molecules ranging from 20–25 nucleotides in length that act mainly as negative regulators of gene expression by promoting the degradation of target mRNAs or repressing their translation [162,163]. Aberrant miRNA expression has been reported in several pathological conditions including cancer, musculoskeletal disorders, painful peripheral neuropathies, and fibromyalgia [164–166].

Our previous expression profiling of hEDS/HSD fibroblasts identified 19 dysregulated miRNAs [123]. Here, we report some examples that might offer interesting clues about the possible involvement of miRNAs in the regulation of potential targets and molecular pathways related to inflammation and Wnt signaling that seem to have a role in the disease pathogenesis.

The most overexpressed miRNA in patients' cells was the miR-378-3p, which is considered a modulator of the epithelial-to-mesenchymal transition and is associated with inflammation and fibrosis through the positive modulation of NF-kB and TNFα pathways [167,168]. miRNA-224, which is also up-regulated in hEDS/HSD cells, is associated with the activation of the Wnt/β-catenin signaling through the inhibition of the expression of glycogen synthase kinase 3β and SFRP2, which are known Wnt suppressors [169]. Therefore, it is likely reasonable to assume that the enhanced expression of this miRNA may contribute to the decreased mRNA level of *SFRP2* observed in hEDS/HSD cells, supporting the hypothesis that the aberrant signaling of the Wnt/β-catenin axis might play a role in the disease mechanisms of hEDS/HSD.

We previously demonstrated the involvement of the ILK in the fibroblast-to-myofibroblast switch of hEDS/HSD cells. The ILK acts downstream of the phosphatidylinositol 3-kinase signaling pathway and negatively regulates the action of the glycogen synthase kinase 3β by phosphorylation of a specific serine residue, further strengthening our assumption that Wnt/β-catenin signaling is involved in hEDS/HSD pathogenesis [49,50]. Consistently, patients' fibroblasts show a reduced expression of the miRNA-23a, which is implicated in the Wnt pathway regulation as well, by inhibiting the expression of FDZ5 and FDZ7 receptors [170]. Decreased levels of this miRNA were also found in cerebrospinal fluid and serum of patients with fibromyalgia [171], a painful disorder in differential diagnosis with hEDS/HSD [172], which might suggest a possible involvement of specific miRNA signatures or a common disease pattern in both conditions. In addition, the altered expression of this miRNA in synovial fibroblasts of psoriatic arthritis patients, results in the enhanced expression of pro-inflammatory mediators and matrix degrading enzymes, further promoting joint degeneration and synovial inflammation [173].

Several studies highlighted the contribution of miRNAs in the modulation of the expression of ECM structural proteins and related signaling molecules, thus emphasizing the close relationship between ECM homeostasis and inflammatory pain related conditions [174–176].

As the expression of miRNAs can be modulated to mediate the expression of their target genes, in-depth in vitro studies on a large cohort of patients' cells could provide further evidence on mechanisms of action of miRNAs and their impact on diverse target genes and altered pathways relevant for the pathophysiology of hEDS/HSD, thus offering new perspectives to identify potential molecular therapeutic targets.

13. Conclusions and Perspectives

Transcriptome and in vitro analyses on cEDS, vEDS, and hEDS/HSD dermal fibroblasts expanded the knowledge about molecular mechanisms involved in the pathophysiology of these connective tissue disorders (Figure 1).

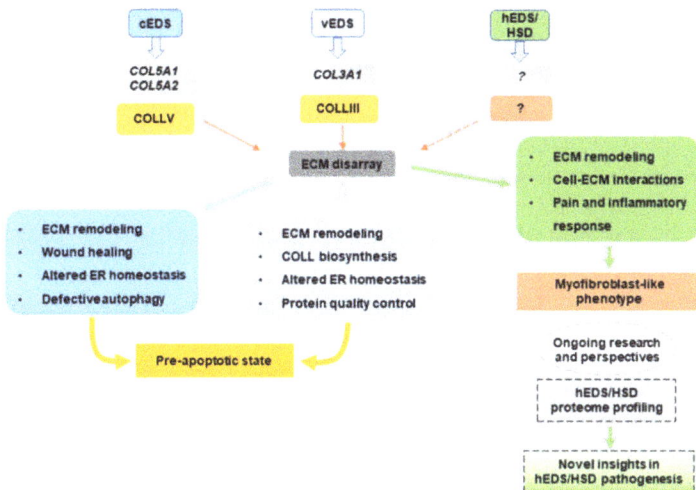

Figure 1. Schematic illustration summarizing the processes likely involved in the pathogenesis of cEDS, vEDS, and hEDS/HSD derived from transcriptome and in vitro studies of patients' skin fibroblasts.

Our findings indicate that these cells share a deregulated expression of many matrix-related genes and a widespread disarray of several ECM structural constituents, thus highlighting the functional relevance of a proper organization and function of the ECM in providing stability to connective tissues. In cEDS and vEDS dermal fibroblasts, the pathological ECM turnover is directly caused by the underlying molecular defect causing abnormal expression of collagen V and collagen III, which, in turn, perturbs key physiological processes critical for collagen processing itself and to maintain cell homeostasis. In the absence of a known genetic etiology, the abnormal ECM organization present in hEDS/HSD cells may be a functional consequence of excessive remodeling due to increased levels of ECM-degrading enzymes and concomitant acquisition of a pro-inflammatory myofibroblast-like phenotype. hEDS/HSD transcriptome profiling for the first time has shed light on different pathobiological aspects of the disease. The dysregulated expression of genes involved in cell-matrix interactions and specific intracellular signaling pathways may have a role in the phenotypic switch of hEDS/HSD cells. Transcriptional changes of different genes and miRNAs involved in molecular pathways related to pain and inflammatory response might provide further clues to dissect the intricate biological events involved in chronic and musculoskeletal pain affecting hEDS/HSD patients. To deepen the knowledge on hEDS/HSD pathophysiology, proteome profiling of patients' cells is currently ongoing to decipher the complex protein network and identify potential bioactive molecules involved in the disease pathogenesis to offer therapeutic options for hEDS/HSD patients.

Author Contributions: Conceptualization: N.C., M.R., M.C.; writing and original draft preparation: N.C., M.R., M.C.; writing, review and editing: N.C., M.R., N.Z., M.C.; supervision: M.C.

Funding: This research received no external funding.

Acknowledgments: The authors wish to thank the patients for their kind availability for these studies and the Fazzo Cusan family for its generous support.

Conflicts of Interest: The authors declare no conflicts of interests.

References

1. Theocharis, A.D.; Manou, D.; Karamanos, N.K. The extracellular matrix as a multitasking player in disease. *FEBS J.* **2019**. [CrossRef] [PubMed]
2. Arseni, L.; Lombardi, A.; Orioli, D. From Structure to Phenotype: Impact of Collagen Alterations on Human Health. *Int. J. Mol. Sci.* **2018**, *19*, 5. [CrossRef] [PubMed]
3. Murphy-Ullrich, J.E.; Sage, E.H. Revisiting the matricellular concept. *Matrix Biol.* **2014**, *37*, 1–14. [CrossRef] [PubMed]
4. Hynes, R.O.; Naba, A. Overview of the matrisome—An inventory of extracellular matrix constituents and functions. *Cold Spring Harb. Perspect. Biol.* **2012**, *4*, a004903. [CrossRef] [PubMed]
5. Naba, A.; Clauser, K.R.; Ding, H.; Whittaker, C.A.; Carr, S.A.; Hynes, R.O. The extracellular matrix: Tools and insights for the "omics" era. *Matrix Biol.* **2016**, *49*, 10–24. [CrossRef] [PubMed]
6. Myllyharju, J.; Kivirikko, K.I. Collagens, modifying enzymes and their mutations in humans, flies and worms. *Trends Genet.* **2004**, *20*, 33–43. [CrossRef] [PubMed]
7. Nyström, A.; Bruckner-Tuderman, L. Matrix molecules and skin biology. *Semin. Cell Dev. Biol.* **2019**, *89*, 136–146. [CrossRef] [PubMed]
8. Gjaltema, R.A.; Bank, R.A. Molecular insights into prolyl and lysyl hydroxylation of fibrillar collagens in health and disease. *Crit. Rev. Biochem. Mol. Biol.* **2017**, *52*, 74–95. [CrossRef] [PubMed]
9. Sorushanova, A.; Delgado, L.M.; Wu, Z.; Shologu, N.; Kshirsagar, A.; Raghunath, R.; Mullen, A.M.; Bayon, Y.; Pandit, A.; Raghunath, M.; et al. The Collagen Suprafamily: From Biosynthesis to Advanced Biomaterial Development. *Adv. Mater.* **2019**, *31*, e1801651. [CrossRef] [PubMed]
10. Malfait, F. Vascular aspects of the Ehlers-Danlos Syndromes. *Matrix Biol.* **2018**, *71–72*, 380–395. [CrossRef] [PubMed]
11. Lenselink, E.A. Role of fibronectin in normal wound healing. *Int. Wound J.* **2015**, *12*, 313–316. [CrossRef] [PubMed]
12. Iozzo, R.V.; Gubbiotti, M.A. Extracellular matrix: The driving force of mammalian diseases. *Matrix Biol.* **2018**, *71–72*, 1–9. [CrossRef] [PubMed]
13. Karamanos, N.K.; Theocharis, A.D.; Neill, T.; Iozzo, R.V. Matrix modeling and remodeling: A biological interplay regulating tissue homeostasis and diseases. *Matrix Biol.* **2019**, *75–76*, 1–11. [CrossRef] [PubMed]
14. Schwartz, M.A. Integrins and extracellular matrix in mechanotransduction. *Cold Spring Harb. Perspect. Biol.* **2010**, *2*, a005066. [CrossRef] [PubMed]
15. Campbell, I.D.; Humphries, M.J. Integrin structure, activation, and interactions. *Cold Spring Harb. Perspect. Biol.* **2011**, *3*, a004994. [CrossRef] [PubMed]
16. Frisch, S.M.; Francis, H. Disruption of epithelial cell matrix interactions induces apoptosis. *J. Cell Biol.* **1994**, *124*, 619–626. [CrossRef] [PubMed]
17. Vlahakis, A.; Debnath, J. The Interconnections between Autophagy and Integrin-Mediated Cell Adhesion. *J. Mol. Biol.* **2017**, *429*, 515–530. [CrossRef]
18. Lock, R.; Debnath, J. Extracellular matrix regulation of autophagy. *Curr. Opin. Cell Biol.* **2008**, *20*, 583–588. [CrossRef]
19. Neill, T.; Schaefer, L.; Iozzo, R.V. Instructive roles of extracellular matrix on autophagy. *Am. J. Pathol.* **2014**, *184*, 2146–2153. [CrossRef]
20. Settembre, C.; Cinque, L.; Bartolomeo, R.; Di Malta, C.; De Leonibus, C.; Forrester, A. Defective collagen proteostasis and matrix formation in the pathogenesis of lysosomal storage disorders. *Matrix Biol.* **2018**, *71–72*, 283–293. [CrossRef]
21. Lamandé, S.R.; Bateman, J.F. Collagen VI disorders: Insights on form and function in the extracellular matrix and beyond. *Matrix Biol.* **2018**, *71–72*, 348–367. [CrossRef]
22. Wilkinson, S. Emerging Principles of Selective ER Autophagy. *J. Mol. Biol.* **2019**. [CrossRef]
23. Bonnans, C.; Chou, J.; Werb, Z. Remodelling the extracellular matrix in development and disease. *Nat. Rev. Mol. Cell Biol.* **2014**, *15*, 786–801. [CrossRef]
24. Frevert, C.W.; Felgenhauer, J.; Wygrecka, M.; Nastase, M.V.; Schaefer, L. Danger-Associated Molecular Patterns Derived From the Extracellular Matrix Provide Temporal Control of Innate Immunity. *J. Histochem. Cytochem.* **2018**, *66*, 213–227. [CrossRef]

25. Genovese, F.; Karsdal, M.A. Protein degradation fragments as diagnostic and prognostic biomarkers of connective tissue diseases: Understanding the extracellular matrix message and implication for current and future serological biomarkers. *Expert Rev. Proteom.* **2016**, *13*, 213–225. [CrossRef]
26. Bhattacharyya, S.; Midwood, K.S.; Yin, H.; Varga, J. Toll-Like Receptor-4 Signaling Drives Persistent Fibroblast Activation and Prevents Fibrosis Resolution in Scleroderma. *Adv. Wound Care* **2017**, *6*, 356–369. [CrossRef]
27. Lamandé, S.R.; Bateman, J.F. Genetic Disorders of the Extracellular Matrix. *Anat. Rec.* **2019**. [CrossRef]
28. Colombi, M.; Dordoni, C.; Chiarelli, N.; Ritelli, M. Differential diagnosis and diagnostic flow chart of joint hypermobility syndrome/Ehlers-Danlos syndrome hypermobility type compared to other heritable connective tissue disorders. *Am. J. Med. Genet. Part C* **2015**, *169*, 6–22. [CrossRef]
29. Bateman, J.F.; Boot-Handford, R.P.; Lamande, S.R. Genetic diseases of connective tissues: Cellular and extra cellular effects of ECM mutations. *Nat. Rev. Genet.* **2009**, *10*, 173–183. [CrossRef]
30. Lu, P.; Takai, K.; Weaver, V.M.; Werb, Z. Extracellular matrix degradation and remodeling in development and disease. *Cold Spring Harb. Perspect. Biol.* **2011**, *3*, a005058. [CrossRef]
31. Karsdal, M.A.; Nielsen, M.J.; Sand, J.M.; Henriksen, K.; Genovese, F.; Bay-Jensen, A.C.; Smith, V.; Adamkewicz, J.I.; Christiansen, C.; Leeming, D.J. Extracellular matrix remodeling: The common denominator in connective tissue diseases. Possibilities for evaluation and current understanding of the matrix as more than a passive architecture, but a key player in tissue failure. *ASSAY Drug Dev. Technol.* **2013**, *11*, 70–92. [CrossRef]
32. Malfait, F.; Francomano, C.; Byers, P.; Belmont, J.; Berglund, B.; Black, J.; Bloom, L.; Bowen, J.M.; Brady, A.F.; Burrows, N.P.; et al. The 2017 international classification of the Ehlers-Danlos syndromes. *Am. J. Med. Genet. C Semin. Med. Genet.* **2017**, *175*, 8–26. [CrossRef]
33. Blackburn, P.R.; Xu, Z.; Tumelty, K.E.; Zhao, R.W.; Monis, W.J.; Harris, K.G.; Gass, J.M.; Cousin, M.A.; Boczek, N.J.; Mitkov, M.V.; et al. Bi-allelic alterations in AEBP1 lead to defective collagen assembly and connective tissue structure resulting in a variant of Ehlers-Danlos syndrome. *Am. J. Hum. Genet.* **2018**, *102*, 696–705. [CrossRef]
34. Hebebrand, M.; Vasileiou, G.; Krumbiegel, M.; Kraus, C.; Uebe, S.; Ekici, A.B.; Thiel, C.T.; Reis, A.; Popp, B. A biallelic truncating AEBP1 variant causes connective tissue disorder in two siblings. *Am. J. Med. Genet. A* **2019**, *179*, 50–56. [CrossRef]
35. Syx, D.; De Wandele, I.; Symoens, S.; De Rycke, R.; Hougrand, O.; Voermans, N.; De Paepe, A.; Malfait, F. Bi-allelic AEBP1 mutations in two patients with Ehlers-Danlos syndrome. *Hum. Mol. Genet.* **2019**, *28*, 1853–1864. [CrossRef]
36. Ritelli, M.; Cinquina, V.; Venturini, M.; Pezzaioli, L.; Formenti, A.M.; Chiarelli, N.; Colombi, M. Expanding the Clinical and Mutational Spectrum of Recessive AEBP1-Related Classical-Like Ehlers-Danlos Syndrome. *Genes.* **2019**, *10*, 135. [CrossRef]
37. Maugeri, A.; De Paepe, A.; Malfait, F. Genetics and testing of Ehlers-Danlos syndrome and of differential diagnostic diseases. In *Ehlers-Danlos Syndrome: A Multidisciplinary Approach*, 1st ed.; Jacobs, J.W.G., Cornelissens, L.J.M., Eds.; IOS Press BV: Amsterdam, The Netherlands, 2018; Chapter 3, pp. 33–46.
38. Damme, T.M.; Syx, D.; Coucke, P.; Symoens, S.; De Paepe, A.; Malfait, F. Genetics of the Ehlers–Danlos syndrome: More than collagen disorders. *Expert Opin. Orphan Drugs* **2015**, *3*, 379–392. [CrossRef]
39. Bowen, J.M.; Sobey, G.J.; Burrows, N.P.; Colombi, M.; Lavallee, M.E.; Malfait, F.; Francomano, C.A. Ehlers-Danlos syndrome, classical type. *Am. J. Med. Genet. C Semin. Med. Genet.* **2017**, *75*, 27–39. [CrossRef]
40. Symoens, S.; Syx, D.; Malfait, F.; Callewaert, B.; De Backer, J.; Vanakker, O.; Coucke, P.; De Paepe, A. Comprehensive molecular analysis demonstrates type V collagen mutations in over 90% of patients with classic EDS and allows to refine diagnostic criteria. *Hum. Mutat.* **2012**, *33*, 1485–1493. [CrossRef]
41. Ritelli, M.; Dordoni, C.; Venturini, M.; Chiarelli, N.; Quinzani, S.; Traversa, M.; Zoppi, N.; Vascellaro, A.; Wischmeijer, A.; Manfredini, E.; et al. Clinical and molecular characterization of 40 patients with classic Ehlers-Danlos syndrome: Identification of 18 COL5A1 and 2 COL5A2 novel mutations. *Orphanet J. Rare Dis.* **2013**, *8*, 58. [CrossRef]
42. Wenstrup, R.J.; Florer, J.B.; Brunskill, E.W.; Bell, S.M.; Chervoneva, I.; Birk, D.E. Type V collagen controls the initiation of collagen fibril assembly. *J. Biol. Chem.* **2004**, *279*, 53331–53337. [CrossRef]
43. Mak, K.M.; Png, C.Y.; Lee, D.J. Type V Collagen in Health, Disease, and Fibrosis. *Anat. Rec.* **2016**, *299*, 613–629. [CrossRef]

44. Zoppi, N.; Gardella, R.; De Paepe, A.; Barlati, S.; Colombi, M. Human fibroblasts with mutations in COL5A1 and COL3A1 genes do not organize collagens and fibronectin in the extracellular matrix, down-regulate α2β1 integrin, and recruit αvβ3 instead of α5β1 integrin. *J. Biol. Chem.* **2004**, *30*, 18157–18168. [CrossRef]
45. Zoppi, N.; Barlati, S.; Colombi, M. FAK-independent αvβ3 integrin-EGFR complexes rescue from anoikis matrix-defective fibroblasts. *Biochim. Biophys. Acta* **2008**, *1783*, 1177–1188. [CrossRef]
46. Zoppi, N.; Ritelli, M.; Colombi, M. Type III and V collagens modulate the expression and assembly of EDA+fibronectin in the extracellular matrix of defective Ehlers-Danlos syndrome fibroblasts. *Biochim. Biophys. Acta* **2012**, *1820*, 1576–1587. [CrossRef]
47. Chiarelli, N.; Carini, G.; Zoppi, N.; Ritelli, M.; Colombi, M. Molecular insights in the pathogenesis of classical Ehlers-Danlos syndrome from transcriptome-wide expression profiling of patients' skin fibroblasts. *PLoS ONE* **2019**, *14*, e0211647. [CrossRef]
48. Viglio, S.; Zoppi, N.; Sangalli, A.; Gallanti, A.; Barlati, S.; Mottes, M.; Colombi, M.; Valli, M. Rescue of migratory defects of Ehlers-Danlos syndrome fibroblasts in vitro by type V collagen but not insulin-like binding protein-1. *J. Investig. Dermatol.* **2008**, *128*, 1915–1919. [CrossRef]
49. Zoppi, N.; Chiarelli, N.; Binetti, S.; Ritelli, M.; Colombi, M. Dermal fibroblast-to-myofibroblast transition sustained by αvβ3 integrin-ILK-Snail1/Slug signaling is a common feature for hypermobile Ehlers-Danlos syndrome and hypermobility spectrum disorders. *Biochim. Biophys. Acta* **2018**, *1864*, 1010–1023. [CrossRef]
50. Zoppi, N.; Chiarelli, N.; Ritelli, M.; Colombi, M. Multifaced Roles of the αvβ3 Integrin in Ehlers-Danlos and Arterial Tortuosity Syndromes' Dermal Fibroblasts. *Int. J. Mol. Sci.* **2018**, *19*, 982. [CrossRef]
51. DeNigris, J.; Yao, Q.; Birk, E.K.; Birk, D.E. Altered dermal fibroblast behavior in a collagen V haploinsufficient murine model of classic Ehlers-Danlos syndrome. *Connect. Tissue Res.* **2016**, *57*, 1–9. [CrossRef]
52. Park, A.C.; Phan, N.; Massoudi, D.; Liu, Z.; Kernien, J.F.; Adams, S.M.; Davidson, J.M.; Birk, D.E.; Liu, B.; Greenspan, D.S. Deficits in Col5a2 Expression Result in Novel Skin and Adipose Abnormalities and Predisposition to Aortic Aneurysms and Dissections. *Am. J. Pathol.* **2017**, *187*, 2300–2311. [CrossRef]
53. Rousselle, P.; Montmasson, M.; Garnier, C. Extracellular matrix contribution to skin wound re-epithelialization. *Matrix Biol.* **2019**, *75–76*, 12–26. [CrossRef]
54. Tracy, L.E.; Minasian, R.A.; Caterson, E.J. Extracellular Matrix and Dermal Fibroblast Function in the Healing Wound. *Adv. Wound Care* **2016**, *5*, 119–136. [CrossRef]
55. Chester, D.; Brown, A.C. The role of biophysical properties of provisional matrix proteins in wound repair. *Matrix Biol.* **2017**, *60–61*, 124–140. [CrossRef]
56. Wang, W.; Li, P.; Li, W.; Jiang, J.; Cui, Y.; Li, S.; Wang, Z. Osteopontin activates mesenchymal stem cells to repair skin wound. *PLoS ONE* **2017**, *12*, e0185346. [CrossRef]
57. Icer, M.A.; Gezmen-Karadag, M. The multiple functions and mechanisms of osteopontin. *Clin. Biochem.* **2018**, *59*, 17–24. [CrossRef]
58. Ross, F.P.; Chappel, J.; Alvarez, J.; Sander, D.; Butler, W.; Farach-Carson, M.; Mintz, K.; Robey, P.G.; Teitelbaum, S.; Cheresh, D. Interactions between the bone matrix proteins osteopontin and bone sialoprotein and the osteoclast integrin alpha v beta 3 potentiate bone resorption. *J. Biol. Chem.* **1993**, *268*, 9901–9907.
59. Giachelli, C.M.; Steitz, S. Osteopontin: A versatile regulator of inflammation and biomineralization. *Matrix Biol.* **2000**, *19*, 615–622. [CrossRef]
60. Liaw, L.; Birk, D.E.; Ballas, C.B.; Whitsitt, J.S.; Davidson, J.M.; Hogan, B.L. Altered wound healing in mice lacking a functional osteopontin gene (spp1). *J. Clin. Investig.* **1998**, *101*, 1468–1478. [CrossRef]
61. Maruhashi, T. Interaction between periostin and BMP-1 promotes proteolytic activation of lysyl oxidase. *J. Biol. Chem.* **2010**, *285*, 13294–13303. [CrossRef]
62. Egbert, M.; Ruetze, M.; Sattler, M.; Wenck, H.; Gallinat, S.; Lucius, R.; Weise, J.M. The matricellular protein periostin contributes to proper collagen function and is downregulated during skin aging. *J. Dermatol. Sci.* **2014**, *73*, 40–48. [CrossRef]
63. González-González, L.; Alonso, J. Periostin: A Matricellular Protein with Multiple Functions in Cancer Development and Progression. *Front. Oncol.* **2018**, *8*, 225. [CrossRef]
64. Li, G.; Jin, R.; Norris, R.A.; Zhang, L.; Yu, S.; Wu, F.; Markwald, R.R.; Nanda, A.; Conway, S.J.; Smyth, S.S.; et al. Periostin mediates vascular smooth muscle cell migration through the integrins alphavbeta3 and alphavbeta5 and focal adhesion kinase (FAK) pathway. *Atherosclerosis* **2010**, *208*, 358–365. [CrossRef]
65. Walker, J.T.; McLeod, K.; Kim, S.; Conway, S.J.; Hamilton, D.W. Periostin as a multifunctional modulator of the wound healing response. *Cell Tissue Res.* **2016**, *365*, 453–465. [CrossRef]

66. Nunomura, S.; Nanri, Y.; Ogawa, M.; Arima, K.; Mitamura, Y.; Yoshihara, T.; Hasuwa, H.; Conway, S.J.; Izuhara, K. Constitutive overexpression of periostin delays wound healing in mouse skin. *Wound Repair Regen.* **2018**, *26*, 6–15. [CrossRef]
67. Lim, P.J.; Lindert, U.; Opitz, L.; Hausser, I.; Rohrbach, M.; Giunta, C. Transcriptome Profiling of Primary Skin Fibroblasts Reveal Distinct Molecular Features Between PLOD1-and FKBP14-Kyphoscoliotic Ehlers-Danlos Syndrome. *Genes* **2019**, *10*, 517. [CrossRef]
68. Penta, K.; Varner, J.A.; Liaw, L.; Hidai, C.; Schatzman, R.; Quertermous, T. Del1 induces integrin signaling and angiogenesis by ligation of alphaVbeta3. *J. Biol. Chem.* **1999**, *274*, 11101–11109. [CrossRef]
69. Shen, W.; Zhu, S.; Qin, H.; Zhong, M.; Wu, J.; Zhang, R.; Song, H. EDIL3 knockdown inhibits retinal angiogenesis through the induction of cell cycle arrest in vitro. *Mol. Med. Rep.* **2017**, *16*, 4054–4060. [CrossRef]
70. Wang, Z.; Boyko, T.; Tran, M.C.; LaRussa, M.; Bhatia, N.; Rashidi, V.; Longaker, M.T.; Yang, G.P. DEL1 protects against chondrocyte apoptosis through integrin binding. *J. Surg. Res.* **2018**, *231*, 1–9. [CrossRef]
71. Hajishengallis, G.; Chavakis, T. DEL-1-Regulated Immune Plasticity and Inflammatory Disorders. *Trends Mol. Med.* **2019**, *25*, 444–459. [CrossRef]
72. Brandt, K.; Grunler, J.; Brismar, K.; Wang, J. Effects of IGFBP-1 and IGFBP-2 and their fragments on migration and IGF-induced proliferation of human dermal fibroblasts. *Growth Horm. IGF Res.* **2015**, *25*, 34–40. [CrossRef]
73. Katayama, H.; Tamai, K.; Shibuya, R.; Nakamura, M.; Mochizuki, M.; Yamaguchi, K.; Kawamura, S.; Tochigi, T.; Sato, I.; Okanishi, T.; et al. Long non-coding RNA HOTAIR promotes cell migration by upregulating insulin growth factor-binding protein 2 in renal cell carcinoma. *Sci. Rep.* **2017**, *7*, 12016. [CrossRef]
74. Ahmad, S.; Bhatia, K.; Kindelin, A.; Ducruet, A.F. The Role of Complement C3a Receptor in Stroke. *Neuromol. Med.* **2019**. [CrossRef]
75. Korkmaz, H.I.; Krijnen, P.A.J.; Ulrich, M.M.W.; de Jong, E.; van Zuijlen, P.P.M.; Niessen, H.W.M. The role of complement in the acute phase response after burns. *Burns* **2017**, *43*, 1390–1399. [CrossRef]
76. Rafail, S.; Kourtzelis, I.; Foukas, P.G.; Markiewski, M.M.; DeAngelis, R.A.; Guariento, M.; Ricklin, D.; Grice, E.A.; Lambris, J.D. Complement deficiency promotes cutaneous wound healing in mice. *J. Immunol.* **2015**, *194*, 1285–1291. [CrossRef]
77. Colombi, M.; Dordoni, C.; Venturini, M.; Ciaccio, C.; Morlino, S.; Chiarelli, N.; Zanca, A.; Calzavara-Pinton, P.; Zoppi, N.; Castori, M.; et al. Spectrum of mucocutaneous, ocular and facial features and delineation of novel presentations in 62 classical Ehlers-Danlos syndrome patients. *Clin. Genet.* **2017**, *92*, 624–631. [CrossRef]
78. Smith, M.; Wilkinson, S. ER homeostasis and autophagy. *Essays Biochem.* **2017**, *61*, 625–635. [CrossRef]
79. Wong, M.Y.; Shoulders, M.D. Targeting defective proteostasis in the collagenopathies. *Curr. Opin. Chem. Biol.* **2019**, *50*, 80–88. [CrossRef]
80. Moon, H.W.; Han, H.G.; Jeon, Y.J. Protein Quality Control in the Endoplasmic Reticulum and Cancer. *Int. J. Mol. Sci.* **2018**, *19*, 3020. [CrossRef]
81. Mogk, A.; Bukau, B.; Kampinga, H.H. Cellular Handling of Protein Aggregates by Disaggregation Machines. *Mol. Cell* **2018**, *69*, 214–226. [CrossRef]
82. Wilkinson, S. ER-phagy: Shaping up and destressing the endoplasmic reticulum. *FEBS J.* **2019**. [CrossRef]
83. Smith, M.D.; Harley, M.E.; Kemp, A.J.; Wills, J.; Lee, M.; Arends, M.; von Kriegsheim, A.; Behrends, C.; Wilkinson, S. CCPG1 is a non-canonical autophagy cargo receptor essential for ER-phagy and pancreatic ER proteostasis. *Dev. Cell* **2018**, *44*, 217–232. [CrossRef]
84. Forrester, A.; De Leonibus, C.; Grumati, P.; Fasana, E.; Piemontese, M.; Staiano, L.; Fregno, I.; Raimondi, A.; Marazza, A.; Bruno, G.; et al. A selective ER-phagy exerts procollagen quality control via a Calnexin-FAM134B complex. *EMBO J.* **2019**, *38*, e99847. [CrossRef]
85. Ishida, Y.; Yamamoto, A.; Kitamura, A.; Lamandé, S.R.; Yoshimori, T.; Bateman, J.F.; Kubota, H.; Nagata, K. Autophagic elimination of misfolded procollagen aggregates in the endoplasmic reticulum as a means of cell protection. *Mol. Biol. Cell* **2009**, *20*, 2744–2754. [CrossRef]
86. Nagahama, M.; Suzuki, M.; Hamada, Y.; Hatsuzawa, K.; Tani, K.; Yamamoto, A.; Tagaya, M. SVIP is a novel VCP/p97-interacting protein whose expression causes cell vacuolation. *Mol. Biol. Cell* **2003**, *14*, 262–273. [CrossRef]
87. Ballar, P.; Zhong, Y.; Nagahama, M.; Tagaya, M.; Shen, Y.; Fang, S. Identification of SVIP as an endogenous inhibitor of endoplasmic reticulum-associated degradation. *J. Biol. Chem.* **2007**, *282*, 33908–33914. [CrossRef]

88. Wang, Y.; Ballar, P.; Zhong, Y.; Zhang, X.; Liu, C.; Zhang, Y.J.; Monteiro, M.J.; Li, J.; Fang, S. SVIP induces localization of p97/VCP to the plasma and lysosomal membranes and regulates autophagy. *PLoS ONE* **2011**, *6*, e24478. [CrossRef]
89. Jia, D.; Wang, Y.Y.; Wang, P.; Huang, Y.; Liang, D.Y.; Wang, D.; Cheng, C.; Zhang, C.; Guo, L.; Liang, P.; et al. SVIP alleviates CCl4-induced liver fibrosis via activating autophagy and protecting hepatocytes. *Cell Death Dis.* **2019**, *10*, 71. [CrossRef]
90. Fung, C.; Lock, R.; Gao, S.; Salas, E.; Debnath, J. Induction of autophagy during extracellular matrix detachment promotes cell survival. *Mol. Biol. Cell* **2008**, *19*, 797–806. [CrossRef]
91. Kawano, S.; Torisu, T.; Esaki, M.; Torisu, K.; Matsuno, Y.; Kitazono, T. Autophagy promotes degradation of internalized collagen and regulates distribution of focal adhesions to suppress cell adhesion. *Biol. Open* **2017**, *6*, 1644–1653. [CrossRef]
92. Byers, P.H.; Belmont, J.; Black, J.; De Backer, J.; Frank, M.; Jeunemaitre, X.; Johnson, D.; Pepin, M.; Robert, L.; Sanders, L.; et al. Diagnosis, natural history, and management in vascular Ehlers-Danlos syndrome. *Am. J. Med. Genet. C Semin. Med. Genet.* **2017**, *175*, 40–47. [CrossRef]
93. Ritelli, M.; Rovati, C.; Venturini, M.; Chiarelli, N.; Cinquina, V.; Castori, M.; Colombi, M. Application of the 2017 criteria for vascular Ehlers-Danlos syndrome in 50 patients ascertained according to the Villefranche nosology. *Clin. Genet.* under review.
94. Pepin, M.; Schwarze, U.; Superti-Furga, A.; Byers, P.H. Clinical and genetic features of Ehlers-Danlos syndrome type IV, the vascular type. *N. Engl. J. Med.* **2000**, *342*, 673–680. [CrossRef]
95. Smith, L.T.; Schwarze, U.; Goldstein, J.; Byers, P.H. Mutations in the COL3A1 gene result in the Ehlers-Danlos syndrome type IV and alterations in the size and distribution of the major collagen fibrils of the dermis. *J. Investig. Dermatol.* **1997**, *108*, 241–247. [CrossRef]
96. Liu, X.; Wu, H.; Byrne, M.; Krane, S.; Jaenisch, R. Type III collagen is crucial for collagen I fibrillogenesis and for normal cardiovascular development. *Proc. Natl. Acad. Sci. USA* **1997**, *94*, 1852–1856. [CrossRef]
97. Chiarelli, N.; Carini, G.; Zoppi, N.; Ritelli, M.; Colombi, M. Transcriptome analysis of skin fibroblasts with dominant negative COL3A1 mutations provides molecular insights into the etiopathology of vascular Ehlers-Danlos syndrome. *PLoS ONE* **2018**, *13*, e0191220. [CrossRef]
98. Sengle, G.; Sakai, L.Y. The fibrillin microfibril scaffold: A niche for growth factors and mechanosensation? *Matrix Biol.* **2015**, *47*, 3–12. [CrossRef]
99. Schiavinato, A.; Keene, D.R.; Wohl, A.P.; Corallo, D.; Colombatti, A.; Wagener, R.; Paulsson, M.; Bonaldo, P.; Sengle, G. Targeting of EMILIN-1 and EMILIN-2 to fibrillin microfibrils facilitates their incorporation into the extracellular matrix. *J. Investig. Dermatol.* **2016**, *136*, 1150–1160. [CrossRef]
100. Boregowda, R.; Paul, E.; White, J.; Ritty, T.M. Bone and soft connective tissue alterations result from loss of fibrillin-2 expression. *Matrix Biol.* **2008**, *27*, 661–666. [CrossRef]
101. Brinckmann, J.; Hunzelmann, N.; Kahle, B.; Rohwedel, J.; Kramer, J.; Gibson, M.A.; Hubmacher, D.; Reinhardt, D.P. Enhanced fibrillin-2 expression is a general feature of wound healing and sclerosis: Potential alteration of cell attachment and storage of TGF-beta. *Lab. Investig.* **2010**, *90*, 739–752. [CrossRef]
102. D'Hondt, S.; Guillemyn, B.; Syx, D.; Symoens, S.; De Rycke, R.; Vanhoutte, L.; Toussaint, W.; Lambrecht, B.N.; De Paepe, A.; Keene, D.R.; et al. Type III collagen affects dermal and vascular collagen fibrillogenesis and tissue integrity in a mutant Col3a1 transgenic mouse model. *Matrix Biol.* **2018**, *70*, 72–83. [CrossRef]
103. Ishikawa, Y.; Bächinger, H.P. A substrate preference for the rough endoplasmic reticulum resident protein FKBP22 during collagen biosynthesis. *J. Biol. Chem.* **2014**, *289*, 18189–18201. [CrossRef]
104. Baumann, M.; Giunta, C.; Krabichler, B.; Rüschendorf, F.; Zoppi, N.; Colombi, M.; Bittner, R.E.; Quijano-Roy, S.; Muntoni, F.; Cirak, S.; et al. Mutations in FKBP14 cause a variant of Ehlers-Danlos syndrome with progressive kyphoscoliosis, myopathy, and hearing loss. *Am. J. Hum. Genet.* **2012**, *90*, 201–216. [CrossRef]
105. Soares Moretti, A.; Martins Laurindo, F.R. Protein disulfide isomerases: Redox connections in and out of the endoplasmic reticulum. *Arch. Biochem. Biophys.* **2017**, *617*, 106–119. [CrossRef]
106. Szegezdi, E.; Macdonald, D.C.; Ni Chonghaile, T.; Gupta, S.; Samali, A. Bcl-2 family on guard at the ER. *Am. J. Physiol.-Cell Physiol.* **2009**, *296*, C941–C953. [CrossRef]
107. Sisinni, L.; Pietrafesa, M.; Lepore, S.; Maddalena, F.; Condelli, V.; Esposito, F.; Landriscina, M. Endoplasmic Reticulum Stress and Unfolded Protein Response in Breast Cancer: The Balance between Apoptosis and Autophagy and Its Role in Drug Resistance. *Int. J. Mol. Sci.* **2019**, *20*, 857. [CrossRef]

108. Boot-Handford, R.P.; Briggs, M.D. The unfolded protein response and its relevance to connective tissue diseases. *Cell Tissue Res.* **2010**, *339*, 197–211. [CrossRef]
109. Hughes, A.; Oxford, A.E.; Tawara, K.; Jorcyk, C.L.; Oxford, J.T. Endoplasmic Reticulum Stress and Unfolded Protein Response in Cartilage Pathophysiology; Contributing Factors to Apoptosis and Osteoarthritis. *Int. J. Mol. Sci.* **2017**, *18*, 665. [CrossRef]
110. Tinkle, B.; Castori, M.; Berglund, B.; Cohen, H.; Grahame, R.; Kazkaz, H.; Levy, H. Hypermobile Ehlers-Danlos syndrome (a.k.a. Ehlers-Danlos syndrome type III and Ehlers-Danlos syndrome hypermobility type): Clinical description and natural history. *Am. J. Med. Genet. C Semin. Med. Genet.* **2017**, *175*, 48–69. [CrossRef]
111. Castori, M.; Tinkle, B.; Levy, H.; Grahame, R.; Malfait, F.; Hakim, A. A framework for the classification of joint hypermobility and related conditions. *Am. J. Med. Genet. C Semin. Med. Genet.* **2017**, *175*, 148–157. [CrossRef]
112. Schalkwijk, J.; Zweers, M.C.; Steijlen, P.M.; Dean, W.B.; Taylor, G.; van Vlijmen, I.M.; van Haren, B.; Miller, W.L.; Bristow, J. A recessive form of the Ehlers-Danlos syndrome caused by tenascin-X deficiency. *N. Engl. J. Med.* **2001**, *345*, 1167–1175. [CrossRef]
113. Merke, D.P.; Chen, W.; Morissette, R.; Xu, Z.; Van Ryzin, C.; Sachdev, V.; Hannoush, H.; Shanbhag, S.M.; Acevedo, A.T.; Nishitani, M.; et al. Tenascin-X haploinsufficiency associated with Ehlers-Danlos syndrome in patients with congenital adrenal hyperplasia. *J. Clin. Endocrinol. Metab.* **2013**, *98*, 379–387. [CrossRef]
114. Morissette, R.; Chen, W.; Perritt, A.F.; Dreiling, J.L.; Arai, A.E.; Sachdev, V.; Hannoush, H.; Mallappa, A.; Xu, Z.; McDonnell, N.B.; et al. Broadening the Spectrum of Ehlers Danlos Syndrome in Patients with Congenital Adrenal Hyperplasia. *J. Clin. Endocrinol. Metab.* **2015**, *100*, 1143–1152. [CrossRef]
115. Syx, D.; Symoens, S.; Steyaert, W.; De Paepe, A.; Coucke, P.J.; Malfait, F. Ehlers-Danlos Syndrome, Hypermobility Type, Is Linked to Chromosome 8p22-8p21.1 in an Extended Belgian Family. *Dis. Markers* **2015**, *2015*, 828970. [CrossRef]
116. Copetti, M.; Morlino, S.; Colombi, M.; Grammatico, P.; Fontana, A.; Castori, M. Severity classes in adults with hypermobile Ehlers-Danlos syndrome/hypermobility spectrum disorders: A pilot study of 105 Italian patients. *Rheumatology* **2019**, kez029. [CrossRef]
117. Castori, M.; Dordoni, C.; Valiante, M.; Sperduti, I.; Ritelli, M.; Morlino, S.; Chiarelli, N.; Celletti, C.; Venturini, M.; Camerota, F.; et al. Nosology and inheritance pattern(s) of joint hypermobility syndrome and Ehlers-Danlos syndrome, hypermobility type: A study of intrafamilial and interfamilial variability in 23 Italian pedigrees. *Am. J. Med. Genet. A* **2014**, *164*, 3010–3020. [CrossRef]
118. Ritelli, M.; Chiarelli, N.; Zoppi, N.; Dordoni, C.; Quinzani, S.; Traversa, M.; Venturini, M.; Calzavara-Pinton, P.; Colombi, M. Insights in the etiopathology of galactosyltransferase II (GalT-II) deficiency from transcriptome-wide expression profiling of skin fibroblasts of two sisters with compound heterozygosity for two novel B3GALT6 mutations. *Mol. Genet. Metab. Rep.* **2014**, *20*, 1–15. [CrossRef]
119. Zoppi, N.; Chiarelli, N.; Cinquina, V.; Ritelli, M.; Colombi, M. GLUT10 deficiency leads to oxidative stress and non-canonical $\alpha v \beta 3$ integrin-mediated TGFβ signaling associated with extracellular matrix disarray in arterial tortuosity syndrome skin fibroblasts. *Hum. Mol. Genet.* **2015**, *24*, 6769–6787. [CrossRef]
120. Manzoni, C.; Kia, D.A.; Vandrovcova, J.; Hardy, J.; Wood, N.W.; Lewis, P.A.; Ferrari, R. Genome, transcriptome and proteome: The rise of omics data and their integration in biomedical sciences. *Brief. Bioinform.* **2018**, *19*, 286–302. [CrossRef]
121. Casamassimi, A.; Federico, A.; Rienzo, M.; Esposito, S.; Ciccodicola, A. Transcriptome Profiling in Human Diseases: New Advances and Perspectives. *Int. J. Mol. Sci.* **2017**, *18*, 1652. [CrossRef]
122. Ideozu, J.E.; Zhang, X.; McColley, S.; Levy, H. Transcriptome Profiling and Molecular Therapeutic Advances in Cystic Fibrosis: Recent Insights. *Genes* **2019**, *10*, 180. [CrossRef]
123. Chiarelli, N.; Carini, G.; Zoppi, N.; Dordoni, C.; Ritelli, M.; Venturini, M.; Castori, M.; Colombi, M. Transcriptome-wide expression profiling in skin fibroblasts of patients with joint hypermobility syndrome/Ehlers-Danlos syndrome hypermobility type. *PLoS ONE* **2016**, *11*, e0161347. [CrossRef]
124. Jansen, K.A.; Atherton, P.; Ballestrem, C. Mechanotransduction at the cell-matrix interface. *Semin. Cell Dev. Biol.* **2017**, *71*, 75–83. [CrossRef]
125. Freedman, B.R.; Bade, N.D.; Riggin, C.N.; Zhang, S.; Haines, P.G.; Ong, K.L.; Janmey, P.A. The (dys) functional extracellular matrix. *Biochim. Biophys. Acta* **2015**, *1853*, 3153–3164. [CrossRef]
126. Kalluri, R.; Neilson, E.R. Epithelial-mesenchymal transition and its implications for fibrosis. *J. Clin. Investig.* **2003**, *112*, 1776–1784. [CrossRef]

127. Lamouille, S.; Xu, J.; Derynck, R. Molecular mechanisms of epithelial-mesenchymal transition. *Nat. Rev. Mol. Cell Biol.* **2014**, *15*, 178–196. [CrossRef]
128. McCrea, P.D.; Maher, M.T.; Gottardi, C.J. Nuclear signaling from cadherin adhesion complexes. *Curr. Top. Dev. Biol.* **2015**, *112*, 129–196.
129. Hinz, B.; Gabbiani, G. Cell-matrix and cell-cell contacts of myofibroblasts: Role in connective tissue remodeling. *Thromb. Haemost.* **2003**, *90*, 993–1002.
130. Michalik, M.; Wójcik-Pszczoła, K.; Paw, M.; Wnuk, D.; Koczurkiewicz, P.; Sanak, M.; Pękala, E.; Madeja, Z. Fibroblast-to-myofibroblast transition in bronchial asthma. *Cell. Mol. Life Sci.* **2018**, *75*, 3943–3961. [CrossRef]
131. Hinz, B.; McCulloch, C.A.; Coelho, N.M. Mechanical regulation of myofibroblast phenoconversion and collagen contraction. *Exp. Cell Res.* **2019**, *379*, 119–128. [CrossRef]
132. Pakshir, P.; Hinz, B. The big five in fibrosis: Macrophages, myofibroblasts, matrix, mechanics, and miscommunication. *Matrix Biol.* **2018**, *68–69*, 81–93. [CrossRef]
133. Mullenbrock, S.; Liu, F.; Szak, S.; Hronowski, X.; Gao, B.; Juhasz, P.; Sun, C.; Liu, M.; McLaughlin, H.; Xiao, Q.; et al. Systems Analysis of Transcriptomic and Proteomic Profiles Identifies Novel Regulation of Fibrotic Programs by miRNAs in Pulmonary Fibrosis Fibroblasts. *Genes* **2018**, *9*, 588. [CrossRef]
134. Van Caam, A.; Vonk, M.; van den Hoogen, F.; van Lent, P.; van der Kraan, P. Unraveling SSc Pathophysiology; The Myofibroblast. *Front. Immunol.* **2018**, *9*, 2452. [CrossRef]
135. Watanabe, T.; Baker Frost, D.A.; Mlakar, L.; Heywood, J.; da Silveira, W.A.; Hardiman, G.; Feghali-Bostwick, C. A Human Skin Model Recapitulates Systemic Sclerosis Dermal Fibrosis and Identifies COL22A1 as a TGFβ Early Response Gene that Mediates Fibroblast to Myofibroblast Transition. *Genes* **2019**, *10*, 75. [CrossRef]
136. Działo, E.; Tkacz, K.; Błyszczuk, P. Crosstalk between the TGF-β and WNT signalling pathways during cardiac fibrogenesis. *Acta Biochim. Pol.* **2018**, *65*, 341–349. [CrossRef]
137. Cruciat, C.M.; Niehrs, C. Secreted and transmembrane wnt inhibitors and activators. *Cold Spring Harb. Perspect. Biol.* **2013**, *5*, a015081. [CrossRef]
138. Carthy, J.M.; Garmaroudi, F.S.; Luo, Z.; McManus, B.M. Wnt3a induces myofibroblast differentiation by upregulating TGF-β signaling through SMAD2 in a β-catenin-dependent manner. *PLoS ONE* **2011**, *6*, e19809. [CrossRef]
139. Blyszczuk, P.; Müller-Edenborn, B.; Valenta, T.; Osto, E.; Stellato, M.; Behnke, S.; Glatz, K.; Basler, K.; Lüscher, T.F.; Distler, O.; et al. Transforming growth factor-β-dependent Wnt secretion controls myofibroblast formation and myocardial fibrosis progression in experimental autoimmune myocarditis. *Eur. Heart J.* **2017**, *38*, 1413–1425. [CrossRef]
140. Abonia, J.P.; Wen, T.; Stucke, E.M.; Grotjan, T.; Griffith, M.S.; Kemme, K.A.; Collins, M.H.; Putnam, P.E.; Franciosi, J.P.; Von Tiehl, K.F.; et al. High prevalence of eosinophilic esophagitis in patients with inherited connective tissue disorders. *J. Allergy Clin. Immunol.* **2013**, *132*, 378–386. [CrossRef]
141. Fikree, A.; Grahame, R.; Aktar, R.; Farmer, A.D.; Hakim, A.J.; Morris, J.K.; Knowles, C.H.; Aziz, Q. A prospective evaluation of undiagnosed joint hypermobility syndrome in patients with gastrointestinal symptoms. *Clin. Gastroenterol. Hepatol.* **2013**, *27*, 569–579. [CrossRef]
142. Morgan, A.W.; Pearson, S.B.; Davies, S.; Gooi, H.C.; Bird, H.A. Asthma and airway collapse in two heritable disorders of connective tissue. *Ann. Rheum. Dis.* **2007**, *66*, 1369–1373. [CrossRef]
143. Rodgers, K.R.; Gui, J.; Dinulos, M.B.; Chou, R.C. Ehlers-Danlos syndrome hypermobility type is associated with rheumatic diseases. *Sci. Rep.* **2017**, *7*, 39636. [CrossRef]
144. Grether-Beck, S.; Felsner, I.; Brenden, H.; Kohne, Z.; Majora, M.; Marini, A.; Jaenicke, T.; Rodriguez-Martin, M.; Trullas, C.; Hupe, M.; et al. Urea uptake enhances barrier function and antimicrobial defense in humans by regulating epidermal gene expression. *J. Investig. Dermatol.* **2012**, *132*, 1561–1572. [CrossRef]
145. Nagahara, M.; Waguri-Nagaya, Y.; Yamagami, T.; Aoyama, M.; Tada, T.; Inoue, K.; Asai, K.; Otsuka, T. TNF-alpha-induced aquaporin 9 in synoviocytes from patients with OA and RA. *Rheumatology* **2010**, *49*, 898–906. [CrossRef]
146. Takeuchi, K.; Hayashi, S.; Matumoto, T.; Hashimoto, S.; Takayama, K.; Chinzei, N.; Kihara, S.; Haneda, M.; Kirizuki, S.; Kuroda, Y.; et al. Downregulation of aquaporin 9 decreases catabolic factor expression through nuclear factor-κB signaling in chondrocytes. *Int. J. Mol. Med.* **2018**, *42*, 1548–1558. [CrossRef]
147. Xu, Y.; Ma, M.; Ippolito, G.C.; Schroeder, H.W.; Carroll, M.C.; Volanakis, J.E. Complement activation in factor D-deficient mice. *Proc. Natl. Acad. Sci. USA* **2001**, *98*, 14577–14582. [CrossRef]

148. Kluzek, S.; Arden, N.K.; Newton, J. Adipokines as potential prognostic biomarkers in patients with acute knee injury. *Biomarkers* **2015**, *20*, 519–525. [CrossRef]
149. Martel-Pelletier, J.; Raynauld, J.P.; Dorais, M.; Abram, F.; Pelletier, J.P. The levels of the adipokines adipsin and leptin are associated with knee osteoarthritis progression as assessed by MRI and incidence of total knee replacement in symptomatic osteoarthritis patients: A post hoc analysis. *Rheumatology* **2016**, *55*, 680–688. [CrossRef]
150. Chandran, V.; Abji, F.; Perruccio, A.V.; Gandhi, R.; Li, S.; Cook, R.J.; Gladman, D.D. Serum-based soluble markers differentiate psoriatic arthritis from osteoarthritis. *Ann. Rheum. Dis.* **2019**, *78*, 796–801. [CrossRef]
151. Lyons, J.J.; Yu, X.; Hughes, J.D.; Le, Q.T.; Jamil, A.; Bai, Y.; Ho, N.; Zhao, M.; Liu, Y.; O'Connell, M.P.; et al. Elevated basal serum tryptase identifies a multisystem disorder associated with increased TPSAB1 copy number. *Nat. Genet.* **2016**, *48*, 1564–1569. [CrossRef]
152. He, Y.W.; Li, H.; Zhang, J.; Hsu, C.L.; Lin, E.; Zhang, N.; Guo, J.; Forbush, K.A.; Bevan, M.J. The extracellular matrix protein mindin is a pattern-recognition molecule for microbial pathogens. *Nat. Immunol.* **2004**, *5*, 88–97. [CrossRef]
153. Jia, W.; Li, H.; He, Y.W. The extracellular matrix protein mindin serves as an integrin ligand and is critical for inflammatory cell recruitment. *Blood* **2005**, *106*, 3854–3859. [CrossRef]
154. Liu, Y.S.; Wang, L.F.; Cheng, X.S.; Huo, Y.N.; Ouyang, X.M.; Liang, L.Y.; Lin, Y.; Wu, J.F.; Ren, J.L.; Guleng, B. The pattern-recognition molecule mindin binds integrin Mac-1 to promote macrophage phagocytosis via Syk activation and NF-κB p65 translocation. *J. Cell. Mol. Med.* **2019**, *23*, 3402–3416. [CrossRef]
155. Clapp, C.; Adán, N.; Ledesma-Colunga, M.G.; Solís-Gutiérrez, M.; Triebel, J.; Martínez de la Escalera, G. The role of the prolactin/vasoinhibin axis in rheumatoid arthritis: An integrative overview. *Cell. Mol. Life Sci.* **2016**, *73*, 2929–2948. [CrossRef]
156. Ledesma-Colunga, M.G.; Adán, N.; Ortiz, G.; Solís-Gutiérrez, M.; López-Barrera, F.; Martínez de la Escalera, G.; Clapp, C. Prolactin blocks the expression of receptor activator of nuclear factor κB ligand and reduces osteoclastogenesis and bone loss in murine inflammatory arthritis. *Arthritis Res. Ther.* **2017**, *19*, 93. [CrossRef]
157. Patil, M.J.; Ruparel, S.B.; Henry, M.A.; Akopian, A.N. Prolactin regulates TRPV1, TRPA1, and TRPM8 in sensory neurons in a sex-dependent manner: Contribution of prolactin receptor to inflammatory pain. *Am. J. Physiol. Endocrinol. Metab.* **2013**, *305*, E1154–E1164. [CrossRef]
158. Basso, L.; Altier, C. Transient Receptor Potential Channels in neuropathic pain. *Curr. Opin. Pharmacol.* **2017**, *32*, 9–15. [CrossRef]
159. Rodríguez-Calvo, R.; Tajes, M.; Vázquez-Carrera, M. The NR4A subfamily of nuclear receptors: Potential new therapeutic targets for the treatment of inflammatory diseases. *Expert Opin. Ther. Targets* **2017**, *21*, 291–304. [CrossRef]
160. Murphy, E.P.; Crean, D. Molecular Interactions between NR4A Orphan Nuclear Receptors and NF-Kb Are Required for Appropriate Inflammatory Responses and Immune Cell Homeostasis. *Biomolecules* **2015**, *5*, 1302–1318. [CrossRef]
161. Banno, A.; Lakshmi, S.P.; Reddy, A.T.; Kim, S.C.; Reddy, R.C. Key Functions and Therapeutic Prospects of Nur77 in Inflammation Related Lung Diseases. *Am. J. Pathol.* **2019**, *189*, 482–491. [CrossRef]
162. Fernandes, J.C.R.; Acuña, S.M.; Aoki, J.I.; Floeter-Winter, L.M.; Muxel, S.M. Long Non-Coding RNAs in the Regulation of Gene Expression: Physiology and Disease. *Noncoding RNA* **2019**, *5*, 17. [CrossRef]
163. Filipowicz, W.; Bhattacharyya, S.N.; Sonenberg, N. Mechanisms of post-transcriptional regulation by microRNAs: Are the answers insight? *Nat. Rev. Genet* **2008**, *9*, 102–114. [CrossRef]
164. Leinders, M.; Üçeyler, N.; Thomann, A.; Sommer, C. Aberrant microRNA expression in patients with painful peripheral neuropathies. *J. Neurol. Sci.* **2017**, *380*, 242–249. [CrossRef]
165. Greco, S.; Cardinali, B.; Falcone, G.; Martelli, F. Circular RNAs in Muscle Function and Disease. *Int. J. Mol. Sci.* **2018**, *19*, 3454. [CrossRef]
166. D'Agnelli, S.; Arendt-Nielsen, L.; Gerra, M.C.; Zatorri, K.; Boggiani, L.; Baciarello, M.; Bignami, E. Fibromyalgia: Genetics and epigenetics insights may provide the basis for the development of diagnostic biomarkers. *Mol. Pain* **2019**, *15*, 1744806918819944. [CrossRef]
167. Kim, J.; Hyun, J.; Wang, S.; Lee, C.; Jung, Y. MicroRNA-378 is involved in hedgehog-driven epithelial-to-mesenchymal transition in hepatocytes of regenerating liver. *Cell Death Dis.* **2018**, *9*, 721. [CrossRef]

168. Zhang, T.; Hu, J.; Wang, X.; Zhao, X.; Li, Z.; Niu, J.; Steer, C.J.; Zheng, G.; Song, G. MicroRNA-378 promotes hepatic inflammation and fibrosis via modulation of the NF-κB-TNFα pathway. *J. Hepatol.* **2019**, *70*, 87–96. [CrossRef]
169. Xiao, Z.; Deng, D.; He, L.; Jiao, H.; Ye, Y.; Liang, L.; Ding, Y.; Liao, W. MicroRNA-224 sustains Wnt/β-catenin signaling and promotes aggressive phenotype of colorectal cancer. *J. Exp. Clin. Cancer Res.* **2016**, *35*, 21.
170. Peng, Y.; Zhang, X.; Feng, X.; Fan, X.; Jin, Z. The crosstalk between microRNAs and the Wnt/β-catenin signaling pathway in cancer. *Oncotarget* **2017**, *8*, 14089–14106. [CrossRef]
171. Bjersing, J.L.; Lundborg, C.; Bokarewa, M.I.; Mannerkorpi, K. Profile of cerebrospinal microRNAs in fibromyalgia. *PLoS ONE* **2013**, *8*, e78762. [CrossRef]
172. Chopra, P.; Tinkle, B.; Hamonet, C.; Brock, I.; Gompel, A.; Bulbena, A.; Francomano, C. Pain management in the Ehlers-Danlos syndromes. *Am. J. Med. Genet. C Semin. Med. Genet.* **2017**, *175*, 212–219. [CrossRef]
173. Wade, S.M.; Trenkmann, M.; McGarry, T.; Canavan, M.; Marzaioli, V.; Wade, S.C.; Veale, D.J.; Fearon, U. Altered expression of microRNA-23a in psoriatic arthritis modulates synovial fibroblast pro-inflammatory mechanisms via phosphodiesterase 4B. *J. Autoimmun.* **2019**, *96*, 86–93. [CrossRef]
174. Tajerian, M.; Clark, J.D. The role of the extracellular matrix in chronic pain following injury. *Pain* **2015**, *156*, 366–370. [CrossRef]
175. Parisien, M.; Samoshkin, A.; Tansley, S.N.; Piltonen, M.H.; Martin, L.J.; El-Hachem, N.; Dagostino, C.; Allegri, M.; Mogil, J.S.; Khoutorsky, A.; et al. Genetic pathway analysis reveals a major role for extracellular matrix organization in inflammatory and neuropathic pain. *Pain* **2019**, *60*, 932–944. [CrossRef]
176. Andersen, H.H.; Duroux, M.; Gazerani, P. MicroRNA as modulators and biomarkers of inflammatory and neuropathic pain conditions. *Neurobiol. Dis.* **2014**, *71*, 159–168. [CrossRef]

© 2019 by the authors. Licensee MDPI, Basel, Switzerland. This article is an open access article distributed under the terms and conditions of the Creative Commons Attribution (CC BY) license (http://creativecommons.org/licenses/by/4.0/).

Article

Transcriptome Profiling of Primary Skin Fibroblasts Reveal Distinct Molecular Features Between *PLOD1*- and *FKBP14*-Kyphoscoliotic Ehlers–Danlos Syndrome

Pei Jin Lim [1], Uschi Lindert [1], Lennart Opitz [2], Ingrid Hausser [3], Marianne Rohrbach [1,*,†] and Cecilia Giunta [1,*,†]

1. Connective Tissue Unit, Division of Metabolism and Children's Research Centre, University Children's Hospital, 8032 Zürich, Switzerland
2. Functional Genomics Center Zurich, University of Zurich/ETH Zurich, Winterthurerstrasse 190, 8057 Zürich, Switzerland
3. Institute of Pathology, Heidelberg University Hospital, 69120 Heidelberg, Germany
* Correspondence: Marianne.Rohrbach@kispi.uzh.ch (M.R.); Cecilia.Giunta@kispi.uzh.ch (C.G.); Tel.: +41-442-667-618 (M.R.); +41-442-667-758 (C.G.)
† These authors contributed equally to this manuscript.

Received: 27 May 2019; Accepted: 4 July 2019; Published: 8 July 2019

Abstract: Kyphoscoliotic Ehlers–Danlos Syndrome (kEDS) is a rare genetic heterogeneous disease clinically characterized by congenital muscle hypotonia, kyphoscoliosis, and joint hypermobility. kEDS is caused by biallelic pathogenic variants in either *PLOD1* or *FKBP14*. *PLOD1* encodes the lysyl hydroxylase 1 enzyme responsible for hydroxylating lysyl residues in the collagen helix, which undergo glycosylation and form crosslinks in the extracellular matrix thus contributing to collagen fibril strength. *FKBP14* encodes a peptidyl-prolyl cis–trans isomerase that catalyzes collagen folding and acts as a chaperone for types III, VI, and X collagen. Despite genetic heterogeneity, affected patients with mutations in either *PLOD1* or *FKBP14* are clinically indistinguishable. We aim to better understand the pathomechanism of kEDS to characterize distinguishing and overlapping molecular features underlying *PLOD1*-kEDS and *FKBP14*-kEDS, and to identify novel molecular targets that may expand treatment strategies. Transcriptome profiling by RNA sequencing of patient-derived skin fibroblasts revealed differential expression of genes encoding extracellular matrix components that are unique between *PLOD1*-kEDS and *FKBP14*-kEDS. Furthermore, we identified genes involved in inner ear development, vascular remodeling, endoplasmic reticulum (ER) stress, and protein trafficking that were differentially expressed in patient fibroblasts compared to controls. Overall, our study presents the first transcriptomics data in kEDS revealing distinct molecular features between *PLOD1*-kEDS and *FKBP14*-kEDS, and serves as a tool to better understand the disease.

Keywords: kyphoscoliotic Ehlers–Danlos Syndrome; EDS type VI; transcriptomics; connective tissue; extracellular matrix; *PLOD1*; *FKBP14*

1. Introduction

According to the 2017 revised Nosology of the Ehlers–Danlos syndrome (EDS) [1], kyphoscoliotic EDS (kEDS, OMIM 225400 and 614557) groups two rare autosomal recessive disorders which are clinically indistinguishable, but genetically distinct as they are caused by pathologic biallelic variants in either procollagen-lysine,2-oxoglutarate 5-dioxygenase 1 (*PLOD1*) or FK506-binding protein 14 (*FKBP14*). *PLOD1* encodes the lysyl hydroxylase 1 (LH1) enzyme, which hydroxylates lysyl residues of Xaa-Lys-Gly tripeptide motif in collagens. Subsequently, the hydroxylated lysyl residues undergo glycosylation with the attachment of galactose or glucosyl-galactose units and after collagen secretion into the extracellular matrix (ECM) they form inter- and intra-molecular crosslinks known to contribute to collagen fibril strength and thus to tissue stability [2]. The lack or loss of function of LH1 leads to underhydroxylation and underglycosylation of lysyl residues in the helical domain of collagen, thereby impairing collagen cross-linking and consequentially causing mechanical instability of the affected connective tissues. Patients deficient in LH1 have an increased ratio of urinary lysyl-pyridinoline to hydroxylysyl-pyridinoline (LP/HP) due to underhydroxylation of collagen lysyl residues. Only recently, pathogenic variants in *FKBP14* have been described in a group of patients with a clinical diagnosis of kEDS but a normal LP/HP ratio [3,4]. *FKBP14* encodes the endoplasmic reticulum (ER)-resident FKBP22 protein, a 22kDa member of the family of FK506-binding peptidyl-prolyl cis–trans isomerases. FKBP22 catalyzes the cis–trans isomerization of prolyl peptide bonds, which is the rate limiting step in procollagen protein folding due to its abundance in proline residues, and has been shown to catalyze the folding of collagen type III [5]. FKBP22 also functions as a chaperone for collagen types III, VI and X where it is thought to prevent premature interactions between collagen during its assembly in the ER [5].

Despite harboring mutations in two genes with different functions, *PLOD1*-kEDS and *FKBP14*-kEDS patients share a common clinical phenotype with major criteria consisting of congenital muscle hypotonia, congenital or early onset kyphoscoliosis, and generalized joint hypermobility; shared minor criteria include skin hyperextensibility, rupture, or aneurysm of medium-sized arteries, easy bruising of skin, and osteopenia or osteoporosis. Gene-specific minor criteria for *PLOD1*-kEDS include skin fragility, while that of *FKBP14*-kEDS include congenital hearing impairment [1]. Pronounced inter- and intrafamilial variability of the clinical presentation have been described in both genetic forms of kEDS [3,4]. In the absence of genetic information, collagen type VI-related myopathies which present with severe neonatal hypotonia, delayed motor development, kyphoscoliosis and joint hypermobility [6–8] represent the major differential diagnosis to kEDS.

Currently, there is no pharmacological treatment for kEDS and the treatment of manifestations aims to alleviate symptoms and focuses on the musculoskeletal and hearing systems [3,4,9]: physical therapy for muscular hypotonia, bracing of unstable joints, corrective surgeries for kyphoscoliosis and prescription of hearing aids for hearing impairment.

The aim of our study is to better understand the pathomechanism of kEDS and to identify overlapping and distinguishing molecular signatures of *PLOD1*-kEDS and *FKBP14*-kEDS. The application of an untargeted general omics approach may serve as a valuable tool to identify novel proteins or pathways involved in the pathogenesis of kEDS that can be pharmacologically targeted to improve the disease symptoms. As a first step towards this goal, by performing RNA sequencing on an in vitro cell culture model of kEDS we obtained and compared the transcriptome profiles of normal control, *PLOD1*-kEDS and *FKBP14*-kEDS patients. Furthermore, the transcriptome profiles of kEDS patients were compared to published transcriptomics data of patients with collagen VI-associated myopathies [10].

2. Materials and Methods

2.1. Subjects and Cell Culture

This study was conducted according to the Declaration of Helsinki for Human Rights and approved by Swiss Ethics (KEK-ZH-Nr. 2019-00811) in the presence of a signed informed consent of the patients or their parents.

As part of the diagnostic workup of kEDS, punch biopsies of the skin for electron microscopy investigations and establishment of fibroblast cultures were previously obtained. The biological material was stored in the Biobank of the Division of Metabolism at the Children's Hospital Zurich. For this study, fibroblasts of three *FKBP14*-kEDS patients, three *PLOD1*-kEDS patients and four healthy controls were used (Table S1). Pathological mutations and clinical findings of the patients are recorded in Table 1. Cells were cultured at 37 °C and 5% CO_2 in Dulbecco's Modified Eagle's Medium (Gibco, 31966-021) supplemented with 10% fetal bovine serum, 100 U/ml penicillin, 100 mg/ml streptomycin, and 0.25 mg/ml Amphotericin B.

Table 1. Gene mutations and clinical presentations of *FKBP14*-kEDS and *PLOD1*-kEDS patients included in this study

Gene Mutations	Clinical Findings				
	Hyperextensible Skin	Kyphoscoliosis	Muscle Hypotonia	Joint Hypermobility	Vascular Abnormality
PLOD1 compound heterozygote c.975+975_1755+?dup/c.1362del p.Glu326_Lys585dup/ p.(Tyr455Thrfs*2) (described as P4 in [11])	+	+ (progressive)	+	+	NR
PLOD1 p.Glu326_Lys585dup homozygous exon 10–16 duplication (described as P1 in [12])	−	+	+	+	−
PLOD1 Leu85Pro homozygous (described as P2 in [12])	−	−	−	+	+ (rupture of artery)
FKBP14 c.362dupC p (Glu122Argfs*7) homozygous (described as P3 in [4])	+	+ (progressive)	+	+	− (#)
FKBP14 c.362dupC p (Glu122Argfs*7) homozygous (described as P1 in [4])	+	+ (progressive)	+	+	−
FKBP14 c.197 + 5_197 + 8del / p.His67* homozygous (described as P4 in [3])	+	+ (progressive)	+	+	NR

The following symbols and abbreviations are used: +, present; −, absent; NR, not reported. (#) A second-degree cousin died of aortic rupture at age 12, suspected *FKBP14* mutation but DNA not available for confirmation. Another second-degree cousin had a dissection of the internal carotid artery at age 50 years [4]. Please refer to the original publications in which the patients were first described in [3,4,11,12] for more detailed clinical characteristics.

2.2. Gene Expression Profiling

Cells were passaged into T75 flasks, fed fresh medium 24 hours after passaging, and RNA was isolated 24 hours later for transcriptome profiling. RNA was harvested using miRNeasy Mini Kit (QIAGEN, 217004) according to the manufacturer's instructions. RNA quality control was performed on an Agilent 2100 Bioanalyzer with RNA integrity number (RIN) values between 9.6 and 10.0. Poly-A purified libraries were prepared using the TruSeq mRNA sample preparation kit (Illumina, 20020595). RNA sequencing was performed on an Illumina HiSeq 4000 instrument at the Functional Genomics Center Zurich. The raw reads were cleaned by removing adapter sequences, trimming low quality ends, and filtering reads with low quality (phred quality <20). Sequence alignment of the resulting high-quality reads to the human genome (build GRCh38.p10) and quantification of transcript expression was carried out using RNA-Seq by Expectation Maximization (RSEM, version 1.3.0) with Ensembl gene models of release 89. A count-based negative binomial model implemented in the software package edgeR (R-version 3.5.1, edgeR 3.24.2) was applied to detect differentially expressed genes (DEGs). DEGs were defined as genes with p-value <0.05 and $\log_2$(fold change) >0.5 or <−0.5. To identify enriched biological processes and cellular components, over-representation enrichment analysis (ORA) using a list of DEGs generated with a more stringent cutoff of p-value <0.01 and gene set enrichment analysis (GSEA) using ranked gene lists were performed with the online toolkit WebGestalt (2017 version) [13,14]. The RNA sequencing data, including raw sequence files for each subject, is available on the European Nucleotide Archive (ENA) and the Gene Expression Omnibus (GEO) database under the accession number PRJEB31335.

2.3. Quantitative RT-PCR for Validation

Candidate genes were selected from the lists of DEGs for validation by quantitative RT-PCR in four independent replicates. RNA was harvested using RNeasy Mini Kit (QIAGEN, 74104) according to manufacturer's instructions, and reverse transcribed to cDNA using the High-Capacity RNA-to-cDNA Kit (Applied Biosysyems, 4387406). cDNA was diluted in RNase-free water to 3 ng/µl for quantitative RT-PCR using Taqman assays (Table 2) on a 7900HT Fast Real-Time PCR System machine (Applied Biosystems). Fold change in gene expression was calculated by the $2^{-\Delta\Delta Ct}$ method with glyceraldehyde-3-phosphate dehydrogenase (*GAPDH*) as an endogenous control.

Table 2. Taqman gene expression assays used for quantitative RT-PCR.

Gene Symbol	Gene Name	Assay ID
GAPDH	glyceraldehyde-3-phosphate dehydrogenase	Hs02758991_g1
ELN	elastin	Hs00355783_m1
POSTN	periostin	Hs01566750_m1
WNT4	Wnt family member 4	Hs01573505_m1
COL15A1	collagen type XV alpha 1	Hs00266332_m1
EFEMP1	EGF-containing Fibulin-like extracellular matrix protein 1	Hs00244575_m1
ALDH1A3	aldehyde dehydrogenase 1 family member A3	Hs00167476_m1
OLFM2	olfactomedin 2	Hs01017934_m1
TM4SF1	transmembrane 4 L six family member 1	Hs01547334_m1
SCAMP5	secretory carrier membrane protein 5	Hs01547727_m1
FGF11	fibroblast growth factor 11	Hs00182803_m1
PLXNA2	plexin A2	Hs00300697_m1
PLEKHA2	Pleckstrin homology domain-containing family A member 2	Hs00952489_m1

3. Results

3.1. Transcriptome Profiling and Differential Expression Analysis

Transcriptome profiling by RNA sequencing identified 298 DEGs in *PLOD1*-kEDS patient-derived fibroblasts compared to controls, of which 139 genes were up-regulated and 159 genes were down-regulated in the patients' cells, as represented in a volcano plot (Figure 1A, Tables S2 and S3). 488 DEGs were identified in *FKBP14*-kEDS patient-derived fibroblasts compared to controls, of which 309 genes were up-regulated and 179 genes were down-regulated (Figure 1B, Tables S4 and S5). The low numbers of DEGs observed in our datasets may be contributed by heterogeneity in multiple factors such as ethnicity, age of biopsy, and sex of patients. Nevertheless, several genes encoding proteins related to ECM composition, protein trafficking, vasculature development, and inner ear development were identified in the lists of DEGs, which can potentially contribute to the pathogenesis as discussed later in this paper.

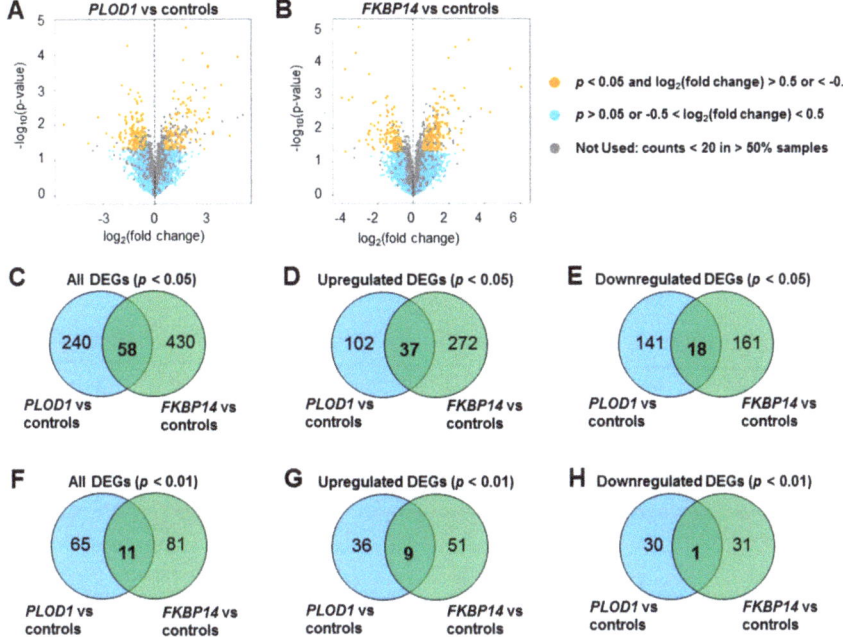

Figure 1. Transcriptome profiles of *PLOD1*-kEDS and *FKBP14*-kEDS patient-derived fibroblasts. Volcano plots of DEGs in (**A**) *PLOD1*-kEDS versus controls and (**B**) *FKBP14*-kEDS versus controls. (**C–E**) Venn diagrams showing number of overlapping and non-overlapping DEGs at $p < 0.05$ (C: all, D: up-regulated, E: down-regulated). (**F–H**) Venn diagrams showing number of overlapping and non-overlapping DEGs at $p < 0.01$ (F: all, G: up-regulated, H: down-regulated).

To further explore the similarities and differences between *PLOD1*-kEDS and *FKBP14*-kEDS patient-derived fibroblasts, we compared DEGs with $p < 0.05$ and $\log_2$ (fold change) >0.5 or <−0.5 in both patient groups, and found 58 overlapping genes (Figure 1C). Of these 58 genes, 37 were up-regulated (Figure 1D) and 18 were down-regulated (Figure 1E) in both *PLOD1*-kEDS and *FKBP14*-kEDS compared to controls. Furthermore, from the more stringent list of DEGs with $p < 0.01$, 11 genes overlapped between *PLOD1*-kEDS and *FKBP14*-kEDS fibroblasts (Figure 1F). Of these 11 DEGs, 9 genes were up-regulated (Figure 1G) in both groups, 1 gene was down-regulated (Figure 1H) in both groups, and 1 gene was up-regulated in *PLOD1*-kEDS but down-regulated in *FKBP14*-kEDS fibroblasts (further discussed in Figure 2A).

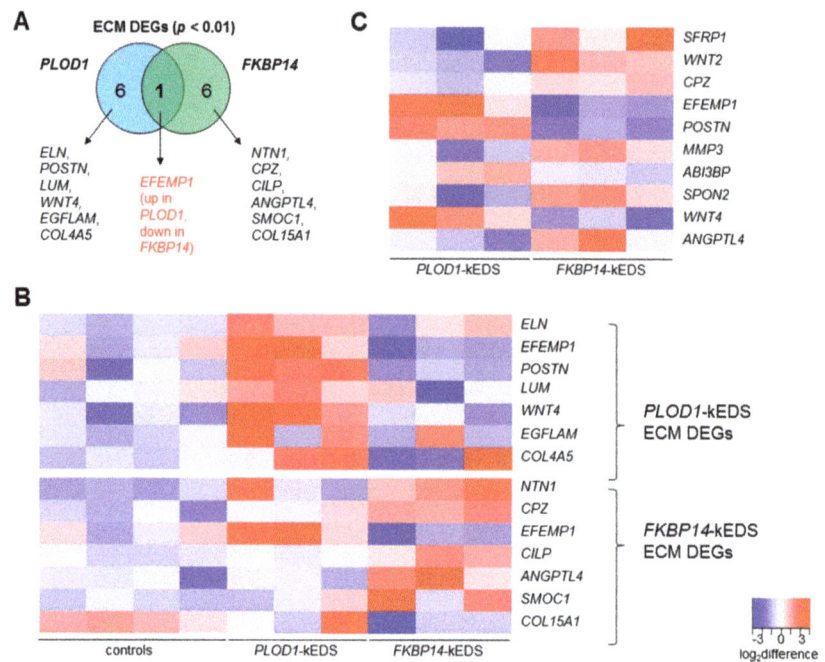

Figure 2. (**A**) Venn diagram depicting unique DEGs contributing to the enrichment of ECM in *PLOD1*-kEDS and *FKBP14*-kEDS patient-derived fibroblasts versus controls. (**B**) Heatmap depicting expression levels of DEGs encoding ECM components per subject. Top panel consists of genes that are differentially regulated in *PLOD1*-kEDS versus controls. Bottom panel consists of genes that are differentially regulated in *FKBP14*-kEDS versus controls. (**C**) Heatmap depicting expression levels of ECM genes that are significantly different between *PLOD1*-kEDS and *FKBP14*-kEDS. The colors indicate the $\log_2$ difference relative to the average expression of all samples within the comparison.

3.2. Gene Ontology and Pathway Analyses

Due to limitations in obtaining patient skin fibroblasts in a rare disease with newly described gene mutations, the sample size in our study was inevitably inadequate to generate a list of DEGs adjusted for false discovery rate (FDR). Instead, using the WebGestalt toolkit, we analyzed the RNA sequencing datasets with two approaches: (i) by setting a more stringent p-value cutoff whereby DEGs with $p < 0.01$ in *PLOD1*-kEDS and *FKBP14*-kEDS patient-derived fibroblasts were used to identify over-represented biological processes and cellular components by ORA and (ii) by performing GSEA in which a list of all detected genes ranked by their average fold change is used to calculate enrichment scores of Kyoto Encyclopedia of Genes and Genomes (KEGG) pathways. Gene ontology terms with an FDR <10% were considered significantly enriched in both approaches.

In the ORA using DEGs with $p < 0.01$, the significantly enriched biological processes and cellular components are summarized in Table 3; DEGs that contribute to these enriched Gene Ontology (GO) terms are listed in Tables S6 and S7. Notably, genes encoding ECM proteins were enriched in the DEGs in both *PLOD1*-kEDS and *FKBP14*-kEDS patient-derived fibroblasts; each patient group was compared to control fibroblasts. However, comparison of DEGs that contributed to over-representation of ECM components demonstrated only one overlapping gene, *EFEMP1*, which was up-regulated in *PLOD1*-kEDS and down-regulated in *FKBP14*-kEDS fibroblasts (Figure 2A). The expression levels of each DEG encoding ECM components in individual control and patient-derived fibroblasts are depicted in a heatmap (Figure 2B). To further investigate the differential expression in ECM-related genes between *PLOD1*-kEDS and *FKBP14*-kEDS, we performed ORA of the DEGs between *PLOD1*-kEDS and *FKBP14*-kEDS fibroblasts with p-value <0.01 and $\log_2$ (fold change) >0.5 or <−0.5 (Table S8) using WebGestalt and observed an enrichment of ECM genes (GO:0005578) with an enrichment ratio of 7.64 and FDR of 6.41×10^{-5}. The expression of these ECM genes that are significantly different between *PLOD1*-kEDS and *FKBP14*-kEDS are summarized in a heatmap (Figure 2C). Hence, the RNA sequencing datasets revealed distinct ECM signatures between *PLOD1*-kEDS and *FKBP14*-kEDS fibroblasts at the transcriptome level.

Table 3. Over-represented gene ontology terms among DEGs in *FKBP14*-kEDS and *PLOD1*-kEDS patient-derived fibroblasts versus controls.

Gene Ontology	Description	Enrichment Ratio	p-value	FDR
FKBP14-kEDS: Biological Process				
GO:0003012	muscle system process	5.52	3.19×10^{-5}	2.42×10^{-2}
GO:0007586	digestion	14.1	1.74×10^{-4}	6.60×10^{-2}
FKBP14-kEDS: Cellular Component				
GO:0005578	proteinaceous extracellular matrix	5.07	3.90×10^{-4}	5.77×10^{-2}
GO:0043235	receptor complex	5.28	8.53×10^{-4}	6.32×10^{-2}
PLOD1-kEDS: Biological Process				
GO:0007219	Notch signaling pathway	8.86	1.26×10^{-5}	9.58×10^{-3}
GO:0007423	sensory organ development	4.77	3.84×10^{-5}	1.46×10^{-2}
GO:0030048	actin filament-based movement	10.7	9.73×10^{-5}	2.02×10^{-2}
GO:0045730	respiratory burst	31.2	1.07×10^{-4}	2.02×10^{-5}
GO:0060326	cell chemotaxis	7.34	1.56×10^{-4}	2.38×10^{-2}
GO:0070098	chemokine-mediated signaling pathway	23.0	2.77×10^{-4}	3.50×10^{-2}
GO:0050920	regulation of chemotaxis	6.74	8.43×10^{-4}	9.16×10^{-2}
GO:0042490	mechanoreceptor differentiation	14.6	1.10×10^{-3}	9.61×10^{-2}
GO:0001655	urogenital system development	4.30	1.14×10^{-3}	9.61×10^{-2}
PLOD1-kEDS: Cellular Component				
GO:0005578	proteinaceous extracellular matrix	5.73	1.77×10^{-4}	2.61×10^{-2}
GO:1990351	transporter complex	6.36	1.03×10^{-3}	5.96×10^{-2}
GO:0044420	extracellular matrix component	7.88	1.56×10^{-3}	5.96×10^{-2}
GO:0031594	neuromuscular junction	12.7	1.61×10^{-3}	5.96×10^{-2}
GO:0043235	receptor complex	4.98	3.03×10^{-3}	8.97×10^{-2}

GSEA of the ranked gene lists from *PLOD1*-kEDS fibroblasts did not identify significantly enriched KEGG pathways. In contrast, KEGG pathways involved in carbohydrate metabolism, namely fructose and mannose metabolism (Figure 3A) as well as glycolysis and gluconeogenesis (Figure 3B), were positively enriched in *FKBP14*-kEDS fibroblasts. Furthermore, genes involved in cell cycle (Figure 3C) and DNA replication (Figure 3D) were over-represented by down-regulated genes in *FKBP14*-kEDS fibroblasts, suggesting alterations in cell proliferation.

The findings from the ORA and GSEA analyses collectively suggests different molecular signatures between *PLOD1*-kEDS and *FKBP14*-kEDS.

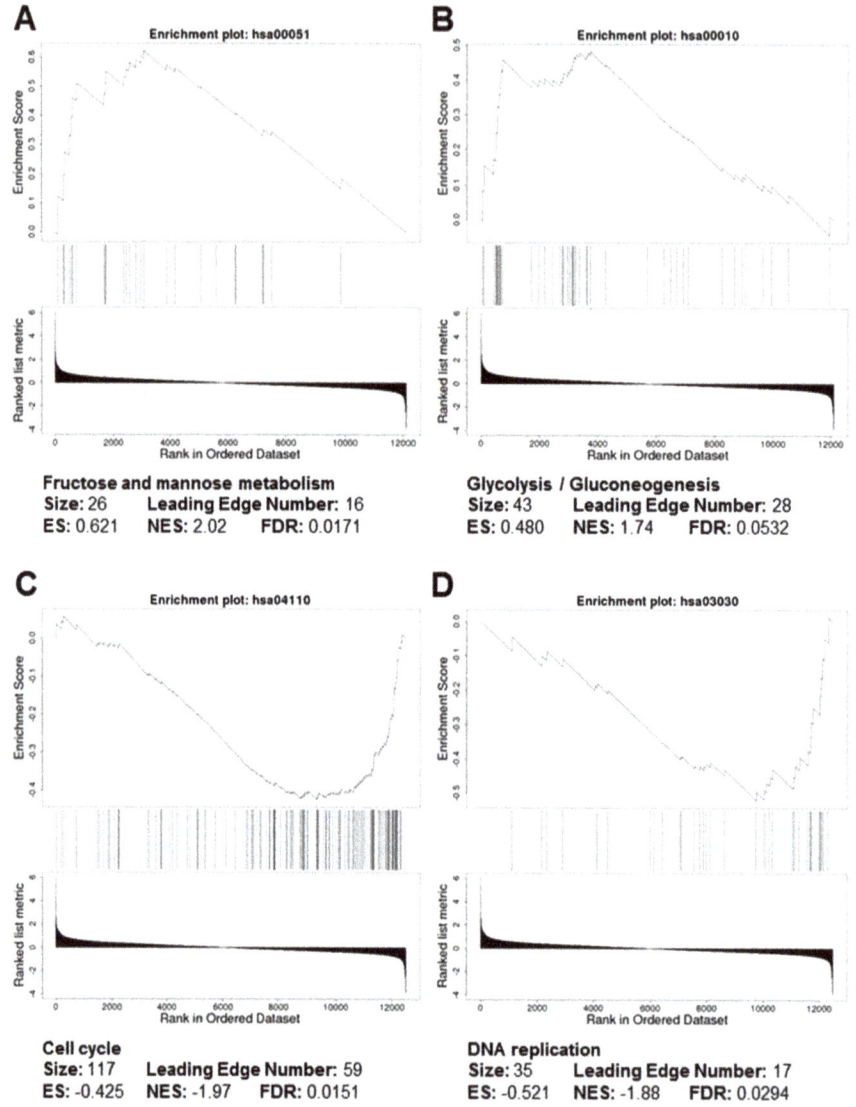

Figure 3. Gene set enrichment analysis (GSEA) plots depicting positive enrichment of (**A**) fructose and mannose metabolism and (**B**) glycolysis / gluconeogenesis and negative enrichment of (**C**) cell cycle and (**D**) DNA replication in *FKBP14*-kEDS versus control fibroblasts.

3.3. Comparison of Transcriptome Profiles between kEDS and Collagen VI-Related Muscular Dystrophies

Due to the overlaps in clinical presentation between collagen VI-related muscular dystrophies (COL6-RD) and kEDS patients, we compared the transcriptome profiles of patient-derived fibroblasts from *PLOD1*-kEDS and *FKBP14*-kEDS (DEGs with $p < 0.01$) with that of COL6-RD patients available through GEO Series accession number GSE103270 [10]. Only one gene, *OLFM2*, was significantly up-regulated in all three patient groups; *EFEMP1* was significantly up-regulated in *PLOD1*-kEDS and COL6-RD but down-regulated in *FKBP14*-kEDS (Figure 4A). While the ECM component, which is biologically relevant to disorders of the connective and muscle tissues, was over-represented in the

DEGs from all three groups (Table 3 and [10]), comparison of DEGs encoding ECM components also revealed only a few overlapping genes among the three patient groups (Figure 4B). Furthermore, the over-representation of muscle system process in *FKBP14*-kEDS fibroblasts (GO:0003012, Table 3) prompted us to compare the genes contributing to this GO term in *FKBP14*-kEDS to the DEGs in COL6-RD fibroblasts. Of the nine genes contributing to the enrichment of muscle system process, only two were significantly differentially regulated in COL6-RD fibroblasts carrying dominant negative (DN) collagen VI mutations; six of these nine genes (three at $p < 0.01$, three at $p < 0.05$) were also differentially regulated in *PLOD1*-kEDS (Figure 4C). These suggest that each disease group have a unique transcriptome profile.

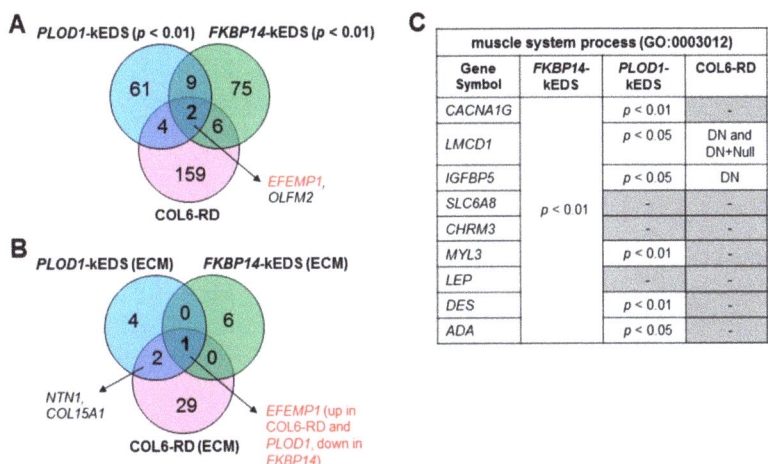

Figure 4. Comparison of transcriptome profiles of *PLOD1*-kEDS, *FKBP14*-kEDS and collagen VI-related muscular dystrophies (COL6-RD) patient-derived fibroblasts. (**A**) Venn diagram depicting number of overlapping and non-overlapping DEGs among three patient groups. (**B**) Venn diagram depicting number of overlapping and non-overlapping DEGs encoding ECM components among three patient groups. (**C**) List of genes contributing to over-representation of muscle system process in *FKBP14*-kEDS versus control fibroblasts and significance of their differential expression in *PLOD1*-kEDS and COL6-RD fibroblasts. DN = dominant negative COL6-RD mutations allowing incorporation of abnormal collagen VI chains; Null = recessive COL6-RD mutations that do not allow incorporation of abnormal chains as described in Butterfield et al., 2017; gene expressions that are non-significantly altered are represented by dashes (-).

3.4. DEGs with Biological Functions that May Contribute to the Pathogenesis of kEDS

From our transcriptomics analysis, we also identified several DEGs with biological functions that may contribute to the pathogenesis of kEDS. We validated the differential expression of these candidate genes by quantitative RT-PCR (Figure 5). How the encoded products of these genes could contribute to the disease pathology is further discussed under Section 4.1 to Section 4.6.

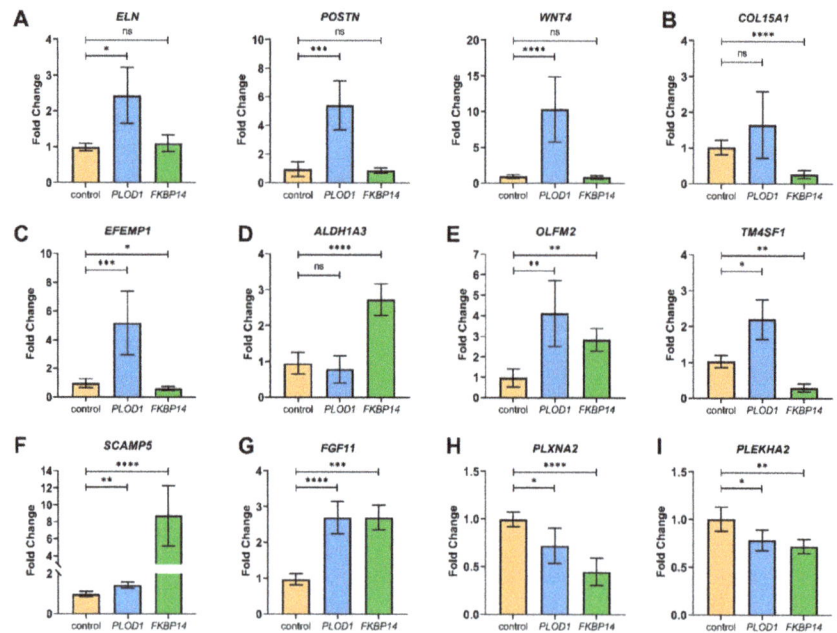

Figure 5. Quantitative RT-PCR was performed to validate the RNA sequencing results. Gene expression levels were measured in four independent replicates per subject, and t-tests were performed (ns = not significant, * = $p < 0.05$, ** $p < 0.005$, *** $p < 0.0005$, **** $p < 0.0001$). Data are expressed as mean ± SEM.

4. Discussion

kEDS represents a paradigm heterogeneous connective tissue disorder in that genetic defects in two different proteins, LH1 and FKBP22, involved in the biosynthesis of collagens lead to a similar clinical phenotype. As an explanation thereof, we hypothesize that *PLOD1* and *FKBP14* mutations might lead to similar alterations of the ECM of connective tissue, either by affecting the same molecular pathways, or by converging the different altered pathways to the same target molecules. To answer this question, we have applied an untargeted omics approach to the identification of common and distinct molecular pathways altered in *FKBP14*-kEDS and *PLOD1*-kEDS.

Our studies were performed in patient-derived primary skin fibroblasts, which are clinically relevant to connective tissue disorders since they are the major cell type producing ECM proteins, and are relatively easy to obtain via skin biopsies. We do, however, acknowledge that clinical characteristics of kEDS patients extend beyond abnormalities in the skin and that the functions of some DEGs may appear more relevant in other ECM-producing cells, including chondrocytes, osteoblasts, and muscle fibroblasts. Nevertheless, the transcriptomics datasets will help to identify candidate DEGs for functional characterization when the other cell types or animal models eventually become available.

Here, we present the outcome of our investigations by transcriptomics of cultured dermal fibroblasts from individuals with *FKBP14*-kEDS and *PLOD1*-kEDS which led to the identification of over-represented biological processes and cellular components. By GSEA, we observed positive enrichment of fructose and mannose metabolism, as well as glycolysis and gluconeogenesis in *FKBP14*-kEDS fibroblasts. Additionally, cell cycle and DNA replication were negatively enriched in *FKBP14*-kEDS fibroblasts, suggesting that cells deficient in FKBP22 may undergo a metabolic switch and alteration in cell proliferation. In vitro studies using mammary epithelial cells have shown that changes in the density of the ECM lead to alterations in glucose metabolism [15]. In another study, up-regulation of glycolysis was observed in healthy skin from the footpad of mice with a dense ECM compared to the abdominal skin with a thinner ECM; glycolysis was also up-regulated in fibrotic skin compared to healthy skin [16]. Reciprocally, perturbations in metabolism can cause alterations in the ECM. In particular, gene and protein expression of ECM components are down-regulated upon suppression of glycolysis [16]. As such, it will be interesting to investigate whether the abnormal deposition of ECM proteins by FKBP22-deficient cells as previously shown [4] cause metabolic changes which can eventually alter cellular bioenergetics and proliferation, or whether alterations in metabolism affects the formation of the ECM by the fibroblasts.

From the transcriptome profiling analysis, we also observed differential expression of several genes encoding products with biological functions that may contribute to the pathogenesis of kEDS (Figure 5). In the following sections, we discuss how the encoded products of these DEGs could contribute to the disease pathology.

4.1. ECM Components

Gene ontology analysis showed significant enrichment of the ECM by DEGs in both *PLOD1*-kEDS and *FKBP14*-kEDS. We validated the up-regulation of *ELN* (encoding elastin), *POSTN* (encoding periostin) and *WNT4* (encoding Wnt family member 4) specifically in *PLOD1*-kEDS (Figure 5A).

Elastin, like collagens, is a structural component of the ECM that is rich in glycine and proline. However, elastin adopts random coil conformations rather than a triple helix structure that give rise to its stretchable properties.

Periostin is preferentially expressed in connective tissues under constant mechanical stress. It accelerates collagen cross-linking through promoting the proteolytic activation of lysyl oxidase (LOX) by bone morphogenetic protein-1 (BMP-1) and serves as a scaffold for collagen and BMP-1 [17]. Periostin also facilitates the formation of ECM meshwork by interacting with other ECM components, including fibronectin and tenascin-C [18]. The protective role of periostin is demonstrated by its induction upon skin injury to promote wound healing [19–21]. Overexpression of periostin has also been described in hypertrophic scar and keloid formation [22] and linked to the pathogenesis of skin fibrosis in systemic sclerosis [23,24]. In contrast, despite elevated *POSTN* expression in the skin fibroblasts (Table 1 and Figure 5A), *PLOD1*-kEDS patients present with poor wound healing and atrophic scarring. A plausible explanation for this observation is the under-hydroxylation of lysine residues in the collagen helical domain due to LH1 deficiency, which impairs collagen cross-linking and fibril assembly, and cannot be compensated for by the activation of LOX by periostin.

The physiological role of WNT4 has been described in the formation of synovial joints [25] and neuromuscular junctions [26]. Up-regulation of *Wnt4* expression during wound healing was demonstrated in a mouse model [27], which could facilitate wound closure by promoting cell migration [28].

Notably, the transcript levels of *ELN*, *POSTN* and *WNT4* can be up-regulated in response to transforming growth factor-beta1 (TGF-β1) stimulation [29–31]. Hence, it remains to be elucidated whether the elevated expression of *ELN*, *POSTN* and *WNT4* is a consequence of enhanced TGF-β signaling in *PLOD1*-kEDS patient fibroblasts, and if so, via which mechanism(s) TGF-β signaling is enhanced.

The down-regulation of *COL15A1* (encoding collagen type XV alpha 1 chain) in *FKBP14*-kEDS was confirmed by quantitative RT-PCR (Figure 5B). Type XV collagen is present widely in the basement membrane zones of cardiac and skeletal myocytes [32]. Studies performed in *Col15a1*-knockout mice demonstrated that deficiency in type XV collagen led to progressive skeletal muscle degeneration and variation in muscle fiber size starting from 13 weeks of age, and increased susceptibility to exercise-induced muscle injury [33]. However, the absence of these abnormal muscular phenotypes in newborn mice suggest that suppressed *COL15A1* expression unlikely explains for the presentation of congenital muscle hypotonia in *FKBP14*-kEDS patients. Nevertheless, rescuing *COL15A1* expression could be a potential therapeutic target to prevent worsening of the muscular phenotype with age in *FKBP14*-kEDS patients.

Interestingly, *EFEMP1* is down-regulated in *FKBP14*-kEDS patient-derived fibroblasts (Figure 5C). *EFEMP1* encodes the EGF-containing fibulin-like extracellular matrix protein, also known as fibulin-3. It is noteworthy that 47% of the *FKBP14*-kEDS patients described [3,4] developed inguinal or umbilical hernias, a phenotype that is also observed in *Efemp1*-knockout mice. Moreover, *Efemp1*-knockout mice have less elastic fibers in the dermis, which also appear more fragmented than elastic fibers in wildtype mice [34]. Therefore, we examined electron micrographs of skin biopsies taken from a control, a *PLOD1*-kEDS patient and two *FKBP14*-kEDS patients and observed fragmentation of elastic fibers in *FKBP14*-kEDS patients but not in the *PLOD1*-kEDS patient (Figure 6). Furthermore, a reduced incorporation of elastin into the network of elastic fibers in *FKBP14*-kEDS patients was observed. These observations, in concordance with the observations by McLaughlin and co-workers in *Efemp1*-knockout mice, suggest that fibulin-3 contributes to the maintenance of ECM integrity and particularly in the formation of elastic fibers. We are currently investigating whether there is indeed a reduction in fibulin-3 protein secretion by *FKBP14*-kEDS fibroblasts in vitro, and whether addition of recombinant fibulin-3 protein can overcome the deficiency and facilitate the formation of elastic fibers in an in vitro model.

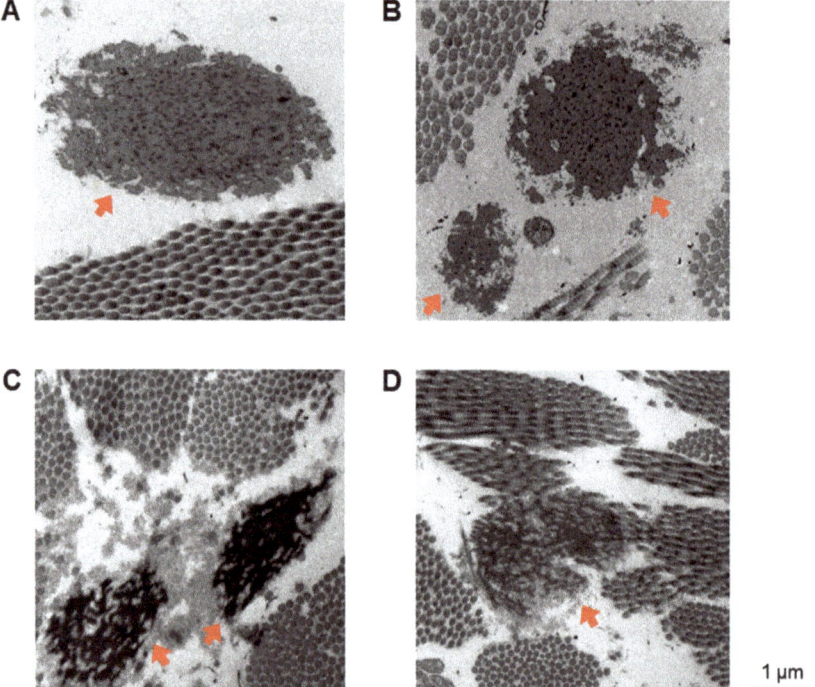

Figure 6. Electron micrographs of skin biopsies from (**A**) a healthy control, (**B**) a *PLOD1*-kEDS patient and (**C,D**) two unrelated *FKBP14*-kEDS patients. Elastic fibers are indicated by red arrows. Electron micrographs were taken at the Center for Microscopy and Image Analysis, University of Zurich with a Philips CM 100, a 100 KV transmission electron microscope equipped with a digital charge-coupled device (CCD) camera for image acquisition.

Contrary to *EFEMP1* down-regulation in *FKBP14*-kEDS, *EFEMP1* was up-regulated in *PLOD1*-kEDS. The overexpression of *EFEMP1* in a chondrogenic cell line in vitro led to suppressed chondrocyte differentiation and dampened expression of aggrecan and types II and X collagen [35]. Thus, whether elevated expression of *EFEMP1* contributes to the pathogenesis of *PLOD1*-kEDS via altering cartilage development remains to be explored, despite the absence of reports on cartilage growth-plate pathologies in both human patients and *Plod1*-knockout mice.

4.2. Inner Ear Development

Sensorineural hearing loss has been reported in 73% of *FKBP14*-kEDS patients [3,4], but not in *PLOD1*-kEDS patients. The transcriptome profiles showed an elevated expression of *ALDH1A3* encoding aldehyde dehydrogenase 1 family member A3 enzyme in *FKBP14*-kEDS patient-derived fibroblasts only, which was confirmed by quantitative RT-PCR (Figure 5D). During normal inner ear canal formation, *Aldh1a3* is repressed by the chromatin remodeler Chd7. Deficiency in Chd7 causes abnormally high levels of *Aldh1a3* expression and inner ear malformation; inner ear development is rescued by the loss of *Aldh1a3* expression in mice deficient in Chd7 [36]. Hence, we postulate a molecular link between enhanced *ALDH1A3* expression and hearing loss in *FKBP14*-kEDS patients that warrants further investigation.

4.3. Vasculature Integrity

We sought to identify DEGs in *PLOD1*-kEDS and *FKBP14*-kEDS that are involved in modulating the vasculature, since rupture or aneurysm of medium-sized arteries have been described in both *PLOD1*-kEDS [12,37] and *FKBP14*-kEDS [3,38,39] and serves as a minor criteria in the diagnosis of kEDS [1]. We observed an up-regulation of *OLFM2* expression (encoding olfactomedin 2) in both *PLOD1*-kEDS and *FKBP14*-kEDS patient-derived fibroblasts (Figure 5E). *OLFM2* is also significantly up-regulated in COL6-RD patient-derived fibroblast [10], although vascular complications involving medium-sized arteries have not been reported in COL6-RD to our knowledge; instead, fenestration and narrow lumens in capillaries have been described in some cases of Ulrich congenital muscular dystrophy [40]. *Olfm2* is up-regulated in balloon-injured arteries and involved in smooth muscle cell phenotypic modulation and vascular remodeling. *Olfm2*-deficient mice are more protected against injury-induced suppression of smooth muscle cell markers and neointimal hyperplasia [41], suggesting that enhanced expression of *OLFM2* may correlate with poorer vascular phenotypes. Notably, *OLFM2* expression was not significantly altered in fibroblasts of vascular EDS patients [42]. Hence, whether elevated *OLFM2* expression contributes to the causation and/or severity of vascular complications in kEDS remains an interesting topic to be explored. The up-regulation of *OLFM2* in *PLOD1*-kEDS, *FKBP14*-kEDS, and COL6-RD also warrants further investigation into whether olfactomedin 2 plays a physiological role in skeletal muscles, which has not yet been described.

In addition, we noted that *TM4SF1* expression was down-regulated in *FKBP14*-kEDS and up-regulated in *PLOD1*-kEDS (Figure 5E). *TM4SF1* encodes the transmembrane-4-L-six-family-1 protein, also known as L6 cell surface antigen, and is highly expressed in the vascular endothelium. In a previous study, knockdown of *TM4SF1* in human umbilical vein endothelial cells (HUVEC cell line) resulted in a senescence phenotype and poor migration in a wound healing assay in vitro. Moreover, the authors showed that the angiogenesis inducer Vascular Endothelial Growth Factor A (VEGF-A) enhanced *Tm4sf1* expression in vivo in mice, while knockdown of *Tm4sf1* blunted the angiogenic effect of VEGF-A, thus demonstrating the importance of TM4SF1 in vessel maturation [43]. As such, it would be interesting to determine if the dysregulation of *TM4SF1* expression contributes to vascular fragility in kEDS.

4.4. Unfolded Protein Response (UPR)

Another striking observation described in the first cohort of *FKBP14*-kEDS patients is the enlargement of ER in skin biopsies [4]. This dilatation of ER is thought to arise from the accumulation of misfolded proteins, in particular of collagens since the rate limiting step in collagen protein folding involves cis–trans isomerization of the prolyl peptide bonds which is catalyzed by FKBP22. ER dilatation coupled with the induction of UPR involving binding immunoglobulin protein (BiP) activation has been demonstrated in several models of connective tissue disorders, including (i) the Osteogenesis

Imperfecta (OI) *Aga2* mouse model harboring *Col1a1* mutations which ablates the conserved C-terminal cysteine C244 residue and introduces additional amino acids into the C-terminal propeptide [44], (ii) in skin fibroblasts isolated from OI patients carrying *COL1A1* mutations in the C-terminal propeptide domain [45], and (iii) in engineered cells with *COL10A1* mutations modelling Schmid metaphyseal chondrodysplasia [46]. However, the transcriptome profiles and gene ontology analyses revealed that genes associated with classical ER associated degradation and UPR, including *CANX*, *CALR*, *HSPA5*, *ATF4*, and *DDIT3*, were not differentially expressed in *FKBP14*-kEDS patient-derived fibroblasts; the same trend was observed in *PLOD1*-kEDS patient-derived fibroblasts (Table S9). The absence of classical UPR activation was also previously observed in other connective tissue disorders, such as in OI patients with mutations in *COL1A1* disrupting triple helix formation [45] and in vascular EDS patients carrying *COL3A1* mutations [42].

Nevertheless, one of the most up-regulated genes in *FKBP14*-kEDS patient-derived fibroblasts was *SCAMP5* encoding the secretory carrier-associated membrane protein 5, which was also induced in *PLOD1*-kEDS patient-derived fibroblasts albeit at a smaller magnitude (Figure 5F). *SCAMP5* expression can be rapidly induced by autophagic stimulation under the control of the master autophagy transcriptional regulator transcription factor EB (TFEB), and its elevated expression was previously described in the striatum of Huntington disease patients [47,48]. Interestingly, SCAMP5 enhances the aggregation of mutant huntingtin by impairing endocytosis [47], yet promotes Golgi fragmentation and unconventional secretion of α-synuclein via exosomes [48]. This warrants for future exploration into the role of SCAMP5 in kEDS patient-derived fibroblasts in relation to ER stress, autophagy and whether it promotes the retention or non-classical secretion of misfolded collagen proteins.

4.5. Bone Remodeling

Osteopenia or osteoporosis is a minor criteria in the diagnosis of kEDS [1], and occurrence of fractures were described in 13% of *FKBP14*-kEDS [3,4]. We observed elevated expression of *FGF11*, which encodes the intracellular fibroblast growth factor 11, in both *PLOD1*-kEDS and *FKBP14*-kEDS (Figure 5G). Bone resorption by osteoclasts is mediated by hypoxic conditions which lead to overexpression of *FGF11* and knockdown of *FGF11* lead to inhibition of bone resorption by osteoclasts in response to hypoxic stimulation [49]. Furthermore, FGF11 protein is strongly expressed in osteoclasts in osteolytic diseases such as rheumatoid synovium and giant cell tumor of bone [49], suggesting that it plays a role in pathological bone resorption. Thus, it will be interesting to determine if FGF11 levels are also elevated in the osteoclasts of kEDS patients, and whether FGF11 is a driver of the bone phenotype in kEDS.

4.6. Others

The expression of *PLXNA2* is down-regulated in *PLOD1*-kEDS and *FKBP14*-kEDS patient fibroblasts (Figure 5H). *PLXNA2* encodes the transmembrane protein plexin-A2, a member of the plexin-A family of semaphorin co-receptors, and is demonstrated to be involved in axon guidance [50]. However, nerve conduction appeared normal in *FKBP14*-kEDS patients [3,4] and *PLOD1*-kEDS patients [12,51]. Nevertheless, the suppression of *PLXNA2* expression was of a higher magnitude in *FKBP14*-kEDS patients, in which sensorineural hearing loss has been reported. Hence, whether diminished *PLXNA2* expression has a causal effect on sensorineural hearing impairment in *FKBP14*-kEDS patients requires further investigation.

In another study, knockdown of *Plxna2* blocked osteoblast differentiation and mineralization in vitro, highlighting the pro-osteogenic role of *Plxna2*. Thus, it may be interesting to investigate whether *PLXNA2* expression is also suppressed in mesenchymal cells of *PLOD1*-kEDS and *FKBP14*-kEDS patients, which may hinder osteoblast differentiation and bone remodeling.

The expression of *PLEKHA2* is diminished in both *PLOD1*-kEDS and *FKBP14*-kEDS patient fibroblasts compared to controls (Figure 5I). *PLEKHA2* expression in dermal fibroblasts of patients with COL6 dominant negative mutations, where mutant type VI collagen chains are incorporated into the collagen fibers, is also significantly lower [10]. *PLEKHA2* encodes pleckstrin homology (PH) domain-containing family A member 2, which is also known as Tandem PH domain-containing protein 2 (TAPP2). TAPP2 is recruited to the plasma membrane via its interaction with phosphatidylinositol-3,4-bisphosphate (PI(3,4)P2) [52,53]. The physiological role of TAPP2 has been well characterized in B-cells, where mice carrying knock-in inactivating mutations of TAPP2 that disrupt its interaction with PI(3,4)P2 develop chronic germinal centers, hyperactive B-cells with higher survival and produce more autoantibodies that worsened with age [54]. Autoimmune phenotypes have, however, not yet been characterized in kEDS patients. Additionally, TAPP2 has also been shown to facilitate the migration of esophageal squamous cell carcinoma cells [55] and malignant B-cells [56] via cytoskeleton reorganization. Hence, further studies on the role of TAPP2 in fibroblast survival, activity, and migration may shed light on its relation to the pathogenesis of kEDS.

5. Conclusions

We have performed the first transcriptomics studies on kEDS which revealed distinct transcriptome signatures between *PLOD1*-kEDS and *FKBP14*-kEDS despite clinical similarities. ECM components are enriched by DEGs in both *PLOD1*-kEDS and *FKBP14*-kEDS, although the genes are unique to each genetic form of kEDS. We aim to validate the candidate DEGs on more patient samples when they become available to strengthen the findings reported here. Proteomics analysis of the ECM component and functional characterization of candidate genes described are currently ongoing and these promise to deepen our understanding of the pathomechanism(s) underlying kEDS.

Supplementary Materials: The following are available online at http://www.mdpi.com/2073-4425/10/7/517/s1, Table S1: Age, sex, and origin of controls and patients, Table S2: List of DEGs that are up-regulated in *PLOD1*-kEDS over controls, Table S3: List of DEGs that are down-regulated in *PLOD1*-kEDS over controls, Table S4: List of DEGs that are up-regulated in *FKBP14*-kEDS over controls, Table S5: List of DEGs that are down-regulated in *FKBP14*-kEDS over controls, Table S6: DEGs contributing to over-representation of gene ontology terms in *PLOD1*-kEDS patient-derived fibroblasts, Table S7: DEGs contributing to over-representation of gene ontology terms in *FKBP14*-kEDS patient-derived fibroblasts, Table S8: List of DEGs in *PLOD1*-kEDS over *FKBP14*-kEDS (p-value <0.01 and $\log_2$ ratio >0.5 or <−0.5), Table S9: Genes involved in ER associated degradation and unfolded protein response.

Author Contributions: Conceptualization: C.G., M.R., U.L., L.O., P.J.L.; data analysis: U.L., P.J.L., L.O., I.H.; manuscript preparation: P.J.L.; manuscript reviewing: C.G., M.R., L.O.

Funding: This work has been supported by the Swiss National Science Foundation Grant No. 31003A-173183 to CG and MR, the Vontobel Foundation and the Armin & Jeannine Kurz Foundation.

Acknowledgments: The authors wish to thank Mike Pope (Ehlers–Danlos Syndrome National Diagnostic Service, London), Matthias Baumann, Christine Fauth and Johannes Zschocke (Medical University of Innsbruck) for sharing patient's material, and Daisy Rymen (Children's Hospital Zurich) for her expert opinion on ER associated degradation and UPR, and for critically evaluating the manuscript.

Conflicts of Interest: The authors declare no conflict of interest.

References

1. Malfait, F.; Francomano, C.; Byers, P.; Belmont, J.; Berglund, B.; Black, J.; Bloom, L.; Bowen, J.M.; Brady, A.F.; Burrows, N.P.; et al. The 2017 international classification of the Ehlers-Danlos syndromes. *Am. J. Med. Genet. Part C Semin. Med. Genet.* **2017**, *175*, 8–26. [CrossRef] [PubMed]
2. Kivirikko, K.I.; Myllylä, R. Posttranslational Enzymes in the Biosynthesis of Collagen: Intracellular Enzymes. *Methods Enzymol.* **1982**, *82*, 245–304. [PubMed]
3. Giunta, C.; Baumann, M.; Fauth, C.; Lindert, U.; Abdalla, F.M.; Brady, A.F.; Collins, J.; Dastgir, J.; Donkervoort, S.; Ghali, N.; et al. A cohort of 17 patients with kyphoscoliotic Ehlers-Danlos syndrome caused by biallelic mutations in FKBP14: Expansion of the clinical and mutational spectrum and description of the natural history. *Genet. Med.* **2018**, *20*, 42–54. [CrossRef] [PubMed]
4. Baumann, M.; Giunta, C.; Krabichler, B.; Rüschendorf, F.; Zoppi, N.; Colombi, M.; Bittner, R.E.; Quijano-Roy, S.; Muntoni, F.; Cirak, S.; et al. Mutations in FKBP14 cause a variant of Ehlers-Danlos syndrome with progressive kyphoscoliosis, myopathy, and hearing loss. *Am. J. Hum. Genet.* **2012**, *90*, 201–216. [CrossRef]

5. Ishikawa, Y.; Bächinger, H.P. A substrate preference for the rough endoplasmic reticulum resident protein FKBP22 during collagen biosynthesis. *J. Biol. Chem.* **2014**, *289*, 18189–18201. [CrossRef]
6. Bönnemann, C.G. The collagen VI-related myopathies: Muscle meets its matrix. *Nat. Rev. Neurol.* **2011**, *7*, 379–390. [CrossRef]
7. Kirschner, J.; Hausser, I.; Zou, Y.; Schreiber, G.; Christen, H.J.; Brown, S.C.; Anton-Lamprecht, I.; Muntoni, F.; Hanefeld, F.; Bönnemann, C.G. Ullrich congenital muscular dystrophy: Connective tissue abnormalities in the skin support overlap with Ehlers-Danlos syndromes. *Am. J. Med. Genet.* **2005**, *132*, 296–301. [CrossRef]
8. Yiş, U.; Dirik, E.; Chambaz, C.; Steinmann, B.; Giunta, C. Differential diagnosis of muscular hypotonia in infants: The kyphoscoliotic type of Ehlers-Danlos syndrome (EDS VI). *Neuromuscul. Disord.* **2008**, *18*, 210–214. [CrossRef]
9. Giunta, C.; Rohrbach, M.; Fauth, C.; Baumann, M. *FKBP14-Kyphoscoliotic Ehlers-Danlos Syndrome*; GeneReviews; Adam, M., Ardinger, H., Pagon, R., Eds.; University of Washington: Seattle, WA, USA, 2019.
10. Butterfield, R.J.; Dunn, D.M.; Hu, Y.; Johnson, K.; Bönnemann, C.G.; Weiss, R.B. Transcriptome profiling identifies regulators of pathogenesis in collagen VI related muscular dystrophy. *PLoS ONE* **2017**, *12*, e0189664. [CrossRef]
11. Giunta, C.; Randolph, A.; Steinmann, B. Mutation analysis of the PLOD1 gene: An efficient multistep approach to the molecular diagnosis of the kyphoscoliotic type of Ehlers-Danlos syndrome (EDS VIA). *Mol. Genet. Metab.* **2005**, *86*, 269–276. [CrossRef]
12. Rohrbach, M.; Vandersteen, A.; Yi, U.; Serdaroglu, G.; Ataman, E.; Chopra, M.; Garcia, S.; Jones, K.; Kariminejad, A.; Kraenzlin, M.; et al. Phenotypic variability of the kyphoscoliotic type of Ehlers-Danlos syndrome (EDS VIA): Clinical, molecular and biochemical delineation. *Orphanet J. Rare Dis.* **2011**, *6*, 46. [CrossRef] [PubMed]
13. Wang, J.; Vasaikar, S.; Shi, Z.; Greer, M.; Zhang, B. WebGestalt 2017: A more comprehensive, powerful, flexible and interactive gene set enrichment analysis toolkit. *Nucleic Acids Res.* **2017**, *45*, W130–W137. [CrossRef] [PubMed]
14. Zhang, B.; Kirov, S.; Snoddy, J. WebGestalt: An integrated system for exploring gene sets in various biological contexts. *Nucleic Acids Res.* **2005**, *33*, W741–W748. [CrossRef] [PubMed]
15. Morris, B.A.; Burkel, B.; Ponik, S.M.; Fan, J.; Condeelis, J.S.; Aguire-Ghiso, J.A.; Castracane, J.; Denu, J.M.; Keely, P.J. Collagen Matrix Density Drives the Metabolic Shift in Breast Cancer Cells. *EBioMedicine* **2016**, *13*, 146–156. [CrossRef] [PubMed]
16. Zhao, X.; Psarianos, P.; Ghoraie, L.S.; Yip, K.; Goldstein, D.; Gilbert, R.; Witterick, I.; Pang, H.; Hussain, A.; Lee, J.H.; et al. Metabolic regulation of dermal fibroblasts contributes to skin extracellular matrix homeostasis and fibrosis. *Nat. Metab.* **2019**, *1*, 147–157. [CrossRef]
17. Maruhashi, T.; Kii, I.; Saito, M.; Kudo, A. Interaction between periostin and BMP-1 promotes proteolytic activation of lysyl oxidase. *J. Biol. Chem.* **2010**, *285*, 13294–13303. [CrossRef]
18. Kii, I.; Nishiyama, T.; Li, M.; Matsumoto, K.I.; Saito, M.; Amizuka, N.; Kudo, A. Incorporation of tenascin-C into the extracellular matrix by periostin underlies an extracellular meshwork architecture. *J. Biol. Chem.* **2010**, *285*, 2028–2039. [CrossRef]
19. Nishiyama, T.; Kii, I.; Kashima, T.G.; Kikuchi, Y.; Ohazama, A.; Shimazaki, M.; Fukayama, M.; Kudo, A. Delayed re-epithelialization in periostin-deficient mice during cutaneous wound healing. *PLoS ONE* **2011**, *6*, e18410. [CrossRef]
20. Elliott, C.G.; Wang, J.; Guo, X.; Xu, S.-W.; Eastwood, M.; Guan, J.; Leask, A.; Conway, S.J.; Hamilton, D.W. Periostin modulates myofibroblast differentiation during full-thickness cutaneous wound repair. *J. Cell Sci.* **2012**, *125*, 121–132. [CrossRef]
21. Ontsuka, K.; Kotobuki, Y.; Shiraishi, H.; Serada, S.; Ohta, S.; Tanemura, A.; Yang, L.; Fujimoto, M.; Arima, K.; Suzuki, S.; et al. Periostin, a matricellular protein, accelerates cutaneous wound repair by activating dermal fibroblasts. *Exp. Dermatol.* **2012**, *21*, 331–336. [CrossRef]
22. Zhou, H.M.; Wang, J.; Elliott, C.; Wen, W.; Hamilton, D.W.; Conway, S.J. Spatiotemporal expression of periostin during skin development and incisional wound healing: Lessons for human fibrotic scar formation. *J. Cell Commun. Signal.* **2010**, *4*, 99–107. [CrossRef] [PubMed]

23. Yang, L.; Serada, S.; Fujimoto, M.; Terao, M.; Kotobuki, Y.; Kitaba, S.; Matsui, S.; Kudo, A.; Naka, T.; Murota, H.; et al. Periostin facilitates skin sclerosis via PI3K/Akt dependent mechanism in a mouse model of scleroderma. *PLoS ONE* **2012**, *7*, e41994. [CrossRef] [PubMed]
24. Yamaguchi, Y.; Ono, J.; Masuoka, M.; Ohta, S.; Izuhara, K.; Ikezawa, Z.; Aihara, M.; Takahashi, K. Serum periostin levels are correlated with progressive skin sclerosis in patients with systemic sclerosis. *Br. J. Dermatol.* **2013**, *168*, 717–725. [CrossRef] [PubMed]
25. Guo, X.; Day, T.F.; Jiang, X.; Garrett-Beal, L.; Topol, L.; Yang, Y. Wnt/β-catenin signaling is sufficient and necessary for synovial joint formation. *Genes Dev.* **2004**, *18*, 2404–2417. [CrossRef] [PubMed]
26. Strochlic, L.; Falk, J.; Goillot, E.; Sigoillot, S.; Bourgeois, F.; Delers, P.; Rouvière, J.; Swain, A.; Castellani, V.; Schaeffer, L.; et al. Wnt4 participates in the formation of vertebrate neuromuscular junction. *PLoS ONE* **2012**, *7*, e29976. [CrossRef] [PubMed]
27. Okuse, T.; Chiba, T.; Katsuumi, I.; Imai, K. Differential expression and localization of WNTs in an animal model of skin wound healing. *Wound Repair Regen.* **2005**, *13*, 491–497. [CrossRef] [PubMed]
28. Prunskaite-Hyyryläinen, R.; Skovorodkin, I.; Xu, Q.; Miinalainen, I.; Shan, J.; Vainio, S.J. Wnt4 coordinates directional cell migration and extension of the müllerian duct essential for ontogenesis of the female reproductive tract. *Hum. Mol. Genet.* **2015**, *25*, 1059–1073. [CrossRef] [PubMed]
29. Kucich, U.; Rosenbloom, J.C.; Abrams, W.R.; Rosenbloom, J. Transforming growth factor-β stabilizes elastin mRNA by a pathway requiring active smads, protein kinase c-δ, and p38. *Am. J. Respir. Cell Mol. Biol.* **2002**, *26*, 183–188. [CrossRef]
30. Colwell, A.S.; Krummel, T.M.; Longaker, M.T.; Lorenz, H.P. Wnt-4 expression is increased in fibroblasts after TGF-beta1 stimulation and during fetal and postnatal wound repair. *Plast. Reconstr. Surg.* **2006**, *117*, 2297–2301. [CrossRef]
31. Chen, G.; Nakamura, I.; Dhanasekaran, R.; Iguchi, E.; Tolosa, E.J.; Romecin, P.A.; Vera, R.E.; Almada, L.L.; Miamen, A.G.; Chaiteerakij, R.; et al. Transcriptional Induction of Periostin by a Sulfatase 2-TGFb1-SMAD Signaling Axis Mediates Tumor Angiogenesis in Hepatocellular Carcinoma. *Cancer Res.* **2017**, *77*, 632–645. [CrossRef]
32. Muona, A.; Eklund, L.; Vaisanen, T.; Pihlajaniemi, T. Developmentally regulated expression of type XV collagen correlates with abnormalities in Col15a1-/-mice. *Matrix Biol.* **2002**, *21*, 89–102. [CrossRef]
33. Eklund, L. Lack of type XV collagen causes a skeletal myopathy and cardiovascular defects in mice. *Proc. Natl. Acad. Sci. USA* **2001**, *98*, 1194–1199. [CrossRef]
34. McLaughlin, P.J.; Bakall, B.; Choi, J.; Liu, Z.; Sasaki, T.; Davis, E.C.; Marmorstein, A.D.; Marmorstein, L.Y. Lack of fibulin-3 causes early aging and herniation, but not macular degeneration in mice. *Hum. Mol. Genet.* **2007**, *16*, 3059–3070. [CrossRef] [PubMed]
35. Wakabayashi, T.; Matsumine, A.; Nakazora, S.; Hasegawa, M.; Iino, T.; Ota, H.; Sonoda, H.; Sudo, A.; Uchida, A. Fibulin-3 negatively regulates chondrocyte differentiation. *Biochem. Biophys. Res. Commun.* **2010**, *391*, 1116–1121. [CrossRef] [PubMed]
36. Yao, H.; Hill, S.F.; Skidmore, J.M.; Sperry, E.D.; Swiderski, D.L.; Sanchez, G.J.; Bartels, C.F.; Raphael, Y.; Scacheri, P.C.; Iwase, S.; et al. CHD7 represses the retinoic acid synthesis enzyme ALDH1A3 during inner ear development. *JCI Insight* **2018**, *3*, 97440. [CrossRef] [PubMed]
37. Gok, E.; Goksel, O.S.; Alpagut, U.; Dayioglu, E. Spontaneous brachial pseudo-aneurysm in a 12-year-old with kyphoscoliosis-type ehlers-danlos syndrome. *Eur. J. Vasc. Endovasc. Surg.* **2012**, *44*, 482–484. [CrossRef]
38. Murray, M.L.; Yang, M.; Fauth, C.; Byers, P.H. FKBP14-related Ehlers-Danlos syndrome: Expansion of the phenotype to include vascular complications. *Am. J. Med. Genet. Part A* **2014**, *164*, 1750–1755. [CrossRef]
39. Dordoni, C.; Ciaccio, C.; Venturini, M.; Calzavara-Pinton, P.; Ritelli, M.; Colombi, M. Further delineation of FKBP14-related Ehlers-Danlos syndrome: A patient with early vascular complications and non-progressive kyphoscoliosis, and literature review. *Am. J. Med. Genet. Part A* **2016**, *170*, 2031–2038. [CrossRef]
40. Niiyama, T.; Higuchi, I.; Hashiguchi, T.; Suehara, M.; Uchida, Y.; Horikiri, T.; Shiraishi, T.; Saitou, A.; Hu, J.; Nakagawa, M.; et al. Capillary changes in skeletal muscle of patients with Ullrich's disease with collagen VI deficiency. *Acta Neuropathol.* **2003**, *106*, 137–142. [CrossRef]
41. Shi, N.; Li, C.X.; Cui, X.B.; Tomarev, S.I.; Chen, S.Y. Olfactomedin 2 Regulates Smooth Muscle Phenotypic Modulation and Vascular Remodeling Through Mediating Runt-Related Transcription Factor 2 Binding to Serum Response Factor. *Arterioscler. Thromb. Vasc. Biol.* **2017**, *37*, 446–454. [CrossRef]

42. Chiarelli, N.; Carini, G.; Zoppi, N.; Ritelli, M.; Colombi, M. Transcriptome analysis of skin fibroblasts with dominant negative COL3A1 mutations provides molecular insights into the etiopathology of vascular Ehlers-Danlos syndrome. *PLoS ONE* **2018**, *13*, 1–24. [CrossRef] [PubMed]
43. Shih, S.C.; Zukauskas, A.; Li, D.; Liu, G.; Ang, L.H.; Nagy, J.A.; Brown, L.F.; Dvorak, H.F. The l6 protein TM4SF1 is critical for endothelial cell function and tumor angiogenesis. *Cancer Res.* **2009**, *69*, 3272–3277. [CrossRef]
44. Lisse, T.S.; Thiele, F.; Fuchs, H.; Hans, W.; Przemeck, G.K.H.; Abe, K.; Rathkolb, B.; Quintanilla-Martinez, L.; Hoelzlwimmer, G.; Helfrich, M.; et al. ER stress-mediated apoptosis in a new mouse model of Osteogenesis imperfecta. *PLoS Genet.* **2008**, *4*, e7. [CrossRef] [PubMed]
45. Chessler, S.D.; Byers, P.H. BiP binds type I procollagen pro?? chains with mutations in the carboxyl-terminal propeptide synthesized by cells from patients with osteogenesis imperfecta. *J. Biol. Chem.* **1993**, *268*, 18226–18233. [PubMed]
46. Wilsoni, R.; Freddi, S.; Chan, D.; Cheah, K.S.E.; Bateman, J.F. Misfolding of collagen X chains harboring schmid metaphyseal chondrodysplasia mutations results in aberrant disulfide bond formation, intracellular retention, and activation of the unfolded protein response. *J. Biol. Chem.* **2005**, *280*, 15544–15552. [CrossRef] [PubMed]
47. Noh, J.Y.; Lee, H.; Song, S.; Kim, N.S.; Im, W.; Kim, M.; Seo, H.; Chung, C.W.; Chang, J.W.; Ferrante, R.J.; et al. SCAMP5 links endoplasmic reticulum stress to the accumulation of expanded polyglutamine protein aggregates via endocytosis inhibition. *J. Biol. Chem.* **2009**, *284*, 11318–11325. [CrossRef] [PubMed]
48. Yang, Y.; Qin, M.; Bao, P.; Xu, W.; Xu, J. Secretory carrier membrane protein 5 is an autophagy inhibitor that promotes the secretion of α-synuclein via exosome. *PLoS ONE* **2017**, *12*, e0180892. [CrossRef]
49. Knowles, H.J. Hypoxia-Induced Fibroblast Growth Factor 11 Stimulates Osteoclast-Mediated Resorption of Bone. *Calcif. Tissue Int.* **2017**, *100*, 382–391. [CrossRef]
50. Mitsogiannis, M.D.; Little, G.E.; Mitchell, K.J. Semaphorin-Plexin signaling influences early ventral telencephalic development and thalamocortical axon guidance. *Neural Dev.* **2017**, *12*, 6. [CrossRef]
51. Kariminejad, A.; Bozorgmehr, B.; Khatami, A.; Kariminejad, M.H.; Giunta, C.; Steinmann, B. Ehlers-Danlos Syndrome Type VI in a 17-Year-Old Iranian Boy with Severe Muscular Weakness—A Diagnostic Challenge? *Iran. J. Pediatr.* **2010**, *20*, 358–362.
52. Marshall, A.J.; Krahn, A.K.; Ma, K.; Duronio, V.; Hou, S. TAPP1 and TAPP2 are targets of phosphatidylinositol 3-kinase signaling in B-cells: Sustained plasma membrane recruitment triggered by the B-cell antigen receptor. *Mol. Cell. Biol.* **2002**, *22*, 5479–5491. [CrossRef] [PubMed]
53. Kimber, W.A.; Trinkle-Mulcahy, L.; Cheung, P.C.F.; Deak, M.; Marsden, L.J.; Kieloch, A.; Watt, S.; Javier, R.T.; Gray, A.; Downes, C.P.; et al. Evidence that the tandem-pleckstrin-homology-domain-containing protein TAPP1 interacts with Ptd(3,4)P2 and the multi-PDZ-domain-containing protein MUPP1 in vivo. *Biochem. J.* **2002**, *361*, 525–536. [CrossRef] [PubMed]
54. Jayachandran, N.; Landego, I.; Hou, S.; Alessi, D.R.; Marshall, A.J. B-cell-intrinsic function of TAPP adaptors in controlling germinal center responses and autoantibody production in mice. *Eur. J. Immunol.* **2017**, *47*, 280–290. [CrossRef] [PubMed]
55. Liu, F.; Ye, F.; Guan, Z.; Zhou, Y.; Ji, F.; Zhang, Q.; Zhang, J.; Zhang, T.; Lu, S. The down-regulation of TAPP2 inhibits the migration of esophageal squamous cell carcinoma and predicts favorable outcome. *Pathol. Res. Pract.* **2017**, *213*, 1556–1562. [CrossRef] [PubMed]
56. Li, H.; Hou, S.; Wu, X.; Nandagopal, S.; Lin, F.; Kung, S.; Marshall, A.J. The Tandem PH Domain-Containing Protein 2 (TAPP2) Regulates Chemokine-Induced Cytoskeletal Reorganization and Malignant B Cell Migration. *PLoS ONE* **2013**, *8*, e57809. [CrossRef] [PubMed]

© 2019 by the authors. Licensee MDPI, Basel, Switzerland. This article is an open access article distributed under the terms and conditions of the Creative Commons Attribution (CC BY) license (http://creativecommons.org/licenses/by/4.0/).

Article

Severe Peripheral Joint Laxity is a Distinctive Clinical Feature of Spondylodysplastic-Ehlers-Danlos Syndrome (EDS)-*B4GALT7* and Spondylodysplastic-EDS-*B3GALT6*

Stefano Giuseppe Caraffi [1], Ilenia Maini [1,2], Ivan Ivanovski [1,3], Marzia Pollazzon [1], Sara Giangiobbe [1], Maurizia Valli [4], Antonio Rossi [4], Silvia Sassi [5], Silvia Faccioli [5], Maja Di Rocco [6], Cinzia Magnani [7], Belinda Campos-Xavier [8], Sheila Unger [8], Andrea Superti-Furga [8,*] and Livia Garavelli [1,*]

[1] Medical Genetics Unit, Maternal and Child Health Department, Azienda USL-IRCCS of Reggio Emilia, 42122 Reggio Emilia, Italy; stefanogiuseppe.caraffi@ausl.re.it (S.G.C.); imaini@ausl.pr.it (I.M.); ivan.ivanovski@ausl.re.it (I.I.); marzia.pollazzon@ausl.re.it (M.P.); sara.giangiobbe@ausl.re.it (S.G.)
[2] Child Neuropsychiatry Unit, Azienda USL of Parma, 43125 Parma, Italy
[3] Department of Surgical, Medical, Dental and Morphological Sciences with interest in Transplant, Oncology and Regenerative Medicine, University of Modena and Reggio Emilia, 42121 Reggio Emilia, Italy
[4] Department of Molecular Medicine, Unit of Biochemistry, University of Pavia, 27100 Pavia, Italy; zuffardi@unipv.it (M.V.); antonio.rossi@unipv.it (A.R.)
[5] Rehabilitation Pediatric Unit, Azienda USL-IRCCS of Reggio Emilia, Reggio Emilia, 42122 Reggio Emilia, Italy; silvia.sassi@ausl.re.it (S.S.); silvia.faccioli@ausl.re.it (S.F.)
[6] Department of Pediatrics, Unit of Rare Diseases, IRCCS Istituto Giannina Gaslini, 16147 Genoa, Italy; majadirocco@gaslini.org
[7] Neonatology and Neonatal Intensive Care Unit, Maternal and Child Department, University of Parma, 43121 Parma, Italy; cinzia.magnani@unipr.it
[8] Division of Genetic Medicine, Centre Hospitalier Universitaire Vaudois (CHUV), University of Lausanne, 1011 Lausanne, Switzerland; belinda.xavier@chuv.ch (B.C.-X.); sheila.unger@chuv.ch (S.U.)
* Correspondence: asuperti@unil.ch (A.S.-F.); livia.garavelli@ausl.re.it (L.G.); Tel.: +41-(0)21-314-92-38 or +41-(0)21-314-00-14 (A.S.-F.); +39-0522-296244 (ext. 295463) (L.G.); Fax: +41-(0)21-314-35-46 (A.S.-F.); +39-0522-296266 (L.G.)

Received: 13 August 2019; Accepted: 10 October 2019; Published: 12 October 2019

Abstract: Variations in genes encoding for the enzymes responsible for synthesizing the linker region of proteoglycans may result in recessive conditions known as "linkeropathies". The two phenotypes related to mutations in genes *B4GALT7* and *B3GALT6* (encoding for galactosyltransferase I and II respectively) are similar, characterized by short stature, hypotonia, joint hypermobility, skeletal features and a suggestive face with prominent forehead, thin soft tissue and prominent eyes. The most outstanding feature of these disorders is the combination of severe connective tissue involvement, often manifesting in newborns and infants, and skeletal dysplasia that becomes apparent during childhood. Here, we intend to more accurately define some of the clinical features of *B4GALT7* and *B3GALT6*-related conditions and underline the extreme hypermobility of distal joints and the soft, doughy skin on the hands and feet as features that may be useful as the first clues for a correct diagnosis.

Keywords: beta-1,3-galactosyltransferase 6 (*B3GALT6*); beta-1,4-galactosyltransferase 7 (*B4GALT7*); spondyloepimetaphyseal dysplasia with joint laxity (SEMDJL1; SEMDJL-Beighton type); Ehlers–Danlos syndrome (EDS); spondylodysplastic Ehlers–Danlos syndrome (spEDS); spEDS-*B3GALT6*; spEDS-*B4GALT7*; extreme laxity of distal joints; soft; doughy skin on the hands and feet

1. Introduction

The extracellular matrix (ECM) of connective tissues like cartilage, bone, tendon, ligaments and skin is a complex interacting arrangement of proteins, glycoproteins and proteoglycans (PGs) that forms the scaffold of the human body and also constitutes the vital environment supporting cell growth and differentiation. PGs are macromolecules that are complex in their structures and represent an important component of both the ECM and of cell membranes. PGs have mechanical and osmotic functions but are strongly involved in signaling pathways controlling cell-to-cell interactions as well as cell growth and differentiation and tissue repair [1–3]. PGs are composed of a core protein and one or more glycosaminoglycan (GAG) chains that are long polysaccharides consisting of repeating disaccharide units. On the basis of the composition of these units, the PGs are subdivided into heparan sulfate (HS) PGs and chondroitin sulfate (CS)/dermatan sulfate (DS) PGs [4]. The biosynthesis of the GAG side chain begins with the creation of a common tetrasaccharide, a so called "linker", covalently linked to the hydroxyl group of serine residues on the core protein (O-glycosylation).

The linker region is synthesized through a coordinated mechanism involving specific glycosyltransferases. It starts with the transfer of a xylose residue by xylosyltransferase I or II (encoded by XYLT1, OMIM *608124, and XYLT2, OMIM *608125), followed by the addition of two galactose residues by galactosyltransferase I, encoded by B4GALT7 (OMIM 604327) and galactosyltransferase II encoded by B3GALT6 (OMIM *615291) (Figure 1). Subsequently there will be a transfer of glucuronic acid by glucuronosyltransferase I encoded by B3GAT3 (OMIM *606374). The so-called "linkeropathies" are recessive conditions caused by variations in genes encoding the enzymes responsible for the synthesis of the linker region of PGs.

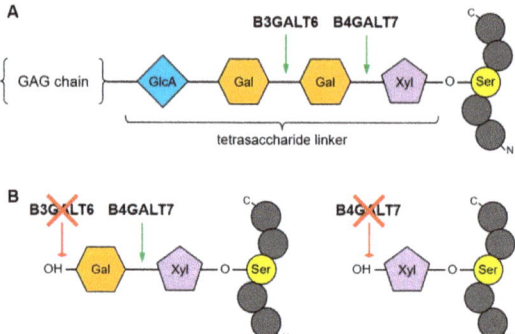

Figure 1. **A.** Galactosyltransferases I (B4GALT7) and II (B3GALT6) catalyze the addition of galactose units at specific positions of the tetrasaccharide region linking Serine residues on the protein backbone to glycosaminoglycan (GAG) chains. Xyl = Xylose, Gal = Galactose, GlcA = Glucuronic Acid. **B.** Functional defects of B4GALT6 or B4GALT7 inhibit addition of one or both Gal monosaccharides, respectively, and prevent GAG chain polymerization.

Variants of the XYLT1 gene are responsible for skeletal dysplasias, including Baratela–Scott syndrome (OMIM 615777, formerly erroneously called Desbuquois dysplasia type 2) [5–10], variants of XYLT2 are the cause of the spondylo-ocular syndrome (OMIM 605822) [11–13], while variants of B3GAT3 cause a Larsen-like syndrome and a more severe skeletal dysplasia (OMIM 245600) [14–20].

In particular, variants in the B4GALT7 and B3GALT6 genes are the cause of a combined skeletal and connective tissue phenotype.

A B4GALT7-related condition was initially described as "progeroid type 1" Ehlers–Danlos syndrome (EDS), due to the phenotype characterized by premature aging and loose elastic skin in one reported patient [21–30]. Subsequently, Cartault et al. in 2015 described a unique homozygous B4GALT7 mutation, p.Arg270Cys, in a large cohort of patients affected by the so-called Larsen of Reunion Island syndrome [31]. This contributed to define the currently known phenotype of *B4GALT7*-related EDS, with round flat face, proptosis, short stature, hypotonia, radioulnar synostosis, osteopenia, hyperextensible skin and joint hypermobility (OMIM #130070) [24,25,31–33].

B3GALT6 variants cause a single disorder that was alternatively described as spondyloepimetaphyseal dysplasia (SEMD) with joint hypermobility (SEMDJL1 or SEMDJL Beighton type; OMIM #271640) in the "Nosology and classification of genetic skeletal disorders: 2015 revision" [34–43] or as severe EDS-like disorder (OMIM #615349) because of the striking joint laxity and muscular hypotonia in infancy [44,45]. Affected patients generally present clinical features of connective tissue weakness in infancy, and subsequently develop signs of skeletal dysplasia; sometimes these signs are already present at birth. Some patients develop life-threatening complications, such as aortic dilatation, aneurysms and cervical spine instability.

Due to the mixed phenotype, with signs of EDS and signs of skeletal dysplasia, the *B4GALT7* and *B3GALT6*-related syndromes have been called "spondylodysplastic EDS" (spEDS) in the recently revised EDS classification of 2017 [46–48].

Hereby we present the clinical features of 3 patients with spEDS. Two are novel patients: Pt. 1 with *B4GALT7*-related EDS, and Pt. 3 with a severe *B3GALT6*-related EDS. Pt. 2 was originally reported as *B3GALT6*-related SEMDJL by Nakajima et al. in 2013, but is included here because his clinical description was largely unpublished and his condition has undergone a significant evolution. Through the description of these patients and a review of previous literature reports, we contribute to a more accurate definition of the clinical features associated with *B4GALT7*- and *B3GALT6*-related syndromes. In particular, we point out the extreme distal joint hypermobility and skin hyperextensibility (more evident in hands and feet) these two conditions have in common, and the peculiar radiological signs of each syndrome. The combination of these features is the main indicator for clinicians to suspect these ultra-rare conditions.

2. Materials and Methods

2.1. Metabolic Labeling of Fibroblast Cultures and PG Synthesis Analysis

Skin fibroblasts from the patient and controls were cultured in minimum essential medium (MEM) with 10% fetal calf serum and antibiotics at 37 °C in a humidified atmosphere containing 5% CO_2. Confluent cells in 10-cm^2 petri dishes were preincubated for 4 h with or without 1 mM p-nitrophenyl β-D-xylopyranoside in MEM containing 250 μM cold Na_2SO_4 without FCS in 5% CO_2 at 37°C. Cells were then labeled with 150 μCi/ml $Na_2[^{35}S]O_4$ in the same medium for 24 h, as described previously [49]. At the end of the labeling period, an equal volume of 100 mM sodium acetate buffer, pH 5.8, containing 8 M urea, 4% Triton X-100, 20 mM ethylenediaminetetraacetic acid, 20 mM N-ethylmaleimide, and 1mM phenylmethanesulfonyl fluoride was added to the medium. The cell layer was harvested in 50 mM sodium acetate buffer, pH 5.8, containing 2 M urea, 2% Triton X-100, and an aliquot was used for protein content determination with the bicinchoninic acid Protein Assay (Pierce), while the remainder was added to the medium. Samples were loaded on 1 ml DEAE Sephacel™ columns; after column washing with 50 mM sodium acetate buffer, pH 6.0, 8 M urea, 0.15 M NaCl, 0.5% Triton X-100 and proteinase inhibitors, PGs were eluted with 1 M NaCl in the same buffer, recovered by precipitation with 9 volumes of ethanol. The pellet was washed with 70% ethanol and then solubilized in water; PGs were quantified by measuring the ^{35}S-activity using a liquid scintillation counter and normalized to the protein content.

Labeled PGs synthesized by cells in absence of p-nitrophenyl β-D-xylopiranoside and purified as described above, were lyophilized, dissolved in 4 M guanidinium chloride, 50 mM sodium acetate buffer, pH 6.0, 0.5% Triton X-100, and analyzed by Size Exclusion Chromatography. Samples were then loaded on a Superose 6 10/300GL column (GE Healthcare) and eluted in the same buffer at 0.2 ml/min. Fractions of 0.4 ml were collected and ^{35}S-activity was measured by scintillation counting [50].

2.2. Molecular Genetic Testing

Genomic DNA of each patient was isolated from peripheral blood mononuclear cells and analyzed using a capture-based IonAmpliSeq custom panel on an Ion Torrent series S5 instrument (Thermo Fisher Scientific). Results were filtered for the *B4GALT7* and *B3GALT6* genes. Identified variants were confirmed by PCR amplification and subsequent bidirectional Sanger sequencing, using primers encompassing the entire exon and intron–exon boundaries comprising the variants in the case of *B4GALT7*, or the entire coding sequence and surrounding untranslated regions in the case of *B3GALT6* single exon. In parallel, DNA samples from the respective parents were analyzed by Sanger sequencing in order to evaluate parental segregation. Primer sequences are available upon request.

3. Patients and Results

3.1. First Patient

3.1.1. Clinical Report

The boy is the first child of healthy, non-consanguineous parents. The pregnancy was unremarkable and the mother was not exposed to any known teratogens during pregnancy. Fetal movements were described as normal. He was born by cesarean delivery at week 38 with birth weight of 2520 g (3rd–10th centile), length 43 cm (<3rd centile), and head circumference of 32 cm (3rd–10th centile). Apgar scores were 9/10. He presented with mesomelic shortening of upper limbs, ulnar deviation of fingers and left hip dysplasia. Cerebral and renal ultrasound were normal, while echocardiography showed atrial septal defect. The baby was discharged from hospital at the age of 6 days, with diagnosis of "arthrogryposis with ulnar deviation of fingers, left hip dysplasia, and congenital heart disease", and later he was fitted with a finger extension device and was given pelvic bimalleolar hip plaster for left hip dysplasia. He was seen at our Medical Genetics Unit at 3 months of age (Figure 2): length and weight, calculated by taking into account the hip plaster, were respectively 51 cm (3rd–10th centile) and 4900 g (3rd centile), and head circumference was 41 cm (25th centile). The physical examination showed: round flat face, mild proptosis, slightly blue sclerae (Figure 2A–D), joint hypermobility especially evident in the hands, mesomelic shortening of the arms, abnormal prono-supination of the left upper limbs, often flexed, short hands and feet with soft skin, bilateral ulnar deviation of fingers, 2nd finger clinodactyly, 2nd–5th finger camptodactyly on the left, and 3rd and 4th finger camptodactyly on the right (Figure 2F–G).

During the clinical follow-up, we noticed an improvement in the finger alterations, but a worsening of the cutaneous hyperextensibility and joint hypermobility. At the age of 3 years and 7 months, he presented with round face, blue sclerae, proptosis (Figure 2D), soft hyperextensible skin and extreme joint hypermobility (Beighton score: 5) in particular of hands and feet, with significant improvement in finger ulnar deviation, clinodactyly, camptodactyly and overlapping toes (Figure 2H–K). Height and weight were 87 cm (<3rd centile) and 9920 g (<3rd centile), head circumference was 49 cm (10th centile). Span was 82 cm and span/height ratio was 0.94. Early psychomotor development was delayed at least in part owing to his orthopedic conditions. He gained head control at 3 months, was able to sit at 3 years and walk aided at the age of 4 years. He uttered his first words at 2 years, but his language skills improved rapidly and significantly thereafter: at 5 years, psychological evaluation reported an IQ of 105.

The skeletal X-rays demonstrated wide anterior ribs, bilateral bowed ulna and radius with dislocation/subluxation and radioulnar synostosis, bilateral metaphyseal widening of the radius, bilateral short and dysmorphic 2nd finger middle phalanx, short first metacarpal (Figure 3A–F).

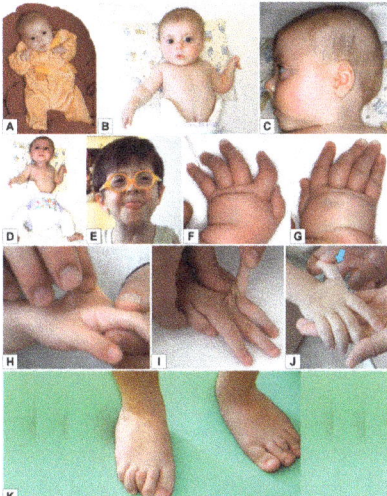

Figure 2. Patient 1 **A**) Age 3 months: ulnar deviation of fingers; **B**) Age 3 months: round flat face, mild proptosis, mesomelic shortening of upper limbs, folded skin on forearm; **C**) Age 3 months: flat profile; **D, F–G**) Age 3 months: ulnar deviation of fingers, 2nd finger clinodactyly, 2nd–5th finger camptodactyly on the left hand, 3rd and 4th finger camptodactyly on the right hand; **E**) Age 3 years 7 months: round face, blue sclerae, proptosis; **H–K**) Age 3 years 7 months: soft, hyperextensible skin and extreme joint hypermobility (Beighton score: 5) in particular of hands and feet; significant improvement in finger ulnar deviation, clinodactyly, camptodactyly of the second finger, and overlapping toes.

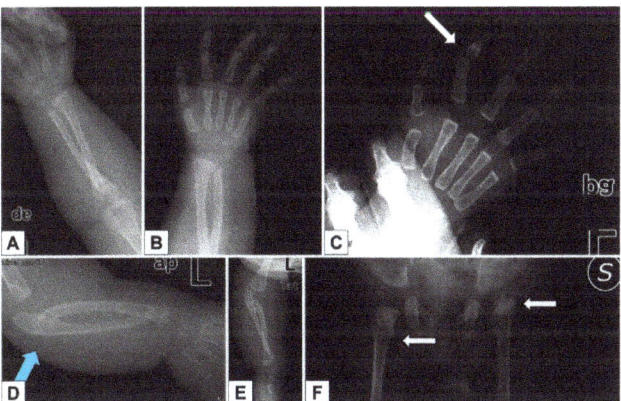

Figure 3. Patient 1 **A–B, D–E**) Bilateral bowing of ulna and radius with dislocation/subluxation and radioulnar synostosis, metaphyseal widening of the radius; **C**) bilateral short and dysmorphic 2nd finger middle phalanx, short first metacarpal; **F**) hip dysplasia.

Electrencephalography, electromyography and brain magnetic resonance imaging were all normal. At the age of 5 years, molecular analysis of *B4GALT7* was performed, showing the presence of two novel pathogenic mutations, indicative of spEDS.

3.1.2. Molecular and Functional Analysis

Karyotype was normal, 46,XY. Our first diagnostic hypothesis was either myopathy or collagen VI pathology, but molecular analysis of the *COL6A1*, *COL6A2*, *COLA6I3*, *RAPSN* genes did not reveal any rare or pathogenic variant. Subsequently, by analyzing a panel of EDS-related genes, two variants were identified in compound heterozygosity in exons 2 and 3 of the *B4GALT7* gene: NM_007255:c.[277dupC];[628C>T], NP_009186:p.[(His93Profs*73)];[(His210Tyr)] (Figure 4).

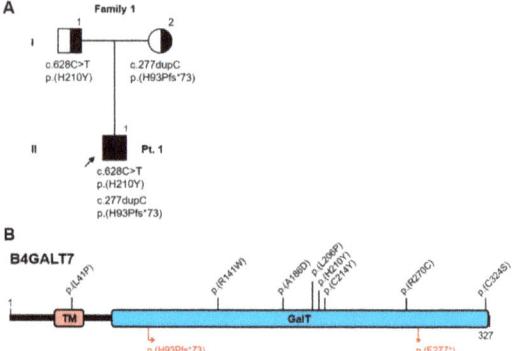

Figure 4. **A)** Pedigree of Family 1; **B)** Schematic representation of the B4GALT7 protein: TM = transmembrane domain, GalT = Galactosyl-transferase catalytic domain. Protein variants reported in the literature are annotated in the cartoon (see Table 1 for references). Top half: missense variants (black). Bottom half: truncating variants (red).

The single nucleotide insertion c.277dupC, inherited from the mother, creates a new reading frame, terminating with a premature stop codon 73 residues downstream, and its transcript is predicted to undergo nonsense-mediated decay (https://nmdprediction.shinyapps.io/nmdescpredictor/). This variant was also described in another patient by Salter et al. in 2016 [32]. The heterozygous C>T transition at nucleotide 628, inherited from the father, results in a non-conservative Histidine to Tyrosine substitution at codon 210 and occurs in a highly conserved residue. This variant has not been reported before, but it is considered deleterious by multiple prediction algorithms (MutationTaster, PolyPhen2, SIFT) and is absent from reference population databases (gnomAD, 1000 Genomes). Based on these criteria and on the clinical features of the patient, we assume that the combination of these variants in trans is causative.

The study of PG synthesis on patient's fibroblasts obtained by skin biopsy did not reveal any significant differences compared to normal controls, neither at basal conditions nor in the presence of beta-xyloside to enhance glycosaminoglycan synthesis. However, the level of PG synthesis in the patient's fibroblasts was at the lower normal limit. Analysis through gel filtration to assess the fraction of whole PGs with respect to glycosaminoglycans did not show any differences. Although these results were not significant *per se*, the slightly reduced PG levels combined with the genetic testing offered some indication that the variants identified in *B4GALT7* may in fact have a deleterious effect on protein function.

Table 1. Patients with *B4GALT7* variants: Review of the literature.

		Hernandez et al. [1979, 1981, 1986][21–23]	Kresse et al. [1987][24]	Faiyaz-Ul-Haque et al. [2004][29]	Guo et al. [2013][30]	Cartault et al. [2015][31]	Arunrut et al. [2016][33]	Salter et al. [2016][32]	Ritelli et al. [2017][31]	Sandler-Wilson et al. [2019][52]	This study, Patient 1	Total (patients with mutation)
Genetics	Diagnosis (Various denominations)	A distinct variant of the EDS	EDS, progeroid type 1	EDS, progeroid type 1	EDS, progeroid type 1	Larsen of Reunion Island syndrome	*B4GALT7*-linkeropathy phenotype	Phenotypic Spectrum of *B4GALT7*	spEDS-*B4GALT7*	spEDS-*B4GALT7*	spEDS-*B4GALT7*	spEDS-*B4GALT7*
	B4GALT7 variants	?	c.[557C>A]; [617T>A]	c.[808C>T]; [808C>T]	c.[122T>C]; [808C>T]	c.[808C>T]; [808C>T]	c.[970T>A]; [970T>A]	c.[277dupC]; [641G>A] c.[421C>T]; [808C>T]	c.[829G>T]; [829G>T]	c.[421C>T]; [808C>T]	c.[277_278insC]; [628C>T]	26/33 Homozygous 7/33 Compound heterozygous
	Variants on protein	?	p.[(A186D)]; [(L206P)]	p.[(R270C)]; [(R270C)]	p.[(L41P)]; [(R270C)]	p.[(R270C)]; [(R270C)]	p.[(C324S)]; [(C324S)]	p.[(H93Pfs*73)]; [(C214Y)] p.[(R141W)]; [(R270C)]	p.[(E277*)]; [(E277*)]	p.[(R141W)]; [(R270C)]	p.[(H93Pfs*73)]; [(H210Y)]	
	Gender	5M	M	1F 1M	M	11M 11F	F	1M 1F	F	1M 1F	M	17M 16F
	Age	8y, 15y, 15y, 16y, 18y	4y 9m	2y, 33y	10y	4y, 46y	5y	3y 6m, 13y	30y	4y, 10y	7y 8m	2y → 46y
Main features	Short stature[a]	4/5	+	2/2	+	22/22	+	2/2	+	2/2	+	33/33
	Radiolunar synostosis[c]	n.a.	+	2/2	+	10/21	-	2/2	-	2/2	+	19/32
	Bowing of limbs[a]	n.a.	+	2/2	+	21/21	-	2/2	-	2/2	+	30/32
	Joint hypermobility especially of the hands[c]	5/5	+	2/2	+	22/22	+	2/2	+	2/2	+	33/33
	Skin hyperextensibility, soft, doughy skin[b]	5/5	+	2/2	+	21/22	+	2/2	+	2/2	+	32/33
Facial dysmorphisms[c]	Progeroid facial appearance	5/5	mild	0/2	-	0/22	-	0/2	-	0/2	-	1/33
	Short face	-	+	2/2	-		+	2/2	-	2/2	+	31/33
	Midface hypoplasia	-	-	2/2	-	22/22	+	2/2	-	2/2	+	30/33
	Narrow mouth	-	+	2/2	-		+	1/2	+	2/2	+	31/33
	Proptosis	-	+	2/2	-		+	2/2	+	2/2	+	32/33
	Cleft palate	-	Bifid uvula	0/2	-	1/22	-	1/2	-	1/2	-	4/33
	Loose skin	-	+	2/2	+	n.a.	n.a.	2/2	+	n.a.	+	8/8

Table 1. Cont.

		Hernandez et al. [1979, 1981, 1986] [21–23]	Kresse et al. [1987][24]	Faiyaz-Ul-Haque et al. [2004][29]	Guo et al. [2013][30]	Cartault et al. [2015][31]	Arunrut et al. [2016][33]	Salter et al. [2016][32]	Ritelli et al. [2017][51]	Sandler-Wilson et al. [2019][52]	This study, Patient 1	Total (patients with mutation)
Other clinical features	Delayed wound healing	5/5	+	n.a.	+	n.a.	+	2/2	+	n.a.	-	7/8
	Cardiovascular abnormalities	1/5 Aortic/Pulmonic Stenosis	n.a.	n.a.	+	n.a.	-	n.a.	-	n.a.	+ -ASD	1/4
	Delayed motor development[b]	5/5	+	2/2	-	n.a.	+	2/2	+	2/2	+	10/11
	Delayed cognitive development[b]	5/5	n.a.	n.a.	Mild learning disabilities	12/22 (learning disabilities)	n.a.	1/2 Severe 1/2 n.a	-	2/2	-	16/29
	Muscle hypotonia[a]	n.a.	+	2/2	mild	n.a.	n.a.	2/2	+	2/2	+	11/11
	Ophthalmological abnormalities[c]	n.a.	-	1/2 Mild esotropia and mild hypertropia	Severe hyperopia, congenital ptosis, intermittent exotropia	5/21 glaucoma 1/21 megalocornea	Nystagmus Iris and optic nerve colobomas Posterior subcapsular cataracts High hyperopia Right-sided ptosis	Severe hypermetropia Small optic nerves Hypermetropia Strabismus	-	2/2 Blue sclerae 1/2 Severe hyperopia	Myopia	13/32
	Osteopenia[b]	n.a.	+	2/2	-	n.a.	n.a.	2/2	+	1/2	+	8/10
	Pes planus[b]	5/5	+	1/2	+	n.a.	+	1/2 1/2 n.a.	+	2/2	+	9/11
	Bilateral elbow contractures or limited elbow movement[c]	n.a.	+	2/2	+	n.a.	n.a.	2/2	-	1/2	+	8/10
	Sensorineural hearing loss	n.a.	-	n.a.	n.a.	n.a.	n.a.	1/2 Conductive hearing loss	+	n.a.	-	2/7
	Other less frequent features	Cryptorchidism 4/5 Inguinal hernia 1/5 Hypogonadism 1/5 Varicose veins 5/5 Multiple nevi 5/5 Dental anomalies 5/5	Dental anomalies: defective and greyish enamel Clavicular exostoses	Yellow discoloration of teeth with defective enamel Mild eventration of the right hemidiaphragm Bilateral equinovarus deformity	Unilateral ptosis	Pectus carinatum 5/22 Bifid thumb 2/22 Scoliosis/Kyphosis 6/22	Pectus carinatum Scoliosis Broad fingertips Subluxation of the distal interphalangeal joints Absence of the pineal gland Prominent scalp veins	Irregular and fragile dentition Scoliosis Bilateral patellar dislocation	Bilateral hallux valgus Lymphedema Scoliosis Temporomandibular joint dislocation	1/2 Chest wall deformity 2/2 Coronal cleft vertebrae 2/2 Sagittal craniosynostosis 1/2 Vesicoureteral reflux	Hip dysplasia Ulnar deviation of fingers	

EDS: Ehlers–Danlos syndrome; n.a. not available; ASD: atrial septal defect; [a]Spondylodysplastic EDS (spEDS) major criteria [46]; [b]spEDS minor criteria [46]; [c]spEDS specific minor criteria for B4GALT7 variants [46]. The 5 patients from Hernandez et al. (column 1) had clinical data compatible with spEDS-B4GALT7, but were not considered in the clinical signs totals because it was unclear whether the clinical diagnosis was confirmed by molecular testing or not [21–23].

3.2. Second Patient

3.2.1. Clinical Report

This patient was partially described in Nakajima et al., 2013 as P9 [45]. He is the first child of healthy non-consanguineous parents, born at the 40th week of gestation by vaginal delivery. During pregnancy, no known exposures to potential teratogens were detected. The mother reported normal fetal movements. Prenatal ultrasound showed bilateral megaureter. At birth, weight was 2870 g (10th centile), length was 48 cm (3rd–10th centile) and head circumference measured 34 cm (25th centile). He presented with: joint hypermobility, soft skin, bilateral congenital talipes equinovarus and right hip dysplasia, which was treated with a Pavlik harness. Renal ultrasound showed a mild bilateral caliceal dilatation, while renal scintigraphy was normal. During the first year of life he was diagnosed with early-onset scoliosis, and he began to wear an orthopedic corset from the age of 4 years. Two years later. he was operated for scoliosis after correction with Halo Gravity Traction and he subsequently received three operations for the lengthening of the pins (until the age of 10 years). He suffered from frequent fractures: a fracture of the left femur after a low trauma at the age of 5 years and 6 months and to the right femur at the age of 8 years and 8 months and a micro-fracture of the right proximal fibula in an area of reduced bone density, compatible with an osseous cyst, at the age of 10 years. Furthermore, he had recurrent luxation of the toes.

Early psychomotor development was normal, but he later suffered from motor delay due to his skeletal and large joint anomalies. He gained head control at the age of 2 months and could sit up at the age of 8 months. He could walk unaided at the age of 2 years and uttered his first words at the age of 1 year. He currently attends the first year of middle school with a support teacher owing to a diminished ability to concentrate. He does not present with intellectual disability or other behavioral disturbances.

At the age of 12 years and 7 months (Figure 5), height was 109cm (<3rd centile <-6SD), weight 29 kg (<3rd centile), head circumference 54 cm (50th–75th centile), span 114 cm, span/height ratio 1.06. Pubertal stage was A0P1B1, with testicular volume 1–2 ml on the left side and 2–3ml on the right. He had sparse hair, high and prominent forehead, large ears, sparse eyebrows, large deeply-set eyes, blue-gray sclerae, hypoplastic columella and misalignment of teeth, which were small, with abnormal enamel and yellow-brown discoloration (Figure 5A–C). He had thin, pale, extremely soft skin, with a prominent venous patterning on the trunk and limbs, limited elbow extension, skin hyperextensibility and distal joint hypermobility especially of the hands (Beighton scale: 6), long and tapered fingers, normal nails. His feet had hypoplastic nails, short and overlapping toes, hallux valgus (Figure 5D–H).

The skeletal X-rays demonstrated severe kyphoscoliosis, osteopenia, thin metacarpals and phalanges (Figure 6A–H).

Audiometric and ophthalmological evaluations were normal. Abdominal ultrasound showed ptosic kidneys, bilateral pelvic ureteral dilatation and thickening of the bladder walls. Echocardiography demonstrated mild dilatation of the aortic bulb and mild mitral valve prolapse.

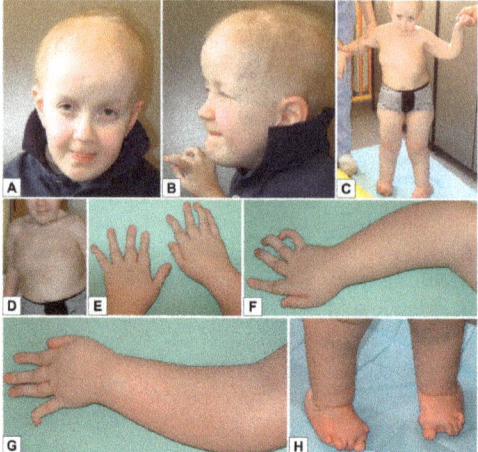

Figure 5. Patient 2 **A–C**) Sparse hair, high and prominent forehead, sparse eyebrows, deeply-set eyes, blue sclerae, hypoplastic columella, large ears; **D–H**) Thin, pale, extremely soft skin, with prominent veins on the trunk and limbs, limited elbow extension, skin hyperextensibility and distal joint hypermobility especially of the hands, long and tapered fingers. Feet: hypoplastic nails, short, overlapping toes, hallux valgus.

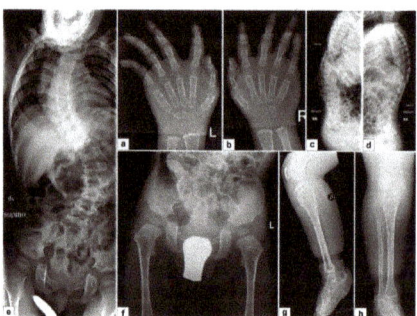

Figure 6. Patient 2 **A–H**) Severe, early onset kyphoscoliosis, osteopenia, thin metacarpals and phalanges.

3.2.2. Molecular Analysis

Sanger sequencing confirmed the two previously reported variants in the single exon of the *B3GALT6* gene [45]: NM_080605:c.[353delA];[925T>A], NP_542172:p.[(Asp118Alafs*160)];[(Ser309Thr)] (Figure 7A,B).

The frameshift variant was inherited from the father and the missense variant from the mother. Both variants can be considered pathogenic, according to the predicted effect on the protein, the consistency of the clinical phenotypes, and their rarity in reference population databases such as gnomAD (MAF=0.0026% and absent, respectively). Variant c.353delA has also been found in our third patient (reported below), while variant c.925T>A had already been reported in other patients in the literature [45,48]. Since this gene is encoded by a single exon, its frameshift variants, identified in various patients, are expected to escape nonsense-mediated mRNA decay. They probably exert a pathogenic effect by leading to an unstable transcript or an unstable/non-functional protein, as suggested by experimental evidence [48].

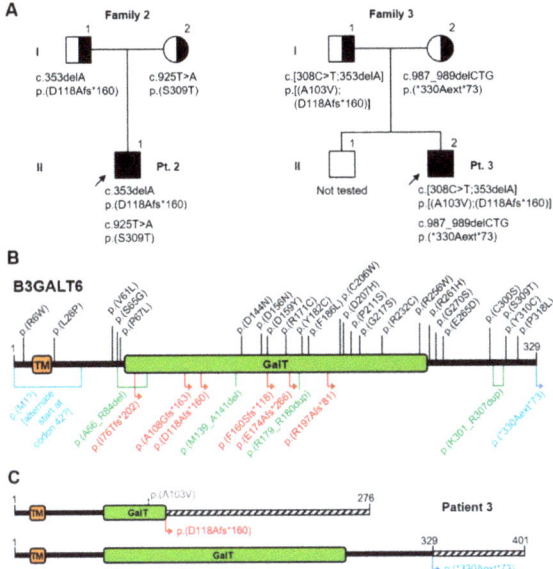

Figure 7. **A)** Pedigree of Families 2 and 3; **B)** Schematic representation of the B3GALT6 protein: TM = transmembrane domain, GalT = Galactosyl-transferase catalytic domain. Protein variants reported in the literature are annotated in the cartoon (see Table 2 for references). Top half: missense variants (black). Bottom half: frameshift variants (red), in-frame indels (green), other length-altering variants (blue); **C)** Protein projection of the two alleles from Pt.3; dashed boxes indicate an altered or extended reading frame.

3.3. Third Patient

3.3.1. Clinical Report

He is the second child of non-consanguineous parents, born at 35th week of gestation by caesarean section, owing to the premature rupture of the amniotic sac. During pregnancy, no known exposures to potential teratogens were detected. Prenatal ultrasound indicated bilateral talipes equinovarus. The mother reported normal fetal movements. At birth, weight was 2500 g (50th centile), length was 47 cm (50th–75th centile) and head circumference measured 34 cm (90th centile). Apgar scores were 1':7 5':9. He presented with joint hypermobility, soft skin, adducted thumbs, right talipes equinovarus. In the second year of life he was diagnosed with early-onset scoliosis, and he began to wear an orthopedic corset.

At age 6 years, he was operated for bilateral cryptorchidism and at age 9 years, he underwent a gastrostomy to correct Barrett's oesophagus.

His motor development was delayed. He gained head control at the age of 4–5 months and could sit up at the age of 24 months. He could walk aided at the age of 5 years, he was able to walk alone at the age of 6 years and 6 months and uttered his first words at the age of 1 year. He currently attends the second year of middle school with a support teacher owing to a diminished ability to concentrate. He does not present with intellectual disability or other behavioral disturbances.

Examination at the age of 12 years and 11 months showed (Figure 8A–F): height 123 cm (<<3rd centile), weight 28 kg (<3rd centile), head circumference 52 cm (10th centile), span 113 cm, span/height ratio 0.91. Pubertal stage were A0P1B1, with testicular volume of 1-2 ml bilaterally. He had sparse hair, high and prominent forehead with high hairline, mild bitemporal depression, fairly large ears, blue-gray sclerae, malar hypoplasia, hypoplastic columella, short philtrum, small misaligned teeth in the lower jaw, flat palate with longitudinal median mucous thickening (Figure 8A–B). He had thin, pale, soft skin, with slightly prominent veins on trunk, limited elbow extension, skin hyperextensibility and distal joint hypermobility especially of the hands (Beighton scale: 6), long and tapered fingers with a tendency to ulnar deviation (Figure 8C–E), hypoplastic nails on the feet (Figure 8F).

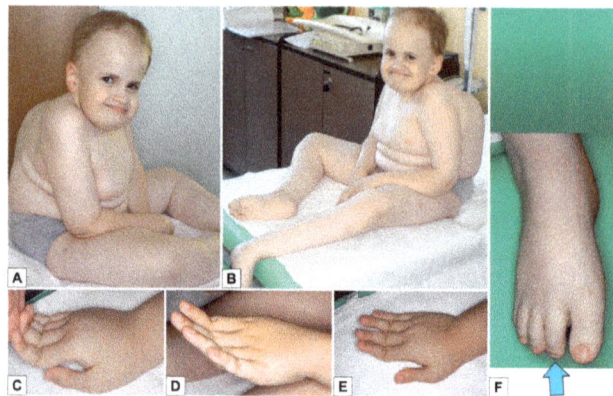

Figure 8. Patient 3 **A,B**) Sparse hair, high and prominent forehead with high hairline, mild bi-temporal depression, fairly large ears, blue-gray sclerae, malar hypoplasia, hypoplastic columella, short philtrum, thin, pale, soft skin with prominent veins on the trunk; **C–E**) Skin hyperextensibility and distal joint hypermobility especially of the hands, long and tapered fingers, with a tendency to ulnar deviation; **F**) Feet: hypoplastic nails.

Skeletal X-rays demonstrated severe kyphoscoliosis, osteopenia (especially in the acetabula, femurs, tibiae and fibulae), and thin metacarpals, metatarsals and phalanges (Figure 9A–F).

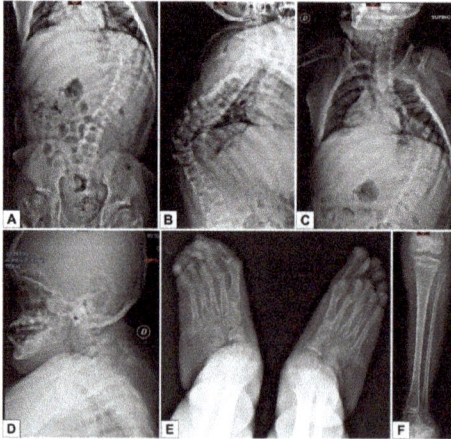

Figure 9. Patient 3 **A–F**) Severe early onset kyphoscoliosis; osteopenia of the acetabula, femur, tibiae and fibulae; thin metacarpals, metatarsals and phalanges.

Audiometric and ophthalmological evaluations were normal. Abdominal ultrasound was normal. Echocardiography demonstrated an atrial septal defect and mild mitralic insufficiency with thickened mitralic valve.

Table 2. Patients with *B3GALT6* variants: Review of the literature.

		Malfait et al. 2013[44]	Nakajima et al. 2013[45]	Sellam et al. 2014[53]	Ritelli et al., 2015[54]	Alazami et al., 2016[8]	Trejo et al.2017[55]	Ben-Mahmoud et al. 2018[56]	Van Damme et al. 2018[48]	This study, Patient 2	This study, Patient 3	Total (patients with mutation)
Genetics	Diagnosis (Various denominations)	EDS-like connective tissue disorder SEMDJL1	B3GALT6 spectrum SEMDJL1 EDS-progeroid form	EDS progeroid type2 SEMDJL1	EDS-like syndrome EDS progeroid type2 SEMDJL1	B3GALT6-phenotype	SEMDJL1	spEDS-B3GALT6 SEMDJL1	spEDS-B3GALT6	spEDS-B3GALT6	spEDS-B3GALT6	spEDS-B3GALT6
	B3GALT6 variants	c.[619G>C]; [619G>C] c.[323_344del]; [619G>C] c.[649G>A]; [649G>A]	9 Compound heterozygous, 1 Heterozygous; see Ref. (Table 3)	?	c.[22TdelT]; [766C>T]	c.[556T>C]; [556T>C] c.[536_541dup]; [536_541dup]	c.[511C>T]; [901_921dup]	c.[618C>G]; [618C>G]	8 Compound heterozygous, 1 Homozygous; see Ref (Table 3)	c.[353delA]; [925T>A]	c.[308C>T;353delA]; [987_989delCTG]	12/45 homozygous 32/45 compound heterozygous 1/45 heterozygous
	Variants on protein	p.[(D207H)]; [(D207H)] p.[(A108Gfs*163)]; [(D207H)] p.[(G217S)]; [(C217S)]	See Figure 7	p.[(D159Y)]; [(E265D)]	p.[(I76Tfs*202)]; [(R256W)]	p.[(F186L)]; [(F186L)] p.[(R179_R180dup)]; [(R179_R180dup)]	p.[(R17C)]; [(K301_K307dup)]	p.[(C206W)]; [(C206W)]	See Figure 7	p.[(D118Afs*160)]; [(S309T)]	p.[(A103V); (D118Afs*160)]; [(*330Aext*73)]	
Main features	Gender	2M 3F	6M 6F	M	2F	1M 4F	3F	1M 2F	7M 5F	M	M	20M 25F
	Age	1y8m→27y	1m→34y	6m	21y, 25y	6w→6y	12y, 15y, 15y	4d→2m	8m→37y	12y 7m	13y 3m	4d→37y
	Short stature [a]	2/3	12/12	+	2/2	3/3	3/3	n.a.	10/10	+	+	35/36
	Kyphoscoliosis (congenital or early onset, progressive) [c]	3/4	12/12	-	2/2	4/5	3/3	n.a.	10/10	+	+	36/39
	Bowing of limbs [a]	2/4	n.a.	+	n.a.	n.a.	3/3	n.a.	10/10	-	-	16/20
	Joint hypermobility especially of the hands [c]	4/4	7/10	n.a.	2/2	5/5	3/3	2/2	10/10	+	+	33/36
	Skin hyperextensibility, soft, doughy skin [b]	4/4	6/10	n.a.	2/2	4/5	3/3	2/2	10/10	+	+	33/38
	Skeletal changes SEMDJL1 [d]	3/4	12/12	n.a.	2/2	2/2	0/1	3/3	9/10	+	+	34/36
Facial dysmorphisms	Prominent forehead	4/4	9/10	+	2/2	4/5	0/3	1/1	8/10	+	+	31/38
	Sparse hair	2/4	3/10	-	0/2	n.a	2/3	0/1	3/3	+	+	12/26
	Midface hypoplasia	2/4	n.a.	+	1/2	4/5	2/3	1/1	8/10	+	+	21/28
	Blue sclerae	3/4	7/10	n.a.	2/2	4/5	1/3	n.a.	6/10	+	+	25/36
	Proptosis	2/4	7/10	+	0/2	n.a.	3/3	n.a.	7/10	-	-	20/32
	Cleft palate	0/4	1/10	-	0/2	n.a.	0/3	-	-	-	-	2/25

Table 2. Cont.

		Malfait et al. 2013[44]	Nakajima et al. 2013[45]	Sellars et al. 2014[53]	Ritelli et al. 2015[54]	Alazami et al. 2016[18]	Trejo et al. 2017[55]	Ben-Mahmoud et al. 2018[56]	Van Damme et al. 2018[48]	This study, Patient 2	This study, Patient 3	Total (patients with mutation)
	Joint hand contractures [c]	2/3	3/12	+	2/2	1/5	2/2	3/3	10/10	-	-	24/40
	Cardiovascular anomalies	n.a.	Mitral regurgitation 1/?	n.a.	2 Mitral valve prolapse	1 Aortic valve stenosis 1 Mitral valve prolapse?	n.a.	n.a.	Aortic root aneurysm 3/8 Cardiac valve anomalies 2/8	Aortic root aneurysm Mitral valve prolapse	Mitral valve prolapse	12/?
	Delayed motor development [b]	4/4	2/?	+	n.a.	4/5	2/2	n.a.	5/9	+	+	17/24
	Delayed cognitive development [b]	2/2	n.a.	n.a.	0/2	4/5	n.a.	n.a.	3/8	-	-	10/20
	Muscle hypotonia [a]	4/4	5/12	+	1/1	2/?	3/3 (1 mild)	1/1	5/9	+	+	25/37
	Ophthalmological anomalies	Myopia 2/4 Retinal detachment1/4	n.a.	Corneal opacity Sclerocornea	0/2	n.a.	n.a	Corneal opacity 3/3	Glaucoma and optic nerve atrophy 1/10 Microcornea 1/10	-	-	7/19
	Osteopenia [b]	4/4	n.a.	+	2/2	3/3	0/3	2/2	8/8	+	+	22/26
	Pes planus [b]	2/2	n.a.	n.a.	2/2	n.a.	n.a	0/2	n.a.	+	+	6/8
	Talipes equinovarus [c]	3/4	4/12	+	n.a	3/5	n.a	2/2	10/10	+	+	23/37
	Peculiar fingers [c]	3/4	7/11	+	2/2	n.a.	n.a	2/2	n.a.	+	+	17/22
	Anomalies of dentition, discoloration of teeth [c]	3/4	n.a.	n.a.	2/2	n.a.	n.a	n.a.	8/9	+	+	13/17
Other clinical features	Less frequent features	Excessive wrinkling of palmar skin (hands and feet) 2/4 Pectus deformity 3/4	Elbow dislocation 9/10 Limited elbow movement 9/11 Carpal synostosis 1/10 Short metacarpals 6/10 Hip dislocation 5/12 Epiphyseal dysplasia of femoral head 4/12	Radioulnar synostosis Early death	2/2 Genu valgus 2/2 Hallux valgus	4/5 Multiple fractures Bilateral dislocated radial head 1/5 Pectus carinatum	3/3 Bilateral radioulnar dislocation 3/3 Hip dysplasia 2/3 Hearing loss 1/3 Pectus carinatum 1/3 Ulnar deviation of the fingers	Contractures of the large joints 2/2 Radioulnar synostosis 2/2 Oligodactyly of the right 3rd finger 1/2 Spontaneous fractures 2/2 Early death 3/3	Sensorineural and conductive hearing loss 1/10 Cervical spine instability 3/7 Laryngeal cleft 1/10 Tracheomalacia 2/10 Spontaneous repeated pneumothoraces 1/10 Chronic respiratory insufficiency 2/10 Pectus carinatum 1/10 Pectus excavatum 1/10 Wilms tumor 1/10 Joint dislocations 10/10 Hip dysplasia 4/6 Fractures 8/9 Hallux valgus 3/10	Prominent superficial veins Limited elbow extension Ptosic kidney, Bilateral caliceal and ureteral dilatation Fractures Right hip dysplasia Recurrent luxation of the toes, Hypoplastic nails Hallux valgus	Prominent superficial veins Barrett's oesophagus Limited elbow extension Bilateral cryptorchidism Hypoplastic nails	

EDS: Ehlers Danlos syndrome; n.a. not available; ASD: atrial septal defect; [a]Spondylodysplastic EDS (spEDS) major criteria [46]; [b]spEDS minor criteria [46]; [c]spEDS specific minor criteria for B3GALT6 variants[46]; [d] Platyspondyly, short ilia, elbow malalignment. Cases from 10 additional families reported by Vorster et al. in 2015 [42] were not listed here due to difficult data interpretation; they include an additional variant, c.235A>C, p.(T79A).

3.3.2. Molecular Analysis

Karyotype was normal, 46XY. Genetic testing revealed three variants in the *B3GALT6* gene: NM_080605:c.[308C>T;353delA];[987_989delCTG], NP_542172:p.[(Ala103Val);(Asp118Alafs*160)]; [(*330Alaext*73)] (Figure 7A–C).

Variant c.353delA, already described as pathogenic (patient 2 and Nakajima et al., 2013 [45]), has been found in *cis* with variant c.308C>T, both inherited from the healthy father. This missense variant is absent from the reference population databases (gnomAD, 1000 genomes), but occurs at a poorly conserved position and is predicted as benign by multiple algorithms (MutationTaster, PolyPhen2, SIFT). It should be classified as a variant of uncertain significance, as there is not enough evidence to suggest whether this variant alone would be tolerated or damaging to protein function, or to indicate a possible contribution to the deleterious effect of the frameshift variant.

Variant c.987_989delCTG, inherited from the healthy mother, disrupts the constitutive stop codon leading to an extended open reading frame, encoding a 72aa longer protein. It is absent from reference population databases (gnomAD, 1000 genomes), and by analogy with the frameshift variants in this gene, it is expected to produce an unstable or non-functional protein. Since it was found in trans with a recognized deleterious mutation, this variant can be considered as likely pathogenic.

4. Discussion

Biallelic variants in the *B4GALT7* and *B3GALT6* genes are responsible for conditions characterized by a combined skeletal and connective tissue phenotype. Their recent classification as spEDS [46] provides a valid assistance for their diagnosis, but the observation of new patients and the study of the clinical signs of those already reported in the literature can further improve our knowledge. In an effort to better define both the similarities and the peculiarities of these syndromes, we reviewed the notable features of our three patients and compared them to those of molecularly confirmed *B4GALT7*- and *B3GALT6*-related cases reported in the literature (Tables 1 and 2) [18,21–24,29–33,44,45,48,51–56]

All three of our patients display a remarkably short stature. Pt. 2 and 3 are more severely affected, in part because of their kyphoscoliosis, with a stature progressing from moderately short (or normal for Pt. 3) at birth to well below the 3rd percentile during childhood. Nearly all patients reported in the literature (65/66 evaluated cases) share this feature, confirming short stature as a main aspect of spEDS.

Joint hypermobility and soft, doughy and hyperextensible skin are some of the most striking characteristics in our patients. These qualities are particularly remarkable in the hands and, to a lesser extent, in the feet, and helped us restrict the diagnosis to an EDS-related condition. Since these features are shared by more than 90% of cases reported in the literature (63/66 and 62/68, respectively), in our opinion their combination should be considered a major diagnostic criterion for spEDS.

Craniofacial features for both *B4GALT7*- and *B3GALT6*-related conditions are quite variable among our patients and the reported individuals for whom either pictures or descriptive data could be evaluated, and do not seem suggestive of a facial gestalt. There are, however, some recurring features that may sometimes help to set the two conditions apart: i. proptosis is more frequent in spEDS-*B4GALT7* (32/33 vs 20/32); ii. midface retrusion has been frequently noted in both conditions, but in spEDS-*B3GALT6* it is often associated with a prominent forehead (31/38), sometimes emphasized by sparse hair (12/26) as in the case of our Pt. 2 and 3, while it would be more appropriate to state that spEDS-*B4GALT7* patients have an entirely flat face, usually short or round (31/33) and with a narrow mouth (31/33). The progeroid aspect of spEDS-*B4GALT7* described in Kresse's child [24] and previously pointed out by Hernandez [21–23] was absent in all other patients, suggesting that it is not strongly associated with mutations in *B4GALT7*. As current terminology indicates, a progeroid facial appearance is the subjective interpretation of a series of facial features which should rather be evaluated individually. Some features such as thin skin or excessive wrinkles may occasionally appear in the description of the cases we reviewed, but never in a combination suggestive of premature aging, and none of these were observed in our patient. Therefore, as already suggested by other authors, we endorse the necessity to remove the term "progeroid type" from this syndrome, because it could be misleading.

The radiological findings are undoubtedly the most significant evidence in distinguishing *B4GALT7*-related and *B3GALT6*-related spEDS: the latter includes more numerous and pronounced elements of skeletal dysplasia, and for this reason has often been classified as such in the past. Radioulnar synostosis has been observed recurrently and almost exclusively in spEDS-*B4GALT7* (19/32 cases, including our patient, vs 3 documented cases of spEDS-*B3GALT6* [53,56]. On the other hand, most of the individuals affected by *B3GALT6*-related spEDS, including our patients, display severe kyphoscoliosis, usually congenital or early onset and progressive (36/39), and several of the skeletal changes associated with SEMDJL1, such as platyspondyly, short iliac bones, elbow dislocation with misalignment of the long bones (33/36). Osteopenia has been noted in both conditions. It has been reported more frequently in *B3GALT6*-related cases (22/26), usually in conjunction with fractures and luxations, as in the case of our patients. In *B4GALT7*-related cases there are significantly fewer records of this feature, but in most reports its absence or presence cannot be verified. Therefore, the actual prevalence of osteopenia is unknown and should probably deserve more consideration when referring spEDS suspects for a radiological exam.

A consequence of skeletal involvement (bowing of limbs) is considered one of the main criteria for suspecting spEDS, and it is possibly more common in the spEDS-*B4GALT7* series (30/32 vs 16/20); in fact, our Pt. 2 and 3 do not have this phenotype. Joint contractures of the upper limbs have been noted for both conditions, though they are usually reported at the elbow in spEDS-*B4GALT7* and at the hands in spEDS-*B3GALT6*. In our experience, it was actually our *B3GALT6*-mutated patients who displayed limited elbow extension, but this is a non-specific clinical sign that is often found in skeletal dysplasia.

Other clinical features appear to be peculiar of *B3GALT6*-related spEDS. The most notable are talipes equinovarus (23/37 patients, including ours), possibly with hypoplastic nails (cfr EDS classification of 2017 [46] and our patients 2 and 3), and abnormalities of the dentition (13/17 patients, including ours), sometimes with yellow-brown discoloration of the teeth (also observed in our Pt. 2).

Motor delay has been noted in the majority of reports in which developmental milestones were evaluated (27/36), including our three patients. It is most likely related to the recurrent skeletal anomalies (bowing of limbs, scoliosis, hip dysplasia) as well as to the frequent reports of hypotonia, also observed in all of our patients (36/48 evaluated cases). In order to overcome the difficulties in sitting and walking, these children often need assistance in the form of rehabilitation with appropriate physiotherapeutic interventions. Depending on the severity of the musculoskeletal features, some individuals may be able to walk without help at a later age than their peers, while others can only manage walking with help.

Developmental delay appears to be present in only about half of the documented spEDS cases (26/49). In many instances, psychomotor development is only mildly delayed, possibly because of the confounding effect of minor behavioral abnormalities. Of our three patients, Pt. 1 IQ has been demonstrated to be average, while Pt. 2 and 3 require a support teacher during school only because of a diminished ability to concentrate, and actually have good comprehensive and expressive skills.

In summary, among all of these clinical features, the most relevant in order to suspect spEDS, because of their rarity in other syndromic conditions, are:

i. the extreme distal joint hypermobility and soft, hyperextensible skin, particularly of the hands;
ii. the radiological signs, which are the main indicator for discriminating spEDS-*B4GALT7*, associated with radioulnar synostosis, and spEDS-*B3GALT6*, characterized by kyphoscoliosis (congenital or early onset and progressive) and by the skeletal signs of SEMDJL1 (platyspondyly, short iliac bones, elbow dislocation).

Current EDS classification includes a third type of spEDS, caused by biallelic defects in SLC39A13 (OMIM *608735), which encodes a zinc transporter involved in connective tissue development [57]. (Radiological findings suggest a minor skeletal involvement compared to other spEDS, but overall,) the small number of reported families is inadequate to discuss a differential diagnosis with good confidence [58,59].

The differential diagnosis of spEDS-*B3GALT6* may be placed with kyphoscoliotic EDS (kEDS), primarily owing to early onset kyphoscoliosis, but the two conditions are very different from the clinical point of view, because the latter is not characterized by significantly short stature, and the joint hypermobility with dislocations/subluxations involves mainly shoulders, hips and knees rather than hands and feet.

The differential diagnosis may be placed also with musculocontractural EDS (mcEDS), but the latter is characterized by congenital multiple contractures, characteristically adduction-flexion contractures, and typical craniofacial features, which are distinct from spEDS.

The main differential diagnosis of spEDS-*B4GALT7* can be given with arthrochalasia EDS (aEDS), which is characterized by congenital bilateral hip dislocation and laxity of small joints, but generally aEDS also presents subluxation of the knees and dorso-lumbar kyphosis.

A clinician should consider the differential diagnosis between spEDS-*B4GAL7* and the *B3GAT3*-related linkeropathy, mainly because of prominent eyes, short stature with joint laxity and radioulnar synostosis. However, it should be noted that patients with *B3GAT3* variants often have cardiovascular abnormalities, and that the rarity of the cases described in the literature does not allow definite conclusions [14,15]. Additional clinical descriptions of these disorders are required in order to characterize the linkeropathies both individually and as a disease group.

The radiological feature of spEDS may further warrant consideration for differential diagnosis with several other skeletal dysplasias. The main discriminating features of *B4GALT7*- and *B3GALT6*-related conditions, i.e. respectively radioulnar synostosis and scoliosis with early onset and rapid evolution, can be found in various skeletal dysplasias, but are associated with such a remarkable degree of joint hypermobility and soft, hyperextensible skin almost exclusively in spEDS.

5. Conclusions

The most striking aspect of spEDS is the combination of clinical signs affecting both the connective tissue and the skeletal system. The criteria identified in the International Classification of EDS by Malfait et al. (2017) clearly show that there is significant overlap between the clinical features of spEDS-*B4GALT7* and spEDS-*B3GALT6*. This is confirmed and expanded upon by the clinical reports and literature review tables (Tables 1 and 2) presented here.

Overall, although skin and joints are similarly affected in both conditions, *B3GALT6* mutations lead to a more extensive and severe involvement of the skeletal system, with features often found in SEMDJL1 such as kyphoscoliosis, platyspondyly, short iliac bones and elbow disclocation.

Careful observation of the hands, with their very soft and hyperextensible skin and extreme distal joint laxity, in combination with the specific radiological signs, can immediately evoke the suspicion of either spEDS-*B4GALT7* (radioulnar synostosis) or spEDS-*B3GALT6* (severe progressive kyphoscoliosis and other SEMDJL1-like skeletal features).

The extreme hypermobility of distal joints and the soft, doughy skin on the hands and feet are rarely seen in other EDS types (except, to some degree, in kEDS and aEDS) and are a valuable clue for the diagnosis, which should be supported by skeletal radiographs and by molecular analysis.

Once spEDS is suspected, direct Sanger sequencing is still a viable and cost-effective option for molecular analysis, since *B4GALT7* and *B3GALT6* have short coding regions and only a few exons (6 and 1, respectively). This should be considered especially if the radiological signs restrict the hypothesis to either gene. However, particularly when attempting a molecular analysis early on, when some of the distinctive clinical signs have not yet evolved, an NGS panel specific for EDS would be advisable, since it can help confirm the differential diagnosis.

Being able to distinguish between spEDS-*B3GALT6* and spEDS-*B4GALT7* is important for clinicians, because some patients in the *B3GALT6*-related group may develop life-threatening complications such as aortic dilatation, aneurysms and cervical spine instability.

In conclusion, accurate diagnosis will help in excluding other causes of peripheral hypotonia, such as neuromuscular disorders, and allow for appropriate physiotherapeutic interventions.

Author Contributions: L.G. and A.S.-F conceived the study. L.G., I.M., I.I., M.P. M.D.R., C.M. and S.G. performed the clinical diagnosis of the patient, genetic counselling and follow-up; M.V. and A.R. performed biochemical studies in skin fibroblasts; S.S. and S.F. investigated bone health parameters; B.C-X. and S.U. carried out the molecular analyses and interpreted the results; I.I. and S.G.C. researched the literature; L.G, I.I., and S.G.C. prepared the manuscript; A.S.F. edited and coordinated the manuscript. All authors discussed, read, and approved the manuscript.

Funding: This research received no external funding.

Acknowledgments: The authors also wish to thank the patients' family members for their cooperation in providing the medical data and photographs necessary for this publication, as well as the photographers Marco Bonazzi and Luca Valcavi. The authors are grateful for the contribution made by the Fondazione Cassa di Risparmio Manodori of Reggio Emilia.

Conflicts of Interest: All authors declare that there are no conflicts of interest concerning this work.

References

1. Kreuger, J.; Spillmann, D.; Li, J.P.; Lindahl, U. Interactions between heparan sulfate and proteins: The concept of specificity. *J. Cell Biol.* **2006**, *174*, 323–327. [CrossRef] [PubMed]
2. Bishop, J.R.; Schuksz, M.; Esko, J.D. Heparan sulphate proteoglycans fine-tune mammalian physiology. *Nature* **2007**, *446*, 1030–1037. [CrossRef] [PubMed]
3. Couchman, J.R.; Pataki, C.A. An Introduction to Proteoglycans and Their Localization. *J. Histochem. Cytochem.* **2012**, *60*, 885–897. [CrossRef] [PubMed]
4. Bülow, H.E.; Hobert, O. The Molecular Diversity of Glycosaminoglycans Shapes Animal Development. *Annu. Rev. Cell Dev. Biol.* **2006**, *22*, 375–407. [CrossRef] [PubMed]
5. Bui, C.; Huber, C.; Tuysuz, B.; Alanay, Y.; Bole-Feysot, C.; Leroy, J.G.; Mortier, G.; Nitschke, P.; Munnich, A.; Cormier-Daire, V. XYLT1 mutations in desbuquois dysplasia type 2. *Am. J. Hum. Genet.* **2014**, *94*, 405–414. [CrossRef] [PubMed]
6. van Koningsbruggen, S.; Knoester, H.; Bakx, R.; Mook, O.; Knegt, L.; Cobben, J.M. Complete and partial XYLT1 deletion in a patient with neonatal short limb skeletal dysplasia. *Am. J. Med. Genet. Part A* **2016**, *170*, 510–514. [CrossRef]
7. Jamsheer, A.; Olech, E.M.; Kozłowski, K.; Niedziela, M.; Sowińska-Seidler, A.; Obara-Moszyńska, M.; Latos-Bieleńska, A.; Karczewski, M.; Zemojtel, T. Exome sequencing reveals two novel compound heterozygous XYLT1 mutations in a Polish patient with Desbuquois dysplasia type 2 and growth hormone deficiency. *J. Hum. Genet.* **2016**, *61*, 577–583. [CrossRef]
8. Guo, L.; Elcioglu, N.H.; Iida, A.; Demirkol, Y.K.; Aras, S.; Matsumoto, N.; Nishimura, G.; Miyake, N.; Ikegawa, S. Novel and recurrent XYLT1 mutations in two Turkish families with Desbuquois dysplasia, type 2. *J. Hum. Genet.* **2017**, *62*, 447–451. [CrossRef]
9. Schreml, J.; Durmaz, B.; Cogulu, O.; Keupp, K.; Beleggia, F.; Pohl, E.; Milz, E.; Coker, M.; Ucar, S.K.; Nürnberg, G.; et al. The missing "link": An autosomal recessive short stature syndrome caused by a hypofunctional XYLT1 mutation. *Hum. Genet.* **2014**, *133*, 29–39. [CrossRef]
10. OMIM. Available online: https://www.omim.org/ (accessed on 1 August 2019).
11. Munns, C.F.; Fahiminiya, S.; Poudel, N.; Munteanu, M.C.; Majewski, J.; Sillence, D.O.; Metcalf, J.P.; Biggin, A.; Glorieux, F.; Fassier, F.; et al. Homozygosity for frameshift mutations in XYLT2 result in a spondylo-ocular syndrome with bone fragility, cataracts, and hearing defects. *Am. J. Hum. Genet.* **2015**, *96*, 971–978. [CrossRef]
12. Taylan, F.; Yavaş Abalı, Z.; Jäntti, N.; Güneş, N.; Darendeliler, F.; Baş, F.; Poyrazoğlu, Ş.; Tamçelik, N.; Tüysüz, B.; Mäkitie, O. Two novel mutations in XYLT2 cause spondyloocular syndrome. *Am. J. Med. Genet. Part A* **2017**, *173*, 3195–3200. [CrossRef] [PubMed]
13. Taylan, F.; Costantini, A.; Coles, N.; Pekkinen, M.; Héon, E.; Şıklar, Z.; Berberoğlu, M.; Kämpe, A.; Kıykım, E.; Grigelioniene, G.; et al. Spondyloocular Syndrome: Novel Mutations in XYLT2 Gene and Expansion of the Phenotypic Spectrum. *J. Bone Miner. Res.* **2016**, *31*, 1577–1585. [CrossRef] [PubMed]

14. Baasanjav, S.; Al-Gazali, L.; Hashiguchi, T.; Mizumoto, S.; Fischer, B.; Horn, D.; Seelow, D.; Ali, B.R.; Aziz, S.A.A.; Langer, R.; et al. Faulty initiation of proteoglycan synthesis causes cardiac and joint defects. *Am. J. Hum. Genet.* **2011**, *89*, 15–27. [CrossRef] [PubMed]
15. Von Oettingen, J.E.; Tan, W.H.; Dauber, A. Skeletal dysplasia, global developmental delay, and multiple congenital anomalies in a 5-year-old boy-Report of the second family with B3GAT3 mutation and expansion of the phenotype. *Am. J. Med. Genet. Part A* **2014**, *164*, 1580–1586. [CrossRef] [PubMed]
16. Jones, K.L.; Schwarze, U.; Adam, M.P.; Byers, P.H.; Mefford, H.C. A homozygous B3GAT3 mutation causes a severe syndrome with multiple fractures, expanding the phenotype of linkeropathy syndromes. *Am. J. Med. Genet. Part A* **2015**, *167*, 2691–2696. [CrossRef]
17. Budde, B.S.; Mizumoto, S.; Kogawa, R.; Becker, C.; Altmüller, J.; Thiele, H.; Rüschendorf, F.; Toliat, M.R.; Kaleschke, G.; Hämmerle, J.M.; et al. Skeletal dysplasia in a consanguineous clan from the island of Nias/Indonesia is caused by a novel mutation in B3GAT3. *Hum. Genet.* **2015**, *134*, 691–704. [CrossRef]
18. Alazami, A.M.; Al-Qattan, S.M.; Faqeih, E.; Alhashem, A.; Alshammari, M.; Alzahrani, F.; Al-Dosari, M.S.; Patel, N.; Alsagheir, A.; Binabbas, B.; et al. Expanding the clinical and genetic heterogeneity of hereditary disorders of connective tissue. *Hum. Genet.* **2016**, *135*, 525–540. [CrossRef]
19. Yauy, K.; Mau-Them, F.T.; Willems, M.; Coubes, C.; Blanchet, P.; Herlin, C.; Arrada, I.T.; Sanchez, E.; Faure, J.M.; Le Gac, M.P.; et al. B3GAT3-related disorder with craniosynostosis and bone fragility due to a unique mutation. *Genet. Med.* **2018**, *20*, 269–274. [CrossRef]
20. Paganini, C.; Costantini, R.; Superti-Furga, A.; Rossi, A. Bone and connective tissue disorders caused by defects in glycosaminoglycan biosynthesis: A panoramic view. *FEBS J.* **2019**, *286*, 3008–3032. [CrossRef]
21. Hernández, A.; Aguirre-Negrete, M.G.; Liparoli, J.C.; Cantú, J.M. Third case of a distinct variant of the Ehlers–Danlos Syndrome (EDS). *Clin. Genet.* **2008**, *20*, 222–224. [CrossRef]
22. Hernández, A.; Aguirre-Negrete, M.G.; González-Flores, S.; Reynoso-Luna, M.C.; Fragoso, R.; Nazará, Z.; AND, G.T.; Cantú, J.M. Ehlers-Danlos features with progeroid facies and mild mental retardation: Further delineation of the syndrome. *Clin. Genet.* **1986**, *30*, 456–461. [CrossRef] [PubMed]
23. Hernandez, A.; Aguirre-Negrete, M.G.; Ramírez-Soltero, S.; González-Mendoza, A.; Martínez, R.M.; Velázquez-Cabrera, A.; Cantú, J.M. A distinct variant of the Ehlers–Danlos syndrome.pdf. *Clin. Genet.* **1979**, *16*, 335–339. [CrossRef] [PubMed]
24. Kresse, H.; Rosthøj, S.; Quentin, E.; Hollmann, J.; Glössl, J.; Okada, S.; Tønnesen, T. Glycosaminoglycan-free small proteoglycan core protein is secreted by fibroblasts from a patient with a syndrome resembling progeroid. *Am. J. Hum. Genet.* **1987**, *41*, 436–453. [PubMed]
25. Quentin, E.; Gladen, A.; Roden, L.; Kresse, H. A genetic defect in the biosynthesis of dermatan sulfate proteoglycan: Galactosyltransferase I deficiency in fibroblasts from a patient with a progeroid syndrome. *Proc. Natl. Acad. Sci. USA* **1990**, *87*, 1342–1346. [CrossRef] [PubMed]
26. Almeida, R.; Levery, S.B.; Mandel, U.; Kresse, H.; Schwientek, T.; Bennett, E.P.; Clausen, H. Cloning and expression of a proteoglycan UDP-galactose: β-xylose β1,4- galactosyltransferase I. A seventh member of the human β4- galactosyltransferase gene family. *J. Biol. Chem.* **1999**, *274*, 26165–26171. [CrossRef] [PubMed]
27. Okajima, T.; Fukumoto, S.; Furukawat, K.; Urano, T.; Furukawa, K. Molecular basis for the progeroid variant of Ehlers–Danlos syndrome. Identification and characterization of two mutations in galactosyltransferase I gene. *J. Biol. Chem.* **1999**, *274*, 28841–28844. [CrossRef] [PubMed]
28. Rahuel-Clermont, S.; Daligault, F.; Piet, M.-H.; Gulberti, S.; Netter, P.; Branlant, G.; Magdalou, J.; Lattard, V. Biochemical and thermodynamic characterization of mutated β1,4-galactosyltransferase 7 involved in the progeroid form of the Ehlers–Danlos syndrome. *Biochem. J.* **2010**, *432*, 303–311. [CrossRef]
29. Faiyaz-Ul-Haque, M.; Zaidi, S.H.E.; Al-Ali, M.; Al-Mureikhi, M.S.; Kennedy, S.; Al-Thani, G.; Tsui, L.-C.; Teebi, A.S. A novel missense mutation in the galactosyltransferase-I (B4GALT7) gene in a family exhibiting facioskeletal anomalies and Ehlers–Danlos syndrome resembling the progeroid type. *Am. J. Med. Genet.* **2004**, *128*, 39–45. [CrossRef]
30. Guo, M.H.; Stoler, J.; Lui, J.; Nilsson, O.; Bianchi, D.W.; Hirschhorn, J.N.; Dauber, A. Redefining the progeroid form of Ehlers–Danlos syndrome: Report of the fourth patient with B4GALT7 deficiency and review of the literature. *Am. J. Med. Genet. Part A* **2013**, *161*, 2519–2527.

31. Cartault, F.; Munier, P.; Jacquemont, M.L.; Vellayoudom, J.; Doray, B.; Payet, C.; Randrianaivo, H.; Laville, J.M.; Munnich, A.; Cormier-Daire, V. Expanding the clinical spectrum of B4GALT7 deficiency: Homozygous p. R270C mutation with founder effect causes Larsen of Reunion Island syndrome. *Eur. J. Hum. Genet.* **2015**, *23*, 49–53. [CrossRef]
32. Salter, C.G.; Davies, J.H.; Moon, R.J.; Fairhurst, J.; Bunyan, D.; Foulds, N. Further defining the phenotypic spectrum of B4GALT7 mutations. *Am. J. Med. Genet. Part A* **2016**, *170*, 1556–1563. [CrossRef] [PubMed]
33. Arunrut, T.; Sabbadini, M.; Jain, M.; Machol, K.; Scaglia, F.; Slavotinek, A. Corneal clouding, cataract, and colobomas with a novel missense mutation in B4GALT7—A review of eye anomalies in the linkeropathy syndromes. *Am. J. Med. Genet. Part A* **2016**, *170*, 2711–2718. [CrossRef] [PubMed]
34. Beighton, P.; Kozlowski, K. Spondylo-Epi-Metaphyseal dysplasia with joint laxity and severe, progressive kyphoscoliosis. *Skelet. Radiol.* **1980**, *5*, 205–212. [CrossRef] [PubMed]
35. Beighton, P.; Kozlowski, K.; Gericke, G.; Wallis, G.; Grobler, L. Spondylo-epimetaphyseal dysplasia with joint laxity and severe, progressive kyphoscoliosis. A potentially lethal dwarfing disorder. *S. Afr. Med. J.* **1983**, *64*, 772–775. [PubMed]
36. Kozlowski, K.; Beighton, P. Radiographic features of spondylo-epimetaphyseal dysplasia with joint laxity and progressive kyphoscoliosis. In *RöFo-Fortschritte Auf Dem Gebiet der Röntgenstrahlen und der Bildgebenden Verfahren*; Georg Thieme Verlag Stuttgart: New York, NY, USA, 1984; Volume 141, pp. 337–341.
37. Beighton, P.; Gericke, G.; Kozlowski, K.; Grobler, L. The manifestations and natural history of spondylo-epi-metaphyseal dysplasia with joint laxity. *Clin. Genet.* **1984**, *26*, 308–317. [CrossRef] [PubMed]
38. Torrington, M.; Beighton, P. The ancestry of spondyloepimetaphyseal dysplasia with joint laxity (SEMDJL) in South Africa. *Clin. Genet.* **2008**, *39*, 210–213. [CrossRef]
39. Beighton, P. Spondyloepimetaphyseal dysplasia with joint laxity (SEMDJL). *J. Med. Genet.* **1994**, *31*, 136–140. [CrossRef]
40. Christianson, A.L.; Beighton, P. Spondyloepimetaphyseal dysplasia with joint laxity (SEMDJL) in three neonates. *Genet. Couns.* **1996**, *7*, 219–225.
41. Hall, C.M.; Elçioglu, N.H.; Shaw, D.G. A distinct form of spondyloepimetaphyseal dysplasia with multiple dislocations. *J. Med. Genet.* **1998**, *35*, 566–572. [CrossRef]
42. Vorster, A.A.; Beighton, P.; Ramesar, R.S. Spondyloepimetaphyseal dysplasia with joint laxity (Beighton type); mutation analysis in eight affected South African families. *Clin. Genet.* **2015**, *87*, 492–495. [CrossRef]
43. Bonafe, L.; Cormier-Daire, V.; Hall, C.; Lachman, R.; Mortier, G.; Mundlos, S.; Nishimura, G.; Sangiorgi, L.; Savarirayan, R.; Sillence, D.; et al. Nosology and classification of genetic skeletal disorders: 2015 revision. *Am. J. Med. Genet. Part A* **2015**, *167*, 2869–2892. [CrossRef] [PubMed]
44. Malfait, F.; Kariminejad, A.; Van Damme, T.; Gauche, C.; Syx, D.; Merhi-Soussi, F.; Gulberti, S.; Symoens, S.; Vanhauwaert, S.; Willaert, A.; et al. Defective initiation of glycosaminoglycan synthesis due to B3GALT6 mutations causes a pleiotropic Ehlers–Danlos-syndrome-like connective tissue disorder. *Am. J. Hum. Genet.* **2013**, *92*, 935–945. [CrossRef] [PubMed]
45. Nakajima, M.; Mizumoto, S.; Miyake, N.; Kogawa, R.; Iida, A.; Ito, H.; Kitoh, H.; Hirayama, A.; Mitsubuchi, H.; Miyazaki, O.; et al. Mutations in B3GALT6, which encodes a glycosaminoglycan linker region enzyme, cause a spectrum of skeletal and connective tissue disorders. *Am. J. Hum. Genet.* **2013**, *92*, 927–934. [CrossRef] [PubMed]
46. Malfait, F.; Francomano, C.; Byers, P.; Belmont, J.; Berglund, B.; Black, J.; Bloom, L.; Bowen, J.M.; Brady, A.F.; Burrows, N.P.; et al. The 2017 international classification of the Ehlers–Danlos syndromes. *Am. J. Med. Genet. Part C* **2017**, *175*, 8–26. [CrossRef] [PubMed]
47. Brady, A.F.; Demirdas, S.; Fournel-Gigleux, S.; Ghali, N.; Giunta, C.; Kapferer-Seebacher, I.; Kosho, T.; Mendoza-Londono, R.; Pope, M.F.; Rohrbach, M.; et al. The Ehlers–Danlos syndromes, rare types. *Am. J. Med. Genet. Part C* **2017**, *175*, 70–115. [CrossRef] [PubMed]
48. Van Damme, T.; Pang, X.; Guillemyn, B.; Gulberti, S.; Syx, D.; De Rycke, R.; Kaye, O.; de Die-Smulders, C.E.M.; Pfundt, R.; Kariminejad, A.; et al. Biallelic B3GALT6 mutations cause spondylodysplastic Ehlers–Danlos syndrome. *Hum. Mol. Genet.* **2018**, *27*, 3475–3487. [CrossRef]
49. Paganini, C.; Monti, L.; Costantini, R.; Besio, R.; Lecci, S.; Biggiogera, M.; Tian, K.; Schwartz, J.M.; Huber, C.; Cormier-Daire, V.; et al. Calcium activated nucleotidase 1 (CANT1) is critical for glycosaminoglycan biosynthesis in cartilage and endochondral ossification. *Matrix Biol.* **2019**, *81*, 70–90. [CrossRef]

50. Nizon, M.; Huber, C.; De Leonardis, F.; Merrina, R.; Forlino, A.; Fradin, M.; Tuysuz, B.; Abu-Libdeh, B.Y.; Alanay, Y.; Albrecht, B.; et al. Further delineation of CANT1 phenotypic spectrum and demonstration of its role in proteoglycan synthesis. *Hum. Mutat.* **2012**, *33*, 1261–1266. [CrossRef]
51. Ritelli, M.; Dordoni, C.; Cinquina, V.; Venturini, M.; Calzavara-Pinton, P.; Colombi, M. Expanding the clinical and mutational spectrum of B4GALT7-spondylodysplastic Ehlers–Danlos syndrome. *Orphanet J. Rare Dis.* **2017**, *12*, 153. [CrossRef]
52. Sandler-Wilson, C.; Wambach, J.A.; Marshall, B.A.; Wegner, D.J.; McAlister, W.; Cole, F.S.; Shinawi, M. Phenotype and response to growth hormone therapy in siblings with B4GALT7 deficiency. *Bone* **2019**, *124*, 14–21. [CrossRef]
53. Sellars, E.A.; Bosanko, K.A.; Lepard, T.; Garnica, A.; Schaefer, G.B. A newborn with complex skeletal abnormalities, joint contractures, and bilateral corneal clouding with sclerocornea. *Semin. Pediatr. Neurol.* **2014**, *21*, 84–87. [CrossRef] [PubMed]
54. Ritelli, M.; Chiarelli, N.; Zoppi, N.; Dordoni, C.; Quinzani, S.; Traversa, M.; Venturini, M.; Calzavara-Pinton, P.; Colombi, M. Insights in the etiopathology of galactosyltransferase II (GalT-II) deficiency from transcriptome-wide expression profiling of skin fibroblasts of two sisters with compound heterozygosity for two novel B3GALT6 mutations. *Mol. Genet. Metab. Rep.* **2015**, *2*, 1–15. [CrossRef] [PubMed]
55. Trejo, P.; Rauch, F.; Glorieux, F.H.; Ouellet, J.; Benaroch, T.; Campeau, P.M. Spondyloepimetaphysial Dysplasia with Joint Laxity in Three Siblings with B3GALT6 Mutations. *Mol. Syndromol.* **2017**, *8*, 303–307. [CrossRef] [PubMed]
56. Ben-Mahmoud, A.; Ben-Salem, S.; Al-Sorkhy, M.; John, A.; Ali, B.R.; Al-Gazali, L. A B3GALT6 variant in patient originally described as Al-Gazali syndrome and implicating the endoplasmic reticulum quality control in the mechanism of some β3GalT6-pathy mutations. *Clin. Genet.* **2018**, *93*, 1148–1158. [CrossRef] [PubMed]
57. Fukada, T.; Civic, N.; Furuichi, T.; Shimoda, S.; Mishima, K.; Higashiyama, H.; Idaira, Y.; Asada, Y.; Kitamura, H.; Yamasaki, S.; et al. The zinc transporter SLC39A13/ZIP13 is required for connective tissue development; its involvement in BMP/TGF-β signaling pathways. *PLoS ONE* **2008**, *3*, e3642. [CrossRef]
58. Giunta, C.; Elçioglu, N.H.; Albrecht, B.; Eich, G.; Chambaz, C.; Janecke, A.R.; Yeowell, H.; Weis, M.; Eyre, D.R.; Kraenzlin, M.; et al. Spondylocheiro Dysplastic Form of the Ehlers–Danlos Syndrome—An Autosomal-Recessive Entity Caused by Mutations in the Zinc Transporter Gene SLC39A13. *Am. J. Hum. Genet.* **2008**, *82*, 1290–1305. [CrossRef] [PubMed]
59. Dusanic, M.; Dekomien, G.; Lücke, T.; Vorgerd, M.; Weis, J.; Epplen, J.T.; Köhler, C.; Hoffjan, S. Novel Nonsense Mutation in *SLC39A13* Initially Presenting as Myopathy: Case Report and Review of the Literature. *Mol. Syndromol.* **2018**, *9*, 100–109. [CrossRef]

© 2019 by the authors. Licensee MDPI, Basel, Switzerland. This article is an open access article distributed under the terms and conditions of the Creative Commons Attribution (CC BY) license (http://creativecommons.org/licenses/by/4.0/).

Article

Further Defining the Phenotypic Spectrum of *B3GAT3* Mutations and Literature Review on Linkeropathy Syndromes

Marco Ritelli [1], Valeria Cinquina [1], Edoardo Giacopuzzi [2], Marina Venturini [3], Nicola Chiarelli [1] and Marina Colombi [1],*

1. Division of Biology and Genetics, Department of Molecular and Translational Medicine, University of Brescia, 25123 Brescia, Italy
2. Genetics Unit, IRCCS Istituto Centro San Giovanni di Dio Fatebenefratelli, 25125 Brescia, Italy
3. Division of Dermatology, Department of Clinical and Experimental Sciences, Spedali Civili University Hospital, 25123 Brescia, Italy
* Correspondence: marina.colombi@unibs.it; Tel.: +39-030-3717-240; Fax: +39-030-371-7241

Received: 25 June 2019; Accepted: 19 August 2019; Published: 21 August 2019

Abstract: The term linkeropathies (LKs) refers to a group of rare heritable connective tissue disorders, characterized by a variable degree of short stature, skeletal dysplasia, joint laxity, cutaneous anomalies, dysmorphism, heart malformation, and developmental delay. The LK genes encode for enzymes that add glycosaminoglycan chains onto proteoglycans via a common tetrasaccharide linker region. Biallelic variants in *XYLT1* and *XYLT2*, encoding xylosyltransferases, are associated with Desbuquois dysplasia type 2 and spondylo-ocular syndrome, respectively. Defects in *B4GALT7* and *B3GALT6*, encoding galactosyltransferases, lead to spondylodysplastic Ehlers-Danlos syndrome (spEDS). Mutations in *B3GAT3*, encoding a glucuronyltransferase, were described in 25 patients from 12 families with variable phenotypes resembling Larsen, Antley-Bixler, Shprintzen-Goldberg, and Geroderma osteodysplastica syndromes. Herein, we report on a 13-year-old girl with a clinical presentation suggestive of spEDS, according to the 2017 EDS nosology, in whom compound heterozygosity for two *B3GAT3* likely pathogenic variants was identified. We review the spectrum of *B3GAT3*-related disorders and provide a comparison of all LK patients reported up to now, highlighting that LKs are a phenotypic continuum bridging EDS and skeletal disorders, hence offering future nosologic perspectives.

Keywords: linkeropathies; *B3GAT3*; Larsen-like syndrome; *B4GALT7*; *B3GALT6*; spondylodysplastic Ehlers-Danlos syndrome; *XYLT1*; *XYLT2*; Desbuquois dysplasia; spondylo-ocular syndrome

---

1. Introduction

Ehlers-Danlos syndrome (EDS) comprises a clinically variable and genetically heterogeneous group of heritable connective tissue disorders (HTCDs) sharing the triad of (generalized) joint hypermobility, cutaneous abnormalities, and internal organ/vascular fragility and dysfunctions. The 2017 EDS nosology recognizes 13 different clinical subtypes and 19 causal genes mainly encoding fibrillar collagens, collagen-modifying proteins, or processing enzymes [1]. A fourteenth subtype has been recently associated with biallelic variants in *AEBP1*, which encodes the aortic carboxypeptidase-like protein (ACLP) associating with collagens in the extracellular matrix (ECM) [2–5]. In addition to the clinical classification, the 2017 EDS nosology introduced a pathogenetic scheme that regrouped the EDS subtypes into seven functional classes (i.e., disorders) of (a) collagen primary structure and collagen processing, (b) collagen folding and crosslinking, (c) structure and function of the myomatrix, (d) glycosaminoglycan (GAG) biosynthesis, (e) intracellular pathways, (f) the complement pathway, and (g) unresolved forms of EDS. The clinical manifestations of EDS are broad and often overlap other HCTDs including some types of skeletal dysplasias, cutis laxa, hereditary myopathies, and TGFβ-related disorders [6,7]. Hence, intermediate or bridging phenotypes presenting the EDS triad are expected at the boundaries of an evolving nosology.

The multisystemic clinical variability of EDS reflects the numerous functions of collagens and their interactors, among which proteoglycans (PGs) are particularly notable. PGs are structurally complex biomacromolecules that are essential in the development, signaling, and homeostasis of many tissues and organs including bone, cartilage, skeletal muscle, eyes, heart, and skin [8–13]. PGs contain one or more variable GAG chains, which are linear polysaccharides consisting of repeating disaccharide blocks attached to a core protein. Depending on the composition of these blocks, the PG superfamily can be subdivided into two major groups: heparan sulfate (HS) and chondroitin sulfate (CS)/dermatan sulfate (DS) PGs. The biosynthesis of GAG chains starts with the formation of a common so-called tetrasaccharide linker region that is covalently attached to a serine residue of the PG core protein (Figure 1). The linker region synthesis is a stepwise process that involves the action of specific glycosyltransferases. It starts with the transfer of a xylose (Xyl) residue by xylosyltransferases I/II (XylT-I/II encoded by *XYLT1* and *XYLT2*, respectively), followed by the addition of two galactose (Gal) residues by galactosyltransferase type I (GalT-I encoded by *B4GALT7*) and type II (GalT-II encoded by *B3GALT6*). The linker region is completed by the transfer of glucuronic acid (GlcA) catalyzed by glucuronosyltransferase I (GlcAT-I encoded by *B3GAT3*), upon which polymerization of the HS or CS/DS chains begins. HS is formed by the alternating addition of disaccharides of N-acetyl-glucosamine (GlcNAc) and GlcA residues, CS by N-acetyl-galactosamine (GalNAc) and GlcA residues, subsequently modified by several sulfotransferases. The formation of DS requires the epimerization of GlcA toward iduronic acid (IdoA), an event accomplished by dermatan sulfate epimerases (DS-epi1 encoded by *DSE*). This allows dermatan 4-o-sulfotransferase 1 (D4ST1 encoded by *CHST14*) to catalyze the 4-o-sulfation of GalNAc, which prevents back-epimerization of the adjacent IdoA (Figure 1) [8–13].

The importance of the correct initiation of GAG synthesis is exemplified by the identification of biallelic variants in all genes encoding the key enzymes in the linker region synthesis, leading to a spectrum of severe multisystemic disorders (Figure 1), a.k.a., linkeropathies (LKs) [11–13].

Two of these LKs, GalT-I- and GalT-II-deficiency, fit with the EDS spectrum and are recognized in the 2017 EDS nosology as spondylodysplastic EDS (spEDS) that also includes patients with mutations in *SLC39A13* encoding the ZIP13 protein involved in the influx of zinc into the cytosol [1,14]. According to the 2017 nosology, spEDS is suggested by two major criteria, short stature and muscle hypotonia, plus characteristic radiographic abnormalities and at least three other minor criteria (general or gene-specific) [1]. At present, 10 patients with molecularly confirmed *B4GALT7*-spEDS have been reported [14–21], 46 with *B3GALT6*-spEDS [14,22–28], and nine with *SLC39A13*-spEDS [14,29–31]. A further 22 patients, all with the same homozygous *B4GALT7* p.(Arg270Cys) missense variant, have been characterized in the ethnic group called white creoles living on Reunion Island (Larsen of Reunion Island syndrome) [32].

Molecular defects in *XYLT1* have been associated with Desbuquois dysplasia type 2 (DBQD2) with 28 molecularly proven patients reported hitherto [33–40], while *XYLT2* mutations cause the so-called spondylo-ocular syndrome (SOS), which has been described in 20 patients thus far [41–46]. Concerning *B3GAT3*, 25 patients from 12 families have been recognized, but range in severity from mild to severe and resemble Larsen (LR)-, Antley-Bixler (AB)-, Shprintzen-Goldberg (SG)-, and Geroderma osteodysplastica (GO)-like syndromes [26,47–53]. Further downstream in the GAG biosynthetic pathway, mutations in the *CHST14* and *DSE* genes have been reported in musculocontractural EDS (mcEDS) type 1 and 2, respectively [14].

Herein, we report on a 13-year-old girl ascertained as a suspect of EDS. Trio-based exome sequencing (ES) revealed compound heterozygosity for two likely pathogenic variants in *B3GAT3*. We provide a comparison of the patient's clinical features with those of the other LK patients reported so far, thus offering future perspectives for clinical research in this field.

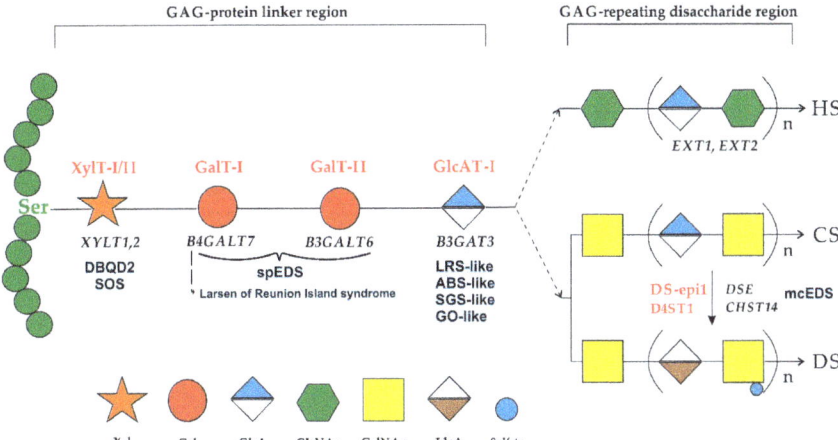

Figure 1. Biosynthetic assembly of the glycosaminoglycan (GAG) backbones of heparan sulfate (HS) and chondroitin sulfate (CS)/dermatan sulfate (DS) chains and related genetic disorders. Each enzyme (in red) and its coding gene (in black) are described near the sugar symbols. After the synthesis of specific core proteins, the synthesis of the GAG-protein linker region is initiated by XylT-I/II, which transfers [Xyl] to the specific Ser residue in the endoplasmic reticulum. The synthesis of the linker region is completed by the consecutive addition of two molecules of [Gal], added by GalT-I/II, followed by the transfer of [GlcA] catalyzed by GlcAT-I in the Golgi. The addition of a [GalNAc] to the linker region commits the growing GAG chain to CS/DS. CS synthesis proceeds with the alternating addition of [GlcA] and [GalNAc] and can be further modified by sulfotransferases. A DS chain is generated after the formation of the chondroitin backbone, when [GlcA] is converted into [IdoA] by DS-epi1, resulting in the formation of the dermatan backbone, where [GalNAc] is sulfated by D4ST1. Alternatively, the addition of the [GlcNAc] to the linker region induces HS biosynthesis. The polymerization of the HS chain is catalyzed by enzymes encoded by the *EXT1* and *EXT2* genes. **Abbreviations**: Ser, Serine; HS, heparin sulfate; CS, chondroitin sulfate; DS, dermatan sulfate; XylT-I/II, xylosyltransferases I/II; GalT-I, galactosyltrasferase I; Galt-II, galactosyltrasferase II; GlcAT-I, glucuronasyltransferase I; DE-epi1; dermatan sulfate epimerases; D4ST1; dermatan 4-o-sulfotransferase 1; Xyl, xylose; Gal, galactose, GlcA; glucuronic acid, GlcNAc, *N*-acetylglucosamine; GalNAc, *N*-acetylgalactisamine; IdoA, iduronic acid; DBQD2, Desbuquois dysplasia type 2, SOS, spondylo-ocular syndrome; spEDS, spondylodysplastic EDS; LRS-like, Larsen-like syndrome; ABS-like, Antley-Bixler-like syndrome; SGS-like, Shprintzen-Goldberg-like syndrome; GO-like, Geroderma osteodysplastica-like; mcEDS, musculocontractural EDS.

2. Patient and Methods

2.1. Ethical Compliance

The patient was evaluated at the specialized outpatient clinic for the diagnosis of EDS and related connective tissue disorders (i.e., the Ehlers-Danlos Syndrome and Inherited Connective Tissue Disorders Clinic, CESED), at the University Hospital Spedali Civili of Brescia. Molecular analysis was achieved in compliance with Italian legislation on genetic diagnostic tests and the patient's parents provided written informed consent for the publication of clinical data and photographs according to the Italian bioethics laws. This study followed the Declaration of Helsinki's principles and was carried out from routine diagnostic activity; a formal ethics review was therefore not requested.

2.2. Amplicon-Based Exome Sequencing

Genomic DNA from the proband and her parents was extracted from peripheral blood leukocytes by standard procedures. Mutational screening was performed by trio-based ES using the Ion Proton platform and the AmpliSeq technology following the manufacturer's recommendations (Thermo Fisher Scientific, South San Francisco, CA, USA). Briefly, whole exome libraries were prepared using the AmpliSeq Exome RDY kit for library preparation. The template preparation of the libraries was performed using the Ion PI Hi-Q OT2 200 kit on the Ion OneTouch 2 starting from 8 µl of the 100 pM libraries. Template preparation and sequencing runs were performed with the Ion PI Hi-Q Sequencing 200 kit. The templated Ion Sphere Particles (ISP) were enriched for positive ISP using the Ion OneTouch ES and sequenced on the Ion Proton with the Ion PI chip v3. Basecalling and sequence alignment against hg19 genome assembly were performed using Ion Torrent Suite software 5.6, and genetic variants were identified using Torrent Suite Variant Caller pipeline 5.6. Variants were decomposed and normalized using the vt tool [54], filtered for quality using GARFIELD-NGS [55], and annotated using ANNOVAR [56]. Variants were filtered according to the following criteria: (i) MAF < 0.01 in 1000G and ExAc v.0.3 populations; (ii) predicted to alter protein product, namely missense, stop-affecting or splice-affecting variants; and (iii) not present in our internal database. Filtered variants were then prioritized based on DANN [57] and M-CAP [58] scores to retain the most likely deleterious variants. RVIS [59] and GDI [60] scores were used to prioritize more intolerant genes. Finally, we only considered variants with a perfect segregation among the parents and proband according to a recessive/de novo model of transmission. To evaluate the putative pathogenicity of the *B3GAT3* variants, we used the following mutation prediction programs: SIFT [61], Mutation Taster [62], CADD [63], PROVEAN [64], GERP++ [65], UMD_prediction [66], LRT [67], Fathmm-MKL [68], VEST [69], and FitCons [70].

2.3. Sanger Sequencing

The *B3GAT3* variants identified by ES (reference sequences: NM_012200.3, NP_036332.2) were confirmed by Sanger sequencing with the BigDye Terminator v1.1 Cycle Sequencing kit on an ABI 3130XL Genetic Analyzer according to the manufacturer's protocols (Thermo Fisher Scientific, South San Francisco, CA, USA) with specific primer sets amplifying exon 3 and exon 4 of *B3GAT3*, respectively (Supplementary Table S1). The sequences were analyzed with Sequencer 5.0 software (Gene Codes Corporation, Ann Arbor, MI, USA) and variants were annotated according to the Human Genome Variation Society (HGVS) nomenclature by using the Alamut Visual software version 2.11 (Interactive Biosoftware, Rouen, France).

3. Results

3.1. Clinical Findings

The proband (LOVD ID #00235371), an Italian 13-year-old girl, was born to non-consanguineous healthy parents and had a healthy sister. Clinical history was remarkable for birth at 41 weeks (height 49 cm, weight 3.2 kg) after induced labor associated with perinatal respiratory distress, anterior ectopic anus, and congenital hip dislocation treated with hip abduction braces, severe neonatal hypotonia, and delayed motor development (delay in walking, first step at three years of age, and acquisition of fine motor skills). She was discharged from the neonatology unit with a diagnosis of generalized joint hypermobility. Medical history further included propensity to develop ecchymoses and surgically treated umbilical hernia. At one year of age, total skeletal x-ray disclosed severe kyphoscoliosis unsuccessfully treated with orthopedic corset, atlantooccipital instability, bilateral radio-ulnar synostosis, abnormalities of the proximal humeral epiphyses, metaphyseal flaring, long and thin bones with widened metaphyses, and delayed bone age. Cervical spine imaging showed significant atlantoaxial and atlanto-occipital instability with flexion and extension. At the age of two, a heart ultrasound revealed an atrial septal defect, which was surgically treated three years later due to tachyarrhythmia and pulmonary hypertension. At ages two and three, respectively, either a clinical diagnosis of SGS or of an unspecified EDS was given. Genetic analyses were not performed. At age four, dual-energy x-ray absorptiometry (DEXA) disclosed severe low bone mineral density for sex and age (z-score < 2 SD) and bisphosphonate treatment was commenced. The progressive kyphoscoliosis and cervical spine instability were surgically treated at age six and 12 without satisfactory improvement. Ophthalmologic evaluation revealed refractive errors such as astigmatism and strabismus. At six years of age, progressive height deficit related to GH deficiency was noticed. GH therapy was started with a discrete outcome, but treatment was interrupted two years later due to the worsening of side effects. Since infancy, the patient suffered from recurrent dislocation of the elbows, shoulders, and knees as well as chronic myalgia of the lower limbs and severe foot pain.

On examination, at 13 years of age, she presented with short stature (height 130 cm; genetic target 169 cm, arm span/height ratio, normal value < 1.05), a weight of 25 kg, severe kyphoscoliosis, short neck, pectus carinatum, genua valga, and muscle hypotonia (Figure 2). Facial dysmorphism included enophthalmos, midface hypoplasia, dolichocephaly, prominent forehead, low-set ears, blue sclerae, downslanting palpebral fissures, long philtrum, narrow palate, and micrognathia. Foot deformities such as sandal gaps, severe pes planovalgus, and clinodactyly of the toes were present (Figure 2). The patient also showed skin hyperextensibility over the neck, elbow, forearm and knees, easy bruising, mild atrophic scarring, generalized joint hypermobility with a Beighton score (BS) of 6/9, clinodactyly of the fifth finger, and long fingers with spatulate distal phalanges. Cognitive development and mentation were normal. DEXA confirmed severe osteopenia and delayed bone age despite bisphosphonates treatment; no fractures were reported. Echocardiogram revealed minimal mitral, tricuspid, and aortic valves insufficiency.

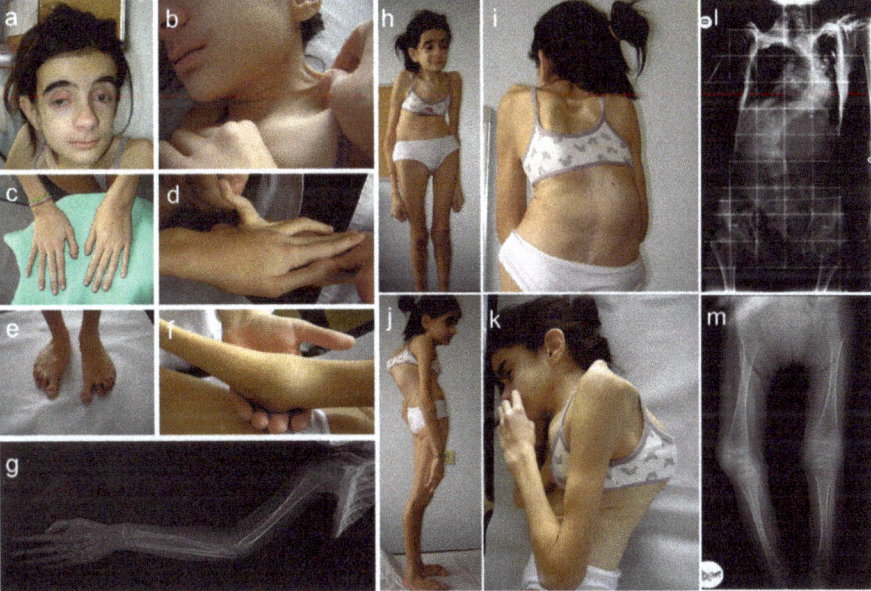

Figure 2. Clinical findings of the patient. Facial dysmorphism: enophthalmos, midface hypoplasia, prominent forehead, micrognathia, low-set ears, and short neck (**a**), skin hyperextensibility over the neck (**b**), long fingers with spatulate distal phalanges and clinodactyly of the fifth finger (**c**), joint laxity of the fifth finger (**d**), foot deformities: sandal gaps, severe pes planovalgus and clinodactyly of the toes (**e**), elbow deformity with reduction in the range of motion (**f**), radioulnar synostosis (**g**), severe kyphoscoliosis (**h–l**), pectus carinatum (**h**), muscle hypotonia (**h–k**), and metaphyseal flaring (**m**).

Overall, the patient fulfilled the minimal criteria suggestive for spEDS according to the 2017 EDS nosology [1], since she presented two major (short stature, muscle hypotonia), four general minor (skin hyperextensibility, foot deformity, delayed motor development, osteopenia), and several gene-specific minor criteria either for *B4GALT7*-spEDS or *B3GALT6*-spEDS (craniofacial dysmorphism, characteristic radiographic findings including severe kyphoscoliosis, radioulnar synostosis, metaphyseal flaring, bilateral elbow deformities, peculiar fingers, and gJHM with recurrent dislocations) (Table 1, Supplementary Table S2). Nevertheless, since the patient also presented some peculiar features not previously reported in spEDS patients such as dolichocephaly, atrial septal defect, and anterior ectopic anus, we performed trio-based ES.

Table 1. Candidate genes after selecting variants.

Gene	Variant	Effect	Inheritance Model	Genotype	Clinvar Phenotype
B3GAT3 (NM_00122200)	c.481C>T (father) c.889C>T (mother)	p.(Arg161Trp) p.(Arg297Trp)	Recessive (comp het)	comp het	Multiple joint dislocations, short stature, craniofacial dysmorphism, and congenital heart defects
BMP8A (NM_181809)	c.333G>T	p.(Met111Ile)	Dominant (de novo)	het	
FES (NM_0002005.3)	c.1778G>A	p.(Arg593Gln)	Dominant (de novo)	het	
NR2F6 (NM_005234)	c.806C>T	p.(Pro269Leu)	Dominant (de novo)	het	
PAK2 (NM_002577)	c.303G>C	p.(Gln101His)	Dominant (de novo)	het	
TRAK1 (NM_001042646)	c.1327G>A	p.(Ser443Gly)	Dominant (de novo)	het	

3.2. Molecular Findings

Summary results of ES are reported in Supplementary Table S3 and Figure S1. After application of the filtering pipeline and prioritization of variants by considering only recessive or de novo variants with perfect segregation among trio members (Supplementary Table S4), six candidate genes were identified (Table 1). Among these, five genes with a de novo variant were excluded, since only *B3GAT3*, associated with multiple joint dislocations, short stature, and craniofacial dysmorphism with or without congenital heart defects (OMIM #245600), was consistent with the patient's phenotype.

In particular, the trio analysis revealed the paternally inherited c.481C>T transition in exon 3, leading to the substitution of a highly conserved and positively charged arginine residue with a larger and neutral tryptophan at position 161 [p.(Arg161Trp)] within the donor substrate binding subdomain of the protein [71,72], and the maternal c.889C>T variant in exon 4, which also resulted in the substitution of an arginine residue with a tryptophan [p.(Arg297Trp)], but within the acceptor substrate binding subdomain (Figures 3A and 4) [71,72].

Both variants are annotated in dbSNPs and have extremely low frequencies in population genomic databases. In particular, the paternal variant was observed in three individuals in GnomAD (rs765246909, 3/251290, no homozygotes, total MAF: C = 0.00001194), and the maternal substitution was also observed in three individuals (rs759636773, 3/251080, no homozygotes, total MAF: C = 0.00001195) (queried on 28 May, 2019). Their putative pathogenicity was estimated through 12 different in silico prediction algorithms that agreed to define p.(Arg161Trp) and p.(Arg297Trp) as high impacting variants (Figure 3B).

By using the InterVar (Clinical Interpretation of Genetic Variants) tool [73], both variants were classified as likely pathogenic (class 4) according to the guidelines of the American College of Medical Genetics and Genomics (ACMG) [74] since (i) both variants are missense substitutions in a gene that has a low rate of benign missense variation and where missense variants are a common mechanism of disease (Table 2); (ii) both variants are located in a critical and well-established functional domain of the protein, i.e., the glycosyltransferase domain; (iii) their extremely low frequency in a publicly available population database; (iv) the multiple lines of computational evidence supporting a deleterious effect; and (v) the patient's phenotype was highly suggestive for a disease with a single genetic etiology.

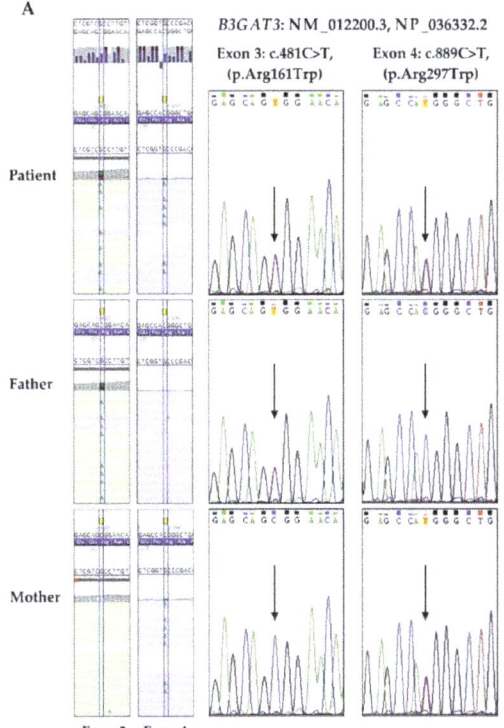

Figure 3. Molecular findings. (A) ES data alignments show the compound heterozygosity of the paternal c.481C>T (p.Arg161Trp) and the maternal c.889C>T (p.Arg297Trp) missense variants. Sanger sequencing confirmed the presence of both variants in the proband (arrows). Healthy parents were heterozygous carriers. Mutations are annotated according to HGVS nomenclature (reference sequences: NM_012200.3, NP_036332.2). (B) In silico predictions of the pathogenicity of the p.(Arg161Trp) and p.(Arg297Trp) missense substitutions by using 12 different algorithms [61–70].

The two variants were submitted to the Leiden Open Variation Database (LOVD variants identifiers: #0000480198 and #0000480199). The functional effect of the missense substitutions on reduced/absent enzymatic activity was not verified due to the unavailability of the patient's fibroblasts.

Table 2. Summary of clinical features of all patients with B3GAT3 variants.

References	Present Patient	[1,2]	[3]	[6]	[8]	[5]	[4,7,8]
Number of patients	n = 1	n = 6	n = 8	n = 1	n=1	n = 1	n = 8
Phenotype	LRS-like	LRS-like	LRS-like	LRS-like	LRS-like	GO-like	ABS/SGS-like
Consanguinity	-	+	+	-	+	+	+
B3GAT3 variant(s) (NM_012200.3)	c.481C>T c.889C>T	c.830G>A homozygous	c.419C>T homozygous	c.1A>G c.671T>A	c.416C>T homozygous	c.245C>T homozygous	c.667G>A homozygous
Protein Change (NP_036332.2)	p.(Arg161Trp) p.(Arg297Trp)	p.(Arg277Gln)	p.(Pro140Leu)	p.(Met1?) p.(Leu224Gln)	p.(Thr139Met)	p.(Pro82Leu)	p.(Gly223Ser)
Skeletal							
Short stature	+	+	+	-	+	+	2/5
Joint hypermobility	+	+	-	+	+	na	1/2
Joint dislocations	+	+	+	+	+	+	3/7
Elbow joint abnormalities	+	+	4/8	na	-	na	+
Multiple fractures	-	-	na	+	-	+	6/8
Kyphoscoliosis	+	-	4/8	+	-	na	1/7
Scoliosis/kyphosis	+	0/1	4/8	+	-	+	2/7
Platyspondyly	-	-	0/2	-	-	+	-
Peculiar fingers (long, slender, tapered, broad, thin, arachnodactyly)	+	+	+	+	+	na	+
Pectus abnormality	+	1/6	-	-	-	na	1/2
Radioulnar synostosis	+	1/1	2/2	-	-	na	7/7
Bowing of limbs	-	na	4/8	+	-	na	0/8
Metaphyseal flaring	+	1/1	na	na	-	na	1/2
Iliac abnormalities	-	1/1	na	na	na	na	1/2
Radial head subluxation or dislocation	-	na	2/2	na	+	na	0/1
Foot deformity	+	+	6/8	-	+	na	+

Table 2. Cont.

References	Present Patient	[1,2]	[3]	[6]	[8]	[5]	[4,7,8]
pes planus	+	1/1	na	-	+	na	na
hallux valgus	+	1/1	6/8	-	+	na	na
club feet	-	0/1	na	-	-	na	+
sandal gap between toes	+	1/1	3/8	-	-	na	1/1
Osteopenia	+	5/6	na	+	+	+	2/2
Cervical spine instability	+	1/1	na	na	na	na	na
Craniofacial							
Midface hypoplasia	+	+	+	+	+	na	7/7
Flat face	-	1/1	na	-	-	na	na
Craniosynostosis	+	+	na	+	na	na	4/7
Frontal bossing	+	1/6	na	+	-	na	3/7
Wide forehead	-	1/1	na	na	-	na	2/2
Blue sclerae	+	0/1	-	+	+	na	5/5
Proptosis or prominent eyes	-	+	na	-	+	na	5/5
Downslanting palpebral fissures	+	3/5	na	+	+	na	1/2
Low-set ears	+	2/5	na	na	na	na	1/1
Depressed nasal bridge	-	5/6	4/8	-	-	na	4/7
Small mouth/microstomia	-	4/6	3/8	-	+	na	2/2
Long upper lip/long philtrum	-/+	na	na	na	+	na	2/2
Cleft palate/bifid uvula	+/-	na	na	+	+	na	1/1
Micrognathia	+	4/6	4/8	na	-	na	0/1
Short and/or webbed neck	+	+	2/8	+	+	na	2/2
Cutaneous							

Table 2. Cont.

References	Present Patient	[1,2]	[3]	[6]	[8]	[5]	[4,7,8]
Skin (hyperextensibility; soft, doughy, thin, translucent skin)	+ (mild)	1/1 skin wrinkling	-	+	-	Cutis laxa	1/2 Cutis laxa
Easy bruising	+	0/1	na	na	na	na	na
Atrophic scarring	+ (mild)	0/1	na	na	na	na	0/1
Other							
Cardiovascular abnormalities	+	6/6	0/3	+	na	+	4/8
Muscle hypotonia	+	0/1	na	+	na	na	4/4
Refractive errors/hypermetropia	+	1/1	na	+	-	na	0/1
Delayed motor development	+	1/1	na	+	+	na	1/1
Delayed cognitive development	-	-	-	-	-	na	1/1
Bone chondromas	-	-	-	-	-	+	-
Anterior ectopic anus	+	-	-	-	-	-	-

Note: +, present in all patients; -, absent in all patients; na, not available; ABS, Antley-Bixler syndrome; GO, Geroderma osteodysplastica; SGS, Shprintzen-Goldberg syndrome.

4. Discussion

The umbrella term LK refers to a group of extremely rare and consequently poorly characterized genetic disorders caused by mutations in genes encoding enzymes responsible for the synthesis of GAG side chains of PGs. Nosologic uncertainty characterizes these disorders, thus contributing to the clinical diagnosis of challenging patients, which is not straightforward at all. Indeed, although the linker region is the identical tetrasaccharide sequence for all PGs in all tissues, biallelic variants in the LK genes are associated with apparently different phenotypes that variably affect the skeletal system and skin, even if remarkable similarities between the different LKs are recognizable [8–13].

In the 2017 EDS nosology, some patients with defects in two out of the five LK genes (i.e., *B4GALT7* and *B3GALT6*) were grouped as spEDS together with those harboring *SLC39A13* mutations, in consideration of the reliable clinical overlap [1], whereas the 22 Larsen of Reunion Island syndrome patients were not included. We have previously suggested that, though some phenotypic variations between Larsen of Reunion Island syndrome and *B4GALT7*-spEDS exist, these conditions should not be considered as different entities [20]. Patients with *B3GAT3*, *XYLT1*, and *XYLT2* mutations were also not classified as spEDS, even though there is a common pathogenic mechanism and numerous shared clinical features. The patient reported in this study corroborates the awareness that LKs are a phenotypic continuum bridging EDS and skeletal disorders. Indeed, the patient was referred to our clinic with a well-founded suspicion of EDS, since she respected the EDS triad and fulfilled the minimal suggestive criteria of spEDS (Supplementary Table S2). In consideration of some peculiar signs not previously associated with spEDS, we performed ES, which revealed compound heterozygosity for two likely pathogenic *B3GAT3* variants.

B3GAT3 is involved in a spectrum of connective tissue and skeletal disorders. Table 2 summarizes the clinical features of the 26 patients from 13 different families, 11 of which were consanguineous with *B3GAT3* mutations reported so far ([26,47–53], present study). Among the *B3GAT3*-related disorders, a LRS-like presentation similar to that of our patient was the most common, but more severe phenotypes resembling ABS, SGS, and GO have also been reported (Table 2).

Historically, Baasanjav et al. [47] first described five patients with short stature, radioulnar synostosis, brachycephaly, and cardiac abnormalities. The authors suggested naming this condition as Larsen-like syndrome, *B3GAT3* type. All patients carried the homozygous c.830G>A, p.(Arg277Gln) missense mutation in the acceptor substrate binding subdomain of the protein. Von Oettingen et al. [48] described a 5-year-old boy with a similar phenotype and the same missense variant. Novel findings were developmental delay, refractive errors, pectus carinatum, atlantoaxial and atlanto-occipital instability, and excessive skin wrinkling. Budde et al. [49] reported eight patients from a large consanguineous family with a LRS-like phenotype without cardiac involvement carrying the c.419C>T, p.(Pro140Leu) pathogenic variant in the donor substrate binding subdomain of the protein. Job et al. [51] described the first compound heterozygous patient, a 6-year-old boy, who presented in addition to the typical LRS-like features, hypotonia, hyperextensible skin, and generalized osteoporosis with multiple fractures. The identified mutations were a null allele (c.1A>G, p.Met1?) and the c.671T>A, p.(Leu224Gln) missense substitution in the acceptor substrate binding subdomain of the protein, respectively. Very recently, Colman et al. [53] characterized a 13-year-old girl with a rather mild phenotype with short stature, short neck, craniofacial dysmorphism, joint hypermobility with dislocation, foot deformities, and mild osteopenia without fractures. The patient was homozygous for the c.416C>T, p.(Thr139Met) missense variant in the donor substrate binding subdomain.

Alazami et al. [26] reported a GO–like syndrome in a patient carrying the homozygous c.245C>T, p.(Pro82Leu) missense variant in the donor substrate binding subdomain of the protein. A detailed clinical description is lacking, but short stature, spondyloepimetaphyseal dysplasia, cutis laxa, generalized osteoporosis with fractures, and several bony chondromas were reported.

Finally, Jones [50], Yauy [52], and Colman et al. [53] described the most severely affected patients reported hitherto, all harboring the same c.667G>A, p.(Gly223Ser) missense mutation in the acceptor substrate binding subdomain of the protein. Jones et al. [50] reported a 12-month-old boy with short stature, hypotonia, global developmental delay, radioulnar synostosis, metaphyseal flaring, craniofacial dysmorphism, sandal gap, bilateral club feet, septal defects, and multiple fractures. Novel findings included blue sclerae, bilateral glaucoma, diaphragmatic hernia, small chest, arachnodactyly, lymphedema, hearing loss, and perinatal cerebral infarction. Colman et al. [53] characterized an infant, who died at the age of 2.5 months and showed cutis laxa, contractures of large and small joints, finger and foot anomalies, short neck, severe asymmetric thorax, dolichocephaly and other facial dysmorphic features, multiple long bones fractures, and bilateral corneal clouding. Likewise, Yauy et al. [52] reported six patients, who all died before one year of age, with craniosynostosis, midface hypoplasia, radioulnar synostosis, multiple neonatal fractures, dislocated joints, joint contractures, and cardiovascular abnormalities. The authors suggested that a B3GAT3-related disorder with craniosynostosis and bone fragility should be considered as a differential diagnosis in the prenatal period for ABS and in the postnatal period for SGS. Indeed, ABS is suspected during pregnancy if ultrasonography shows craniofacial deformities due to craniosynostosis (a hallmark of ABS), midface hypoplasia, bilateral radiohumeral, or radioulnar synostosis [52,75]. In the postnatal period, SGS can also be suspected, based on the association of craniosynostosis and arachnodactyly. Indeed, SGS is characterized by craniosynostosis and distinctive craniofacial features, skeletal anomalies (arachnodactyly, dolichostenomelia, camptodactyly, pes planus, pectus excavatum or carinatum, kyphoscoliosis), cardiovascular anomalies, intellectual deficiency, and brain anomalies [76].

Interestingly, our patient received a diagnosis of SGS in infancy, though there was the absence of several cardinal features, above all intellectual disability and brain abnormalities. Overall, the clinical presentation of our patient is consistent with a moderate-severe LRS-like phenotype, characterized by short stature, hypotonia, joint hypermobility with recurrent dislocations, radioulnar synostosis, peculiar fingers, foot deformities, midface hypoplasia, short neck, and cardiovascular abnormalities. Of note, although severe osteopenia was present, our patient never experienced fractures. Other features observed in our patient such as skin hyperextensibility, pectus carinatum, atlantoaxial and atlanto-occipital instability, severe kyphoscoliosis, and blue sclerae have been rarely reported in other individuals with B3GAT3 mutations, while easy bruising, atrophic scarring, and anterior ectopic anus are novel findings (Table 2). Since most of the initial reports focused on a particular aspect of the phenotype, mostly skeletal features, it remains possible that cutaneous and other systemic features that were present have not been described.

The heterogeneous LK phenotypes, ranging from mild to severe and even lethal presentations, seem to be related to specific B3GAT3 mutations, though there has been a limited number of patients and pathogenic variants described so far (Table 2, Figure 4) [52,53].

In particular, the more severe phenotypes appear to harbor mutations located within the acceptor substrate binding subdomain of the catalytic domain of the protein, whereas more mildly affected patients seem to have mutations in the donor substrate binding subdomain [52,53,77]. The intermediated phenotype of our patient, compound heterozygous for missense variants in both these subdomains (p.(Arg161Trp) in the donor and p.(Arg297Trp) in the acceptor substrate binding subdomain, respectively), corroborates this preliminary genotype–phenotype correlation.

Concerning the p.(Arg161Trp) missense variant, Fondeur-Gelinotte et al. [78] previously demonstrated the importance of this residue in the uridine-diphosphate (UDP)-GlcA donor binding site by site-directed mutagenesis and biochemical analyses. Concerning the consequence of the p.(Arg297Trp) missense variant in the acceptor substrate binding subdomain, in silico modeling using the Hope software [79] suggested that its 3D structure should be severely compromised. Indeed, the wild type residue forms either a hydrogen bond with a glutamic acid at position 295 or a salt bridge with glutamic acid at positions 206 and 295. The differences in size and hydrophobicity between the wild type and mutant residue are predicted to perturb both the hydrogen bond and the ionic interaction, thus likely abolishing the GlcAT-I activity.

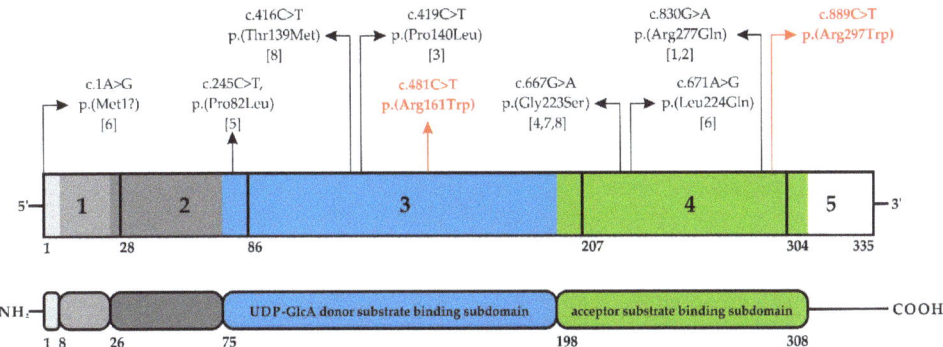

Figure 4. Schematic illustration of the *B3GAT3* structure and glucuronosyltransferase I protein domains. The different *B3GAT3* pathogenic variants found in all patients reported thus far [1-8, present study] are shown over the diagram. The variants identified in this study are in red. Variants are annotated according to HGVS nomenclature (reference sequences: NM_012200.3, NP_036332.2). (**Below**) The protein structure of GlcAT-I is reported with its multiple domains. In different grey scales from the N-terminus (NH_2) to the C-terminus (COOH), the small cytoplasmic domain (residues 1–7), the transmembrane domain (residues 8–25), and the proline-rich stem region (residues 26–74); in blue and green the catalytic domain consisting of the UDP-GlcUA (uridine diphosphate—(β-D-) glucuronic acid) donor substrate binding subdomain (residues 75–197, in blue), and the acceptor substrate binding subdomain (residues 198–308, in green), according to references [71,72,78].

The impression that LKs, even though likely not fully characterized yet, should be considered as a phenotypic continuum emerges from the comparison of all LK patients reported up to now (Table 3), which shows several overlapping clinical features, but also some differences. These data support the idea that the enzymes involved in the biosynthesis of the PG linker region may be part of a larger enzyme complex rather than functioning independently, and that the existing phenotypic disparity could be due to varying levels and spatiotemporal gene expression in different tissues [23,48].

Table 3. Clinical features of linkeropathies.

Genes	B3GAT3	B4GALT7	B3GALT6	SLC39A13	XYLT1	XYLT2
Number of patients	n = 26	n = 32	n = 46	n = 9	n = 28	n = 20
Skeletal						
Short stature	18/23	29/29	36/46	+	+	7/17
Joint hypermobility	10/19	+	37/46	+	13/14	2/5
Joint dislocation	9/25	+	37/46	+	14/14	na
Joint contractures (hands, elbow)	19/24	4/9	30/46	3/9	4/5	na
Low bone density/osteopenia	10/11	6/32	20/46	7/9	2/2	15/15
Multiple fractures	8/18	1/10	21/46	na	na	19/19
Kypho/scoliosis	8/24	7/32	32/46	1/7	10/12	12/13
Platyspondyly	1/20	-	13/36	+	5/9	18/18
Peculiar fingers [a]	25/25	1/30	13/36	7/7	19/20	9/12
Foot deformity [b]	22/25	9/10	24/46	8/8	9/13	12/12
Pectus excavatum/carinatum	3/19	1/1	2/10	na	10/12	4/4
Radioulnar synostosis	11/13	18/31	1/36	-	na	na

Table 3. Cont.

Genes	B3GAT3	B4GALT7	B3GALT6	SLC39A13	XYLT1	XYLT2
Metaphyseal flaring	3/5	4/8	23/46	4/6	4/4	0/1
Monkey-wrench femora	0/4	1/2	na	na	11/12	na
Iliac abnormalities	2/4	-	27/46	-	4/4	2/2
Radial head subluxation or dislocation	3/5	17/31	15/36	3/6	1/1	na
Bowing of limbs	5/19	7/32	13/46	8/8	5/5	na
Advance bone age/carpal ossification	0/10	1/32	5/16	na	13/14	-
Craniofacial						
Midface hypoplasia	24/25	na	8/10	0/6	1/1	na
Flat face	1/4	29/32	22/36	1/1	18/18	na
Craniosynostosis	12/15	6/8	1/36	-	na	na
Frontal bossing	4/12	-	29/46	+	na	na
Wide forehead	2/5	29/32	na	na	1/1	1/1
Blue sclerae	8/17	6/10	30/46	+	5/6	2/6
Proptosis or prominent eyes	12/14	28/30	20/46	+	6/7	na
Wide-spaced eyes	3/7	28/32	-	1/1	1/1	4/4
Low-set ears	2/8	7/10	20/46	na	na	4/4
Depressed nasal bridge	13/20	-	10/46	1/1	21/21	4/4
Small mouth/microstomia	10/19	28/30	-	-	2/2	na
Long upper lip/long philtrum	4/4	-	15/36	na	6/6	1/1
Cleft palate/bifid uvula	4/4	4/31	6/46	3/8	7/16	na
Micrognathia	9/17	3/32	14/46	-	3/3	na
Short and/or webbed neck	13/19	-	-	8/8	5/5	4/4
Abnormal dentition	1/2	6/10	17/46	8/9	3/5	2/14
Ocular						
Refractive errors/hypermetropia	3/5	12/29	1/10	1/9	1/5	14/14
Clouded cornea	2/4	1/30	1/46	-	0/5	na
Cataract	0/26	0/30	na	na	0/5	18/20
Retinal detachment	0/26	Na	na	na	0/5	9/16
Cutaneous						
Hyperextensible, soft, doughy, thin, translucent skin; cutis laxa	4/15	30/32	29/46	+	2/2	1/5
Atrophic scarring	1/5	4/32	10/46	5/7	na	na
Other						
Cardiovascular abnormalities	13/20	-	2/36	-	1/9	7/19
Muscle hypotonia	6/7	10/32	21/46	3/3	2/2	10/10
Delayed motor development	5/5	8/10	12/46	3/6	10/13	6/13
Delayed cognitive development	1/18	19/32	14/46	0/6	17/19	9/17
Deafness	1/3	2/32	2/46	-	2/8	12/20

Note: na, not available; +, present in all patients; -, absent in all patients; [a]: Peculiar fingers including long, slender, tapered, broad, thin, arachnodactyly; [b]: Foot deformity including pes planus, hallux valgus, club feet, sandal gap; B3GAT3: ([26,47–53], present patient); B4GALT7: [15–21,32]; B3GALT6: [22–28], SLC39A13: [29–31], XYLT1: [33–40]; XYLT2: [41–46].

The phenotypic features with a high rate of incidence shared among all LKs include short stature, joint laxity with dislocations, craniofacial dysmorphism (especially prominent forehead/eyes and blue sclerae), pectus abnormalities, peculiar fingers, foot deformities, and to a variable degree hypotonia and developmental delay. Delayed cognitive development is instead more frequently observed in *XYLT1*-related DBQD2, followed by *XYLT2*-related SOS, and *B4GALT7*- and *B3GALT6*-spEDS (Table 3).

While joint hypermobility is a common trait of all LKs, joint contractures are more frequently observed in patients with *B3GAT3* and *B3GALT6* mutations. Low bone mineral density/osteopenia and radiographic abnormalities are also very common, whereas generalized osteoporosis and multiple fractures are more frequently encountered in *B3GALT6*-spEDS and in SOS. Likewise, among the numerous radiographic findings, severe progressive kyphoscoliosis is particularly frequent in *B3GALT6*-spEDS, SOS, and DBQD2. In addition to progressive (kypho)scoliosis, short long bones with a monkey wrench or Swedish key appearance of the femora, and advanced carpal and tarsal ossification are also more frequently observed in DBQD2 (Table 3). Of note, the presence or absence of specific hand anomalies, comprising an extra-ossification center distal to the second metacarpal, delta phalanx, or bifid distal thumb phalanx, together with dislocations of the interphalangeal joints are considered radiographic hallmarks distinguishing DBQD1 (resulting from pathogenic variants in *CANT1*) from DBQD2 [34]. Platyspondyly, a hallmark of "spondylo"-dysplasia, has never been recognized in *B4GALT7*-spEDS and *B3GAT3*-related disorders, except in the patient reported by Alazami et al. [26], whereas it is frequently reported either in *B3GALT6*- and *SLC39A13*-spEDS or in SOS, whereas in DBQD2 it was recognized in five out of the nine investigated patients. Radioulnar synostosis seems more specific for *B3GAT3*- and *B4GALT7*-related disorders, the latter including either *B4GALT7*-spEDS (8/10) or Larsen of Reunion Island syndrome (10/21). Radiographic anomalies that are more common in *B3GALT6*-spEDS when compared to other LKs comprise metaphyseal flaring, iliac abnormalities, radial head subluxation/dislocation, which is also frequent in *B4GALT7*-spEDS, and bowing of limbs, which is shared with *SLC39A13*-spEDS (Table 3).

Among the plethora of craniofacial dysmorphism, midface hypoplasia, craniosynostosis, and short/webbed neck are particularly frequent in *B3GAT3*-related disorders. Wide forehead, hypertelorism, and microstomia are more common in *B4GALT7*-spEDS, whereas frontal bossing, micrognathia, and abnormal dentition characterize *B3GALT6*- and *SLC39A13*-spEDS. Flat face is shared among all spEDS subtypes and DBQD2 (Table 3).

Ocular involvement might facilitate the differential between SOS and other LKs. In particular, the presence of cataract, retinal detachment has been described thus far only in patients with *XYLT2* mutations. Refractive errors/hypermetropia is also frequent in SOS, but has also been found in *B4GALT7*-spEDS and *B3GAT3*-disorders. Apart from eye involvement, deafness is also more frequent in SOS when compared to the other LKs (Table 3).

Marked cutaneous anomalies including hyperextensible, soft, doughy, thin and translucent skin, and atrophic scarring seem to distinguish spEDS patients, party justifying their inclusion within the EDS spectrum. On the other hand, in all the other LKs, several patients (including ours) with skin hyperextensibility and other abnormalities (including even cutis laxa-like features) have been published (Table 3). As above-mentioned for *B3GAT3*, it remains possible that in more than a few patients, the cutaneous involvement was not investigated. Therefore, further reports are needed to define the real incidence of skin abnormalities in LKs.

Concerning cardiovascular involvement, anomalies such as septal defects, aortic valve dysplasia, aortic root and ascending aorta dilatation, and mitral valve prolapse, are more recurrent in *B3GAT3*-related disorders as well as in SOS (Table 3).

5. Conclusions

In summary, our findings expand the *B3GAT3* allelic repertoire, corroborate the emergent genotype-phenotype correlations, and confirm the extended phenotypic range of *B3GAT3* mutations overlapping skeletal dysplasia and soft HCTDs including EDS. Furthermore, we provided a comprehensive overview of the phenotypic features of B3GAT3-related disorders and of all the LKs, thus offering future nosologic perspectives for either EDS or skeletal dysplasias. Given the convergent pathogenic pathway and the important clinical overlap not only among B3GAT3-associated diseases and spEDS (a term that, in our opinion, is unfortunate, since platyspondyly is not present in all subtypes), but also with DBDQ2 and SOS, this group of HCTDs should be considered as a phenotypic continuum and not as distinct entities.. In the genetic era of the classification of disorders, we propose that the term LK is preceded by the specific causal gene. Further reports on additional patients are awaited as well as functional studies on the spatiotemporal expression of the different glycosyltransferases to unravel the molecular mechanisms involved in the pathophysiology of LKs needed to identify potential therapeutic options.

Supplementary Materials: The following are available online at http://www.mdpi.com/2073-4425/10/9/631/s1, Table S1. Primers. Table S2. Clinical features of the present patient and comparison with spEDS and B4GALT7-LRS. Table S3. Summary of exome sequencing results. Figure S1. Coverage distribution. Table S4. Variants with perfect segregation among trio members.

Author Contributions: M.C. and M.R. conceived the study. M.V. and M.C. performed the clinical evaluation of the patient, genetic counselling, and follow-up; M.R., V.C., and E.G. carried out the molecular analyses; M.R., V.C., and N.C. researched the literature; M.R., V.C., and N.C. prepared the manuscript; M.C. edited and coordinated the manuscript. All authors discussed, read, and approved the manuscript.

Funding: No funding was received for this project.

Acknowledgments: The authors wish to thank the patient for her cooperation during the diagnostic process and the Fazzo Cusan family for the generous support.

Conflicts of Interest: All authors declare that there are no conflicts of interest concerning this work.

References

1. Malfait, F.; Francomano, C.; Byers, P.; Belmont, J.; Berglund, B.; Black, J.; Bloom, L.; Bowen, J.M.; Brady, A.F.; Burrows, N.P.; et al. The 2017 international classification of the Ehlers-Danlos syndromes. *Am. J. Med. Genet. C* **2017**, *175*, 8–26. [CrossRef] [PubMed]
2. Blackburn, P.R.; Xu, Z.; Tumelty, K.E.; Zhao, R.W.; Monis, W.J.; Harris, K.G.; Gass, J.M.; Cousin, M.A.; Boczek, N.J.; Mitkov, M.V.; et al. Bi-allelic alterations in *AEBP1* lead to defective collagen assembly and connective tissue structure resulting in a variant of Ehlers-Danlos syndrome. *Am. J. Hum. Genet.* **2018**, *102*, 696–705. [CrossRef] [PubMed]
3. Hebebrand, M.; Vasileiou, G.; Krumbiegel, M.; Kraus, C.; Uebe, S.; Ekici, A.B.; Thiel, C.T.; Reis, A.; Popp, B. A biallelic truncating *AEBP1* variant causes connective tissue disorder in two siblings. *Am. J. Med. Genet. A* **2019**, *179*, 50–56. [CrossRef] [PubMed]
4. Syx, D.; De Wandele, I.; Symoens, S.; De Rycke, R.; Hougrand, O.; Voermans, N.; De Paepe, A.; Malfait, F. Bi-allelic *AEBP1* mutations in two patients with Ehlers-Danlos syndrome. *Hum. Mol. Genet.* **2019**, *28*, 1853–1864. [CrossRef] [PubMed]
5. Ritelli, M.; Cinquina, V.; Venturini, M.; Pezzaioli, L.; Formenti, A.M.; Chiarelli, N.; Colombi, M. Expanding the Clinical and Mutational Spectrum of Recessive *AEBP1*-Related Classical-Like Ehlers-Danlos Syndrome. *Genes* **2019**, *10*, 135. [CrossRef] [PubMed]
6. Castori, M.; Colombi, M. Generalized joint hypermobility, joint hypermobility syndrome and Ehlers-Danlos syndrome, hypermobility type. *Am. J. Med. Genet. C Semin. Med. Genet.* **2015**, *169*, 1–5. [CrossRef] [PubMed]
7. Colombi, M.; Dordoni, C.; Chiarelli, N.; Ritelli, M. Differential diagnosis and diagnostic flow chart of joint hypermobility syndrome/Ehlers-Danlos syndrome hypermobility type compared to other heritable connective tissue disorders. *Am. J. Med. Genet. Part C* **2015**, *169*, 6–22. [CrossRef] [PubMed]
8. Sugahara, K.; Kitagawa, H. Recent advances in the study of the biosynthesis and functions of sulfated glycosaminoglycans. *Curr. Opin. Struct. Biol.* **2000**, *10*, 518–527. [CrossRef]
9. Bülow, H.E.; Hober, O. The molecular diversity of glycosaminoglycans shapes animal development. *Annu. Rev. Cell Dev. Biol.* **2006**, *22*, 375–407. [CrossRef]
10. Esko, J.D.; Kimata, K.; Lindahl, U. Proteoglycans and Sulfated Glycosaminoglycans. In *Essential of Glycobiology*, 2nd ed.; Varki, A., Cummings, R.D., Esko, J.D., Freeze, H.H., Stanley, P., Bertozzi, C.R.,

Hart, G.W., Etzler, M.E., Eds.; Cold Spring Harbor Laboratory Press: Cold Spring Harbor, NY, USA, 2009; Chapter 16.
11. Mizumoto, S.; Ikegawa, S.; Sugahara, K. Human genetic disorders caused by mutations in genes encoding biosynthetic enzymes for sulfated glycosaminoglycans. *J. Biol. Chem.* **2013**, *19*, 10953–10961. [CrossRef]
12. Taylan, F.; Mäkitie, O. Abnormal Proteoglycan Synthesis Due to Gene Defects Causes Skeletal Diseases with Overlapping Phenotypes. *Horm. Metab. Res.* **2016**, *48*, 745–754. [CrossRef] [PubMed]
13. Mizumoto, S.; Kosho, T.; Yamada, S.; Sugahara, K. Pathophysiological Significance of Dermatan Sulfate Proteoglycans Revealed by Human Genetic Disorders. *Pharmaceuticals* **2017**, *10*, 34. [CrossRef] [PubMed]
14. Brady, A.F.; Demirdas, S.; Fournel-Gigleux, S.; Ghali, N.; Giunta, C.; Kapferer-Seebacher, I.; Kosho, T.; Mendoza-Londono, R.; Pope, M.F.; Rohrbach, M.; et al. The Ehlers-Danlos syndromes, rare types. *Am. J. Med. Genet. C* **2017**, *175*, 70–115. [CrossRef] [PubMed]
15. Kresse, H.; Rosthøj, S.; Quentin, E.; Hollmann, J.; Glössl, J.; Okada, S.; Tønnesen, T. Glycosaminoglycan-free small proteoglycan core protein is secreted by fibroblasts from a patient with a syndrome resembling progeroid. *Am. J. Hum. Genet.* **1987**, *41*, 436–453. [PubMed]
16. Faiyaz-Ul-Haque, M.; Zaidi, S.H.; Al-Ali, M.; Al-Mureikhi, M.S.; Kennedy, S.; Al-Thani, G.; Tsui, L.C.; Teebi, A.S. A novel missense mutation in the galactosyltransferase-I (*B4GALT7*) gene in a family exhibiting facioskeletal anomalies and Ehlers-Danlos syndrome resembling the progeroid type. *Am. J. Med. Genet. A* **2004**, *128*, 39–45. [CrossRef] [PubMed]
17. Guo, M.H.; Stoler, J.; Lui, J.; Nilsson, O.; Bianchi, D.W.; Hirschhorn, J.N.; Dauber, A. Redefining the progeroid form of Ehlers-Danlos syndrome: Report of the fourth patient with *B4GALT7* deficiency and review of the literature. *Am. J. Med. Genet. A* **2013**, *161*, 2519–2527. [CrossRef]
18. Salter, C.G.; Davies, J.H.; Moon, R.J.; Fairhurst, J.; Bunyan, D.; DDD Study; Foulds, N. Further defining the phenotypic spectrum of *B4GALT7* mutations. *Am. J. Med. Genet. A* **2016**, *170*, 1556–1563. [CrossRef] [PubMed]
19. Arunrut, T.; Sabbadini, M.; Jain, M.; Machol, K.; Scaglia, F.; Slavotinek, A. Corneal clouding, cataract, and colobomas with a novel missense mutation in B4GALT7-a review of eye anomalies in the linkeropathy syndromes. *Am. J. Med. Genet. A* **2016**, *170*, 2711–2718. [CrossRef] [PubMed]
20. Ritelli, M.; Dordoni, C.; Cinquina, V.; Venturini, M.; Calzavara-Pinton, P.; Colombi, M. Expanding the clinical and mutational spectrum of *B4GALT7*-spondylodysplastic Ehlers-Danlos syndrome. *Orphanet J. Rare Dis.* **2017**, *7*, 153. [CrossRef] [PubMed]
21. Sandler-Wilson, C.; Wambach, J.A.; Marshall, B.A.; Wegner, D.J.; McAlister, W.; Cole, F.S.; Shinawi, M. Phenotype and response to growth hormone therapy in siblings with *B4GALT7* deficiency. *Bone* **2019**, *124*, 14–21. [CrossRef] [PubMed]
22. Malfait, F.; Kariminejad, A.; Van Damme, T.; Gauche, C.; Syx, D.; Merhi-Soussi, F.; Gulberti, S.; Symoens, S.; Vanhauwaert, S.; Willaert, A.; et al. Defective initiation of glycosaminoglycan synthesis due to *B3GALT6* mutations causes a pleiotropic Ehlers-Danlos-syndrome-like connective tissue disorder. *Am. J. Hum. Genet.* **2013**, *92*, 935–945. [CrossRef] [PubMed]
23. Nakajima, M.; Mizumoto, S.; Miyake, N.; Kogawa, R.; Iida, A.; Ito, H.; Kitoh, H.; Hirayama, A.; Mitsubuchi, H.; Miyazaki, O.; et al. Mutations in *B3GALT6*, which encodes a glycosaminoglycan linker region enzyme, cause a spectrum of skeletal and connective tissue disorders. *Am. J. Hum. Genet.* **2013**, *92*, 927–934. [CrossRef] [PubMed]
24. Sellars, E.A.; Bosanko, K.A.; Lepard, T.; Garnica, A.; Schaefer, G.B. A newborn with complex skeletal abnormalities, joint contractures, and bilateral corneal clouding with sclerocornea. *Semin. Pediatr. Neurol.* **2014**, *21*, 84–87. [CrossRef] [PubMed]
25. Ritelli, M.; Chiarelli, N.; Zoppi, N.; Dordoni, C.; Quinzani, S.; Traversa, M.; Venturini, M.; Calzavara-Pinton, P.; Colombi, M. Insights in the etiopathology of galactosyltransferase II (GalT-II) deficiency from transcriptome-wide expression profiling of skin fibroblasts of two sisters with compound heterozygosity for two novel *B3GALT6* mutations. *Mol. Genet. Metab. Rep.* **2014**, *20*, 1–15. [CrossRef] [PubMed]
26. Alazami, A.M.; Al-Qattan, S.M.; Faqeih, E.; Alhashem, A.; Alshammari, M.; Alzahrani, F.; Al-Dosari, M.S.; Patel, N.; Alsagheir, A.; Binabbas, B.; et al. Expanding the clinical and genetic heterogeneity of hereditary disorders of connective tissue. *Hum. Genet.* **2016**, *135*, 525–540. [CrossRef] [PubMed]
27. Honey, E.M. Spondyloepimetaphyseal dysplasia with joint laxity (Beighton type): A unique South African disorder. *S. Afr. Med. J.* **2016**, *106*, S54–S56. [CrossRef] [PubMed]

28. Van Damme, T.; Pang, X.; Guillemyn, B.; Gulberti, S.; Syx, D.; De Rycke, R.; Kaye, O.; De Die-Smulders, C.E.M.; Pfundt, R.; Kariminejad, A.; et al. Biallelic B3GALT6 mutations cause spondylodysplastic Ehlers-Danlos syndrome. *Hum. Mol. Genet.* **2018**, *27*, 3475–3487. [CrossRef] [PubMed]
29. Giunta, C.; Elçioglu, N.H.; Albrecht, B.; Eich, G.; Chambaz, C.; Janecke, A.R.; Yeowell, H.; Weis, M.; Eyre, D.R.; Kraenzlin, M.; et al. Spondylocheiro dysplastic form of the Ehlers-Danlos syndrome—An autosomal-recessive entity caused by mutations in the zinc transporter gene *SLC39A13*. *Am. J. Hum. Genet.* **2008**, *82*, 1290–1305. [CrossRef]
30. Fukada, T.; Civic, N.; Furuichi, T.; Shimoda, S.; Mishima, K.; Higashiyama, H.; Idaira, Y.; Asada, Y.; Kitamura, H.; Yamasaki, S.T.; et al. The zinc transporter *SLC39A13*/ZIP13 is required for connective tissue development; its involvement in BMP/TGF-β signaling pathways. *PLoS ONE* **2008**, *3*, e3642. [CrossRef]
31. Dusanic, M.; Dekomien, G.; Lücke, T.; Vorgerd, M.; Weis, J.; Epplen, J.T.; Köhler, C.; Hoffjan, S. Novel Nonsense Mutation in *SLC39A13* Initially Presenting as Myopathy: Case Report and Review of the Literature. *Mol. Syndromol.* **2018**, *9*, 100–109. [CrossRef]
32. Cartault, F.; Munier, P.; Jacquemont, M.L.; Vellayoudom, J.; Doray, B.; Payet, C.; Randrianaivo, H.; Laville, J.M.; Munnich, A.; Cormier-Daire, V. Expanding the clinical spectrum of *B4GALT7* deficiency: Homozygous p.R270C mutation with founder effect causes Larsen of Reunion Island syndrome. *Eur. J. Hum. Genet.* **2015**, *23*, 49–53. [CrossRef]
33. Schreml, J.; Durmaz, B.; Cogulu, O.; Keupp, K.; Beleggia, F.; Pohl, E.; Milz, E.; Coker, M.; Ucar, S.K.; Nürnberg, G.; et al. The missing "link": An autosomal recessive short stature syndrome caused by a hypofunctional *XYLT1* mutation. *Hum. Genet.* **2014**, *133*, 29–39. [CrossRef]
34. Bui, C.; Huber, C.; Tuysuz, B.; Alanay, Y.; Bole-Feysot, C.; Leroy, J.G.; Mortier, G.; Nitschke, P.; Munnich, A.; Cormier-Daire, V. *XYLT1* mutations in Desbuquois dysplasia type 2. *Am. J. Hum. Genet.* **2014**, *94*, 405–414. [CrossRef]
35. Van Koningsbruggen, S.; Knoester, H.; Bakx, R.; Mook, O.; Knegt, L.; Cobben, J.M. Complete and partial *XYLT1* deletion in a patient with neonatal short limb skeletal dysplasia. *Am. J. Med. Genet. A* **2016**, *170*, 510–514. [CrossRef]
36. Jamsheer, A.; Olech, E.M.; Kozłowski, K.; Niedziela, M.; Sowińska-Seidler, A.; Obara-Moszyńska, M.; Latos-Bieleńska, A.; Karczewski, M.; Zemojtel, T. Exome sequencing reveals two novel compound heterozygous *XYLT1* mutations in a Polish patient with Desbuquois dysplasia type 2 and growth hormone deficiency. *J. Hum. Genet.* **2016**, *61*, 577–583. [CrossRef]
37. Silveira, C.; Leal, G.F.; Cavalcanti, D.P. Desbuquois dysplasia type II in a patient with a homozygous mutation in *XYLT1* and new unusual findings. *Am. J. Med. Genet. A* **2016**, *170*, 3043–3047. [CrossRef]
38. Guo, L.; Elcioglu, N.H.; Iida, A.; Demirkol, Y.K.; Aras, S.; Matsumoto, N.; Nishimura, G.; Miyake, N.; Ikegawa, S. Novel and recurrent *XYLT1* mutations in two Turkish families with Desbuquois dysplasia, type 2. *J. Hum. Genet.* **2017**, *62*, 447–451. [CrossRef]
39. Al-Jezawi, N.K.; Ali, B.R.; Al-Gazali, L. Endoplasmic reticulum retention of xylosyltransferase 1 (*XYLT1*) mutants underlying Desbuquois dysplasia type II. *Am. J. Med. Genet. A* **2017**, *173*, 1773–1781. [CrossRef]
40. LaCroix, A.J.; Stabley, D.; Sahraoui, R.; Adam, M.P.; Mehaffey, M.; Kernan, K.; Myers, C.T.; Fagerstrom, C.; Anadiotis, G.; Akkari, Y.M.; et al. GGC Repeat Expansion and Exon 1 Methylation of *XYLT1* Is a Common Pathogenic Variant in Baratela-Scott Syndrome. *Am. J. Hum. Genet.* **2019**, *104*, 35–44. [CrossRef]
41. Munns, C.F.; Fahiminiya, S.; Poudel, N.; Munteanu, M.C.; Majewski, J.; Sillence, D.O.; Metcalf, J.P.; Biggin, A.; Glorieux, F.; Fassier, F.; et al. Homozygosity for frameshift mutations in *XYLT2* result in a spondylo-ocular syndrome with bone fragility, cataracts, and hearing defects. *Am. J. Hum. Genet.* **2015**, *96*, 971–978. [CrossRef]
42. Taylan, F.; Costantini, A.; Coles, N.; Pekkinen, M.; Héon, E.; Şıklar, Z.; Berberoğlu, M.; Kämpe, A.; Kıykım, E.; Grigelioniene, G.; et al. Spondyloocular Syndrome: Novel Mutations in *XYLT2* Gene and Expansion of the Phenotypic Spectrum. *J. Bone Miner. Res.* **2016**, *31*, 1577–1585. [CrossRef]
43. Taylan, F.; Yavaş Abalı, Z.; Jäntti, N.; Güneş, N.; Darendeliler, F.; Baş, F.; Poyrazoğlu, Ş.; Tamçelik, N.; Tüysüz, B.; Mäkitie, O. Two novel mutations in *XYLT2* cause spondyloocular syndrome. *Am. J. Med. Genet. A* **2017**, *173*, 3195–3200. [CrossRef]
44. Umair, M.; Eckstein, G.; Rudolph, G.; Strom, T.; Graf, E.; Hendig, D.; Hoover, J.; Alanay, J.; Meitinger, T.; Schmidt, H.; et al. Homozygous *XYLT2* variants as a cause of spondyloocular syndrome. *Clin. Genet.* **2018**, *93*, 913–918. [CrossRef]

45. Guleray, N.; Simsek Kiper, P.O.; Utine, G.E.; Boduroglu, K.; Alikasifoglu, M. Intrafamilial variability of *XYLT2*-related spondyloocular syndrome. *Eur. J. Med. Genet.* **2018**, *1769*, 30611–30616. [CrossRef]
46. Kausar, M.; Chew, E.G.Y.; Ullah, H.; Anees, M.; Khor, C.C.; Foo, J.N.; Makitie, O.; Siddiqi, S. A Novel Homozygous Frameshift Variant in *XYLT2* Causes Spondyloocular Syndrome in a Consanguineous Pakistani Family. *Front. Genet.* **2019**, *10*, 144. [CrossRef]
47. Baasanjav, S.; Al-Gazali, L.; Hashiguchi, T.; Mizumoto, S.; Fischer, B.; Horn, D.; Seelow, D.; Ali, B.R.; Aziz, S.A.; Langer, R.; et al. Faulty initiation of proteoglycan synthesis causes cardiac and joint defects. *Am. J. Hum. Genet.* **2011**, *15*, 15–27. [CrossRef]
48. Von Oettingen, J.E.; Tan, W.H.; Dauber, A. Skeletal dysplasia, global developmental delay, and multiple congenital anomalies in a 5-year-old boy-report of the second family with *B3GAT3* mutation and expansion of the phenotype. *Am. J. Med. Genet. A* **2014**, *164*, 1580–1586. [CrossRef]
49. Budde, B.S.; Mizumoto, S.; Kogawa, R.; Becker, C.; Altmüller, J.; Thiele, H.; Rüschendorf, F.; Toliat, M.R.; Kaleschke, G.; Hämmerle, J.M.; et al. Skeletal dysplasia in a consanguineous clan from the island of Nias/Indonesia is caused by a novel mutation in *B3GAT3*. *Hum. Genet.* **2015**, *134*, 691–704. [CrossRef]
50. Jones, K.L.; Schwarze, U.; Adam, M.P.; Byers, P.H.; Mefford, H.C. A homozygous *B3GAT3* mutation causes a severe syndrome with multiple fractures, expanding the phenotype of linkeropathy syndromes. *Am. J. Med. Genet. A* **2015**, *167*, 2691–2696. [CrossRef]
51. Job, F.; Mizumoto, S.; Smith, L.; Couser, N.; Brazil, A.; Saal, H.; Patterson, M.; Gibson, M.I.; Soden, S.; Miller, N.; et al. Functional validation of novel compound heterozygous variants in *B3GAT3* resulting in severe osteopenia and fractures: Expanding the disease phenotype. *BMC Med. Genet.* **2016**, *17*, 86. [CrossRef]
52. Yauy, K.; Tran Mau-Them, F.; Willems, M.; Coubes, C.; Blanchet, P.; Herlin, C.; Taleb Arrada, I.; Sanchez, E.; Faure, J.M.; Le Gac, M.P.; et al. B3GAT3-related disorder with craniosynostosis and bone fragility due to a unique mutation. *Genet. Med.* **2018**, *20*, 269–274. [CrossRef]
53. Colman, M.; Van Damme, T.; Steichen-Gersdorf, E.; Laccone, F.; Nampoothiri, S.; Syx, D.; Guillemyn, B.; Symoens, S.; Malfait, F. The clinical and mutational spectrum of B3GAT3 linkeropathy: Two case reports and literature review. *Orphanet. J. Rare Dis.* **2019**, *14*, 138. [CrossRef]
54. San Lucas, F.A.; Wang, G.; Scheet, P.; Peng, B. Integrated annotation and analysis of genetic variants from next-generation sequencing studies with variant tools. *Bioinformatics* **2012**, *1*, 421–422. [CrossRef]
55. Ravasio, V.; Ritelli, M.; Legati, A.; Giacopuzzi, E. GARFIELD-NGS: Genomic vARiants FIltering by dEep Learning moDels in NGS. *Bioinformatics* **2018**, *1*, 3038–3040. [CrossRef]
56. Wang, K.; Li, M.; Hakonarson, H. Annovar: Functional annotation of genetic variants from high-throughput sequencing data. *Nucleic Acids. Res.* **2010**, *38*, 164. [CrossRef]
57. Quang, D.; Chen, Y.; Xie, X. Dann: A deep learning approach for annotating the pathogenicity of genetic variants. *Bioinformatics* **2015**, *31*, 761–763. [CrossRef]
58. Jagadeesh, K.A.; Wenger, A.M.; Berger, M.J.; Guturu, H.; Stenson, P.D.; Cooper, D.N.; Bernstein, J.A.; Bejerano, G. M-CAP eliminates a majority of variants of uncertain significance in clinical exomes at high sensitivity. *Nat. Genet.* **2016**, *48*, 1581–1586. [CrossRef]
59. Petrovski, S.; Wang, Q.; Heinzen, E.L.; Allen, A.S.; Goldstein, D.B. Genic intolerance to functional variation and the interpretation of personal genomes. *PLoS Genet.* **2013**, *9*, e1003709. [CrossRef]
60. Itan, Y.; Shang, L.; Boisson, B.; Patin, E.; Bolze, A.; Moncada-Vélez, M.; Scott, E.; Ciancanelli, M.J.; Lafaille, F.G.; Markle, J.G.; et al. The human gene damage index as a gene-level approach to prioritizing exome variants. *Proc. Natl. Acad. Sci. USA* **2015**, *112*, 13615–13620. [CrossRef]
61. Sim, N.L.; Kumar, P.; Hu, J.; Henikoff, S.; Schneider, G.; Ng, P.C. SIFT web server: Predicting effects of amino acid substitutions on proteins. *Nucleic Acids Res.* **2012**, *40*, W452–W457. [CrossRef]
62. Schwarz, J.M.; Cooper, D.N.; Schuelke, M.; Seelow, D. MutationTaster2: Mutation prediction for the deep-sequencing age. *Nat. Methods* **2014**, *11*, 361–362. [CrossRef]
63. Rentzsch, P.; Witten, D.; Cooper, G.M.; Shendure, J.; Kircher, M. CADD: Predicting the deleteriousness of variants throughout the human genome. *Nucleic Acids Res.* **2018**. [CrossRef]
64. Choi, Y.; Chan, A.P. PROVEAN web server: A tool to predict the functional effect of amino acid substitutions and indels. *Bioinformatics* **2015**, *31*, 2745–2747. [CrossRef]
65. Eugene, V.; Davydov, E.V.; Goode, D.L.; Sirota, M.; Cooper, G.M.; Sidow, A.; Batzoglou, S. Identifying a High Fraction of the Human Genome to be under Selective Constraint Using GERP++. *PLoS Comput. Biol.* **2010**, *6*, e1001025. [CrossRef]

66. Salgado, D.; Desvignes, J.P.; Rai, G.; Blanchard, A.; Miltgen, M.; Pinard, A.; Lévy, N.; Collod-Béroud, G.; Béroud, C. UMD-Predictor: A High-Throughput Sequencing Compliant System for Pathogenicity Prediction of any Human cDNA Substitution. *Hum. Mutat.* **2016**, *35*, 439–446. [CrossRef]
67. Chun, S.; Fay, J.C. Identification of deleterious mutations within three human genomes. *Genom. Res.* **2009**, *19*, 1553–1561. [CrossRef]
68. Shihab, H.A.; Rogers, M.F.; Gough, J.; Mort, M.; Cooper, D.N.; Day, I.N.; Gaunt, T.R.; Campbell, C. An integrative approach to predicting the functional effects of non-coding and coding sequence variation. *Bioinformatics* **2015**, *31*, 1536–1543. [CrossRef]
69. Carter, H.; Douville, C.; Stenson, P.D.; Cooper, D.N.; Karchin, R. Identifying Mendelian disease genes with the Variant Effect Scoring Tool BMC. *Genomics* **2013**, *14*, S3. [CrossRef]
70. Arbiza, L.; Gronau, I.; Aksoy, B.A.; Hubisz, M.J.; Gulko, B.; Keinan, A.; Siepel, A. Genome-wide inference of natural selection on human transcription factor binding sites. *Nat. Genet.* **2013**, *45*, 723–729. [CrossRef]
71. Pedersen, L.C.; Tsuchida, K.; Kitagawa, H.; Sugahara, K.; Darden, T.A.; Negishi, M. Heparan/chondroitin sulfate biosynthesis. Structure and mechanism of human glucuronyltransferase I. *J. Biol. Chem.* **2000**, *277*, 34580–34585. [CrossRef]
72. Pedersen, L.C.; Darden, T.A.; Negishi, M. Crystal structure of β1,3-glucuronyltransferase I in complex with active donor substrate UDP-GlcUA. *J. Biol. Chem.* **2002**, *277*, 21869–21873. [CrossRef]
73. Li, Q.; Wang, K. InterVar: Clinical Interpretation of Genetic Variants by the 2015 ACMG-AMP Guidelines. *Am. J. Hum. Genet.* **2017**, *100*, 267–280. [CrossRef]
74. Richards, S.; Aziz, N.; Bale, S.; Bick, D.; Das, S.; Gastier-Foster, J.; Grody, W.W.; Hegde, M.; Lyon, E.; Spector, E.; et al. ACMG Laboratory Quality Assurance Committee. Standards and guidelines for the interpretation of sequence variants: A joint consensus recommendation of the American College of Medical Genetics and Genomics and the Association for Molecular Pathology. *Genet. Med.* **2015**, *17*, 405–424. [CrossRef]
75. Oldani, E.; Garel, C.; Bucourt, M.; Carbillon, L. Prenatal diagnosis of Antley-Bixler syndrome and POR deficiency. *Am. J. Case.* **2015**, *16*, 882–885. [CrossRef]
76. Greally, M.T. Shprintzen-Goldberg Syndrome. In *GeneReviews®*; Adam, M.P., Ardinger, H.H., Pagon, R.A., Wallace, S.E., Bean, L.J.H., Stephens, K., Amemiya, A., Eds.; University of Washington: Seattle, WA, USA, 2006; pp. 1993–2019.
77. Ouzzine, M.; Gulberti, S.; Levoin, N.; Netter, P.; Magdalou, J.; Fournel-Gigleux, S. The donor substrate specificity of the human β 1,3-glucuronosyltransferase I toward UDP-glucuronic acid is determined by two crucial histidine and arginine residues. *J. Biol. Chem.* **2002**, *227*, 25439–25445. [CrossRef]
78. Fondeur-Gelinotte, M.; Lattard, V.; Oriol, R.; Mollicone, R.; Jacquinet, J.C.; Mulliert, G.; Gulberti, S.; Netter, P.; Magdalou, J.; Ouzzine, M.; et al. Phylogenetic and mutational analyses reveal key residues for UDP-glucuronic acid binding and activity of β 1,3-glucuronosyltransferase I (GlcAT-I). *Protein Sci.* **2006**, *15*, 1667–1678. [CrossRef]
79. Venselaar, H.; Te Beek, T.A.; Kuipers, R.K.; Hekkelman, M.L.; Vriend, G. Protein structure analysis of mutations causing inheritable diseases. An e-Science approach with life scientist friendly interfaces. *BMC Bioinform.* **2010**, *11*, 548. [CrossRef]

© 2019 by the authors. Licensee MDPI, Basel, Switzerland. This article is an open access article distributed under the terms and conditions of the Creative Commons Attribution (CC BY) license (http://creativecommons.org/licenses/by/4.0/).

Article

The Connective Tissue Disorder Associated with Recessive Variants in the *SLC39A13* Zinc Transporter Gene (Spondylo-Dysplastic Ehlers–Danlos Syndrome Type 3): Insights from Four Novel Patients and Follow-Up on Two Original Cases

Camille Kumps [1], Belinda Campos-Xavier [1], Yvonne Hilhorst-Hofstee [2], Carlo Marcelis [3], Marius Kraenzlin [4], Nicole Fleischer [5], Sheila Unger [1] and Andrea Superti-Furga [1,*]

[1] Division of Genetic Medicine, Lausanne University Hospital (CHUV), 1011 Lausanne, Switzerland; camille.kumps@chuv.ch (C.K.); belinda.xavier@chuv.ch (B.C.-X.); sheila.unger@chuv.ch (S.U.)
[2] Department of Clinical Genetics, Leiden University Medical Centre, 2333 ZA Leiden, The Netherlands; y.hilhorst-hofstee@lumc.nl
[3] Department of Human Genetics, Radboud University Nijmegen Medical Center, 6525 GA Nijmegen, The Netherlands; carlo.marcelis@radboudumc.nl
[4] Clinic for Endocrinology, Diabetes & Metabolism, University Hospital Basel, 4031 Basel, Switzerland; marius.kraenzlin@unibas.ch
[5] FDNA Inc., Boston, MA 02111, USA; nicole@fdna.com
* Correspondence: asuperti@unil.ch

Received: 2 April 2020; Accepted: 10 April 2020; Published: 14 April 2020

Abstract: Recessive loss-of-function variants in *SLC39A13*, a putative zinc transporter gene, were first associated with a connective tissue disorder that is now called "Ehlers–Danlos syndrome, spondylodysplastic form type 3" (SCD-EDS, OMIM 612350) in 2008. Nine individuals have been described. We describe here four additional affected individuals from three consanguineous families and the follow up of two of the original cases. In our series, cardinal findings included thin and finely wrinkled skin of the hands and feet, characteristic facial features with downslanting palpebral fissures, mild hypertelorism, prominent eyes with a paucity of periorbital fat, blueish sclerae, microdontia, or oligodontia, and—in contrast to most types of Ehlers–Danlos syndrome—significant short stature of childhood onset. Mild radiographic changes were observed, among which platyspondyly is a useful diagnostic feature. Two of our patients developed severe keratoconus, and two suffered from cerebrovascular accidents in their twenties, suggesting that there may be a vascular component to this condition. All patients tested had a significantly reduced ratio of the two collagen-derived crosslink derivates, pyridinoline-to-deoxypyridinoline, in urine, suggesting that this simple test is diagnostically useful. Additionally, analysis of the facial features of affected individuals by DeepGestalt technology confirmed their specificity and may be sufficient to suggest the diagnosis directly. Given that the clinical presentation in childhood consists mainly of short stature and characteristic facial features, the differential diagnosis is not necessarily that of a connective tissue disorder and therefore, we propose that *SLC39A13* is included in gene panels designed to address dysmorphism and short stature. This approach may result in more efficient diagnosis.

Keywords: Ehlers–Danlos syndrome; SLC39A13; dysmorphology; short stature; connective tissue; DeepGestalt technology

1. Introduction

The Ehlers–Danlos syndrome (EDS) is a group of disorders that affect connective tissues in the skin, ligaments, joints, blood vessels, and other organs. Defects in connective tissue cause a wide range of clinical manifestations, from mildly loose joints to life-threatening conditions, such as arterial bleeds. From the original description as a disorder of hyperelastic skin and lax joints at the beginning of the 20th century [1], a first classification with four main types was proposed in 1970 [2]; molecular advances have allowed the recognition of many distinct disorders that, although different from the classic EDS types described by Beighton, have been given the moniker of "Ehlers–Danlos syndrome" as a reflection of the presence of connective tissue fragility. Thus, the recent version of EDS classification has been expanded to include a wide range of disorders, including skeletal dysplasia [3]. However, the clinical criteria remain relatively non-specific, and clinical diagnosis can be difficult.

A connective tissue disorder associated with recessive biallelic variants in *SLC39A13*, and a mouse knock-out model for the same gene, were described in 2008 by two separate but collaborating groups [4,5]. The features noted in these eight patients were postnatal-onset short stature, protuberant eyes with bluish sclerae and down-slanting palpebral fissures, thin and moderately hyperelastic skin with bruisability, and hands with finely wrinkled palms, tapering fingers, thenar atrophy, and moderate hypermobility of the small joints. Skeletal radiographs showed a moderate degree of platyspondyly with irregular vertebral end plates as well as minor epimetaphyseal changes in appendicular bones. A reduced molar ratio of pyridinoline-to-deoxypyridinoline in urine was observed, indicating reduced collagen lysyl hydroxylation. The latter finding is typically observed in collagen lysyl hydroxylase deficiency (EDS type VI-A), but in that condition, there is severe muscular hypotonia from birth, progressive kyphoscoliosis, and normal height (apart from kyphoscoliosis). Additionally, there were no pathogenic mutations in *PLOD1*; instead, the patients reported by Giunta et al. (2008) were homozygous for an in-frame 9-bp deletion in *SLC39A13*, c.483_491del (p.F162_164del), while those reported by Fukada et al. (2008) were homozygous for the *SLC39A13* variant c.221G>A (p.G74D). The novel condition was given the name of "spondylo-cheiro-dysplastic EDS" for the distinguishing features of the hand and the platyspondyly [4,5]. Later, the EDS nosology has used the name "spondylodysplastic EDS" for a group of three conditions, B3GALT6 deficiency (better known as spondyloepimetaphyseal dysplasia with joint laxity, Beighton type), B4GALT7 deficiency, and SLC39A13 deficiency (the original spondylo-cheiro-dysplastic type) [3]. Of note, these three conditions are also included in the 2019 revision of the skeletal dysplasia nosology [6]. SLC39A13 deficiency is rare and following the initial eight affected individuals, only a single case report has been published [7]. We present a follow up of two of the original patients [5] as well as clinical, radiological, and genetic findings on four new patients from three families.

2. Materials and Methods

All affected individuals and/or their legal representatives in the study gave their informed consent to the use of their clinical data, as well as for molecular studies in a diagnostic context (see below).

2.1. Molecular Analysis

Molecular studies for patients 1 and 2 were done in the Lausanne laboratory as described [5]. Molecular studies in patients 3 to 6 were done for diagnostic purposes with appropriate informed consent from patients and their guardians. The studies were done using routine diagnostic sequencing procedures in certified diagnostic laboratories: Analysis for patients 3, 5, and 6 was done in Lausanne, while the analysis of patient 4 was done in Nijmegen.

2.2. Analysis of Pyridinium Crosslink Products by HPLC

Spot urine samples were acid hydrolyzed and analyzed by reverse-phase HPLC as previously described [4].

2.3. Analysis of Facial Features

Anonymized frontal facial photographs of individuals with confirmed biallelic *SLC39A13* variants were used to capture the facial gestalt of SLC39A13 deficiency and to compare it to individuals who are clinically normal. All images were fully de-identified through the use of the DeepGestalt image analysis technology [8]. De-identified data of images from 8 patients (the six patients reported here, plus two of the patients described by Giunta et al. (2008) whose photographs had been submitted to us for clinical consultation) were uploaded to the Face2Gene Research app [9] and matched to controls by age, sex, and ethnicity to produce artificial composite images. Because of the small number of patients, the comparison was run twice [10] with two different sets of matched controls. The comparison and separation quality between the three groups was evaluated by measuring the area under the curve (AUC) of the receiver operating characteristic (ROC) curve. To estimate the statistical power of DeepGestalt in distinguishing affected individuals from controls, a cross-validation scheme was used, including a series of binary comparisons between all groups. For these binary comparisons, the data was split randomly multiple times into training sets and test sets. Each such set contained half of the samples from the group, and this random process was repeated 10 times [8,10]. The results of the binary comparisons were reported both numerically and graphically.

2.4. Clinical Reports

2.4.1. Individuals 1 and 2 (reported in part by Fukada et al., 2008 [5])

The parents, of Portuguese origin, were of average stature and clinically unremarkable. They were not knowingly related, although molecular workup showed that they shared a common haplotype harboring the *SLC39A13* pathogenic variant [5]. The affected children, a boy and a girl, were born at term from uncomplicated pregnancies and were of normal size and weight at birth but showed progressive short stature beginning in the second half of the first year of life. Among the clinical signs in early childhood were muscular hypotonia and soft skin, leading to the diagnostic suspicion of the Ehlers–Danlos syndrome (EDS). During childhood, the main clinical signs and features were thin, fragile, but not hyperelastic skin that bruised easily and was particularly thin on the hands and feet; varicose veins; moderate joint laxity; blueish or greyish sclerae; down-slanting palpebral fissures; and the absence of one or more teeth in permanent dentition. Both had astigmatism in childhood. Radiographic examination revealed moderate osteopenia, flattened or biconcave vertebral bodies with flaky irregularity of the endplates as well as mild dysplastic changes at the metaphyses of long bones and of the phalanges. At the time of the first report, they were 28 years old and 145 cm tall (individual 1, male); and 20 years old and 135 cm tall (individual 2, female). Their body proportions were normal, indicating that the platyspondyly was accompanied by shortening of the long bones. Their skin remained thin and fragile, and the subcutaneous fat tissue was sparse. Both individuals had marked venous varicosities on their legs and feet. Individual 1 suffered from a cerebral hemorrhage posteriorly to the left putamen at age 22 years, from which he recovered completely. He has successfully completed higher education.

Subsequent to the initial publication in 2008, individual 1 has had no further complications; he is professionally active, has married, and his wife has given birth to a healthy child. Individual 2, his younger sister, had obtained higher education degrees at age 22 and 25 years. At age 25 years, cerebral vascular imaging was obtained, and no abnormalities were observed. At age 26 years, two weeks after the interruption of hormonal contraception prescribed because of irregular cycles, she suffered from arterial thromboembolism that caused cerebral ischemia with right arm paresis and Broca's aphasia. Several days after initiation of aspirin therapy, she developed cerebral hemorrhage leading to complete right hemiparesis. Fortunately, she was able to partially recover over four years on physiotherapy but still has partial function of the right hand and myoclonus on the right arm. She received speech/language therapy; however, she still has some degree of dysphasia. Nevertheless, she was able to return to her previous work.

2.4.2. Individual 3

This is a girl born by caesarean section at 33 weeks of gestation from a dizygotic twin pregnancy. Both the patient and her twin sister were 2500 g at birth. The patient developed progressive short stature over her first years of life, while her twin sister grew normally. Motor and intellectual development were normal. The parents are first cousins of Turkish origin and measure 148 cm and 170 cm with normal proportions. An elder brother had normal growth and body proportions. During her teenage years, she developed severe keratoconus with no complications so far. She complains about back pain and her skin bruises easily, with hematomas and blood blisters at sites of friction. Radiographic examination of the spine at age 15 revealed platyspondyly of the thoracic and lumbar vertebrae with mild anterior beaking. On physical examination at age 23, she had short stature. There was moderate diffuse joint laxity most marked on the finger joints. The skin on her hands is thin and wrinkled and she has tapering fingers with narrow end phalanges. Facial features include downslanting palpebral fissures, a flat face, protuberant eyes with reduced periocular tissue, greyish sclerae, and a small mouth.

2.4.3. Individual 4

This boy was the fourth child from healthy parents of Turkish descent. The parents were second cousins and measured 165 cm (father) and 158 cm (mother). His three older sibs have normal growth. He was born after an uneventful term pregnancy with a birthweight of 3500g. He developed progressive short stature over his first years of life and was first seen by a pediatrician at age 8 1/2 years. Height was 113.4cm (−3.8 SDS) and weight was 23.5kg (−1.5 SDS). Psychomotor development was normal. He complained of knee and ankle pain after exercise. He had mild weakness of the hands, making it difficult to open bottles. He had no clear dysmorphic features, but his palpebral fissures were down-slanting and he had oligodontia (missing five teeth). He has had no eye complication so far. He had a short trunk and stands with lumbar hyperlordosis. He had loose skin on hands and feet, moderate hyperlaxity of fingers and toes, valgus deformity of the ankles, and bilateral pes planus. Radiological examination showed platyspondyly of the thoracic and lumbar spine with mild anterior beaking and mild metaphyseal irregularities especially of the distal ulna. He first received a clinical diagnosis of spondylometaphyseal dysplasia. Molecular analysis of *SHOX*, *FGFR3*, *TRPV4*, and *LTBP3* were negative, but further studies showed a biallelic missense change in *SLC39A13* (Table 1). Additional urine analysis showed a low ratio of pyridinoline-to-deoxypyridinoline (Figure 3).

Table 1. Overview of the molecular results.

	Pat. 1 *	Pat. 2 *	Pat. 3	Pat. 4	Pat. 5	Pat. 6
SLC39A13 (NM_001128225.3)	c.G221A/c.G221A	c.G221A/c.G221A	c.483_491delCTTCCTGGC/ c.483_491delCTTCCTGGC	c.793G>A/c.793G>A	c.1019delT/c.1019delT	c.1019delT/c.1019delT
Protein (NP_001121697.2)	p.Gly74Asp	p.Gly74Asp	p.Phe162Ala164del	p.Asp265Asn	p.Leu340ProfsTer23	p.Leu340ProfsTer23
Frequency in gnomAD **	absent	absent	extremely low, AF = 0.000009	absent	absent	absent
Predictors of pathogenicity on protein ***	pathogenic	pathogenic	pathogenic	pathogenic	pathogenic	pathogenic

Notes: * these two siblings have been reported by us before (Fukada et al., 2008); ** as assessed in March 2020 on the gnomAD database (https://gnomad.broadinstitute.org); *** as assessed in March 2020 on the VARSOME integrative website (https://varsome.com).

2.4.4. Individuals 5 and 6

The family originates from Syria and access to early medical records is impossible. The parents, who are first cousins, are clinically healthy and measure 150 cm in height (mother) and 160 cm (father). The proband (patient 6; Figure 1) is a 9-year-old boy referred for evaluation for small stature. We found that his elder sister is similarly affected (patient 5), while a younger sister is clinically normal. Individual 6 was born at term after an uneventful pregnancy, although IUGR was noted and the mother reports reduced fetal movements throughout the pregnancy. Birth data were not recorded, but the mother remembers his length to be reduced. At birth, the mother noted the he had no spontaneous movements, had edema on both feet, and increased palmar creases on his hands and feet. Throughout childhood, psychomotor development was slightly delayed, but intelligence and behavior at age 9 years were normal. At this age, his growth parameters were height 111 cm (−3.9 SD), weight 19 kg (−3.3 SD), and head circumference 51.5 cm (−0.78 SD). He had down-slanting palpebral fissures, protruding eyes, hypertelorism, and a large forehead. An ophthalmologic evaluation revealed bilateral myopia and marked keratoconus. Hearing was normal. Four teeth of the permanent dentition were found to be missing. His elbow joints appeared prominent, but he had no other signs of joint deformity or scoliosis. He had hyperlaxity of small joints and complained occasionally of joint pain particularly in his knees. The skin was thin and wrinkled, particularly on the palmar side of his hands and feet, and the thenar muscles were underdeveloped. Radiographic studies showed small epiphyses and posterior subluxation of the radii and mild platyspondyly. Routine blood tests were unremarkable. Analysis of *SHOX* done prior to referral was negative.

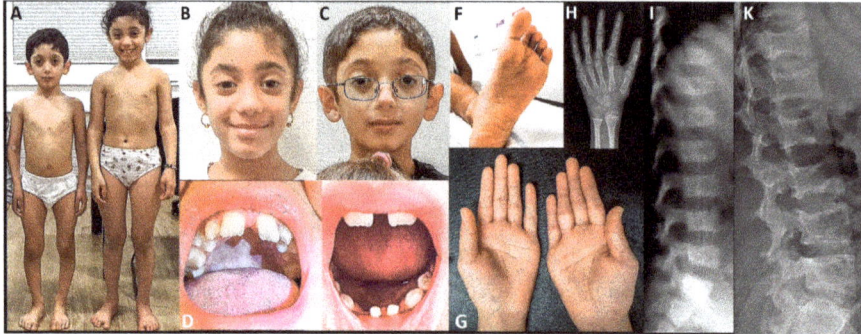

Figure 1. Clinical and radiographic findings. Individuals 5 (panels **A** and **B**) and 6 (panels **A** and **C**) at the age of eight and ten years. Panels **D** and **E**: missing upper incisives, and missing lower incisives with persistence of primary teeth in individuals 6 at age 9 years. Panel **F**: foot of individual 6 showing finely wrinkled skin and abnormally deep furrows. Panel **G**: Palmar aspects of the hands of individual 5 showing thin skin with increased number of fine wrinkles and mild thenar and hypothenar atrophy. Panel **H** shows the hand radiograph of individual 5 at age ten years. There is mild diaphyseal overconstriction of radius and ulna, metacarpals, and phalanges; bone maturation is roughly appropriate; overall, the changes are mild and non-diagnostic. Panels **I** and **K**: lateral spine radiographs of individual 3. At age 6 years, there is moderate platyspondyly. At age 15, the end plates have a marked concave conformation.

Individual 5 is the older sister of individual 6. According to the birth date declared at immigration, she would be 10 years old. However, we doubt the accuracy of this date and suspect that she is older than the stated age. Birth parameters have not been recorded. Her growth parameters were as follows: height 131.5 cm (−1 SD for age 10), weight 29 kg (−0.81 SD for age 10), and head circumference 51.5 cm (−0.55 SD for age 10). Shortly after she was referred, she had menarche and her growth has slowed down significantly. She had protruding eyes and a large forehead, but her facial appearance was less conspicuous than that of her brother or other affected individuals. She had myopia, and four incisor teeth were missing. Intelligence and hearing were normal. She had long tapering fingers and her interphalangeal joints were hyperextensible. The skin was thin and wrinkled, particularly on the palmar side of her hands and feet.

3. Results

Figure 1 shows some clinically relevant clinical and radiographic features in SLC39A13-deficient individuals.

Table 1 shows the genetic variants identified in *SLC39A13*. Of note, the variants were homozygous in all patients, underlining the rarity of the disorder.

Figure 2 shows the results of the automated DeepGestalt technology. Both comparisons (*SLC39A13* group vs. control group 1, and *SLC39A13* group vs. control group 2) gave highly significant results (AUC= 0.991 p = 0.006 and AUC= 0.986 p = 0.004).

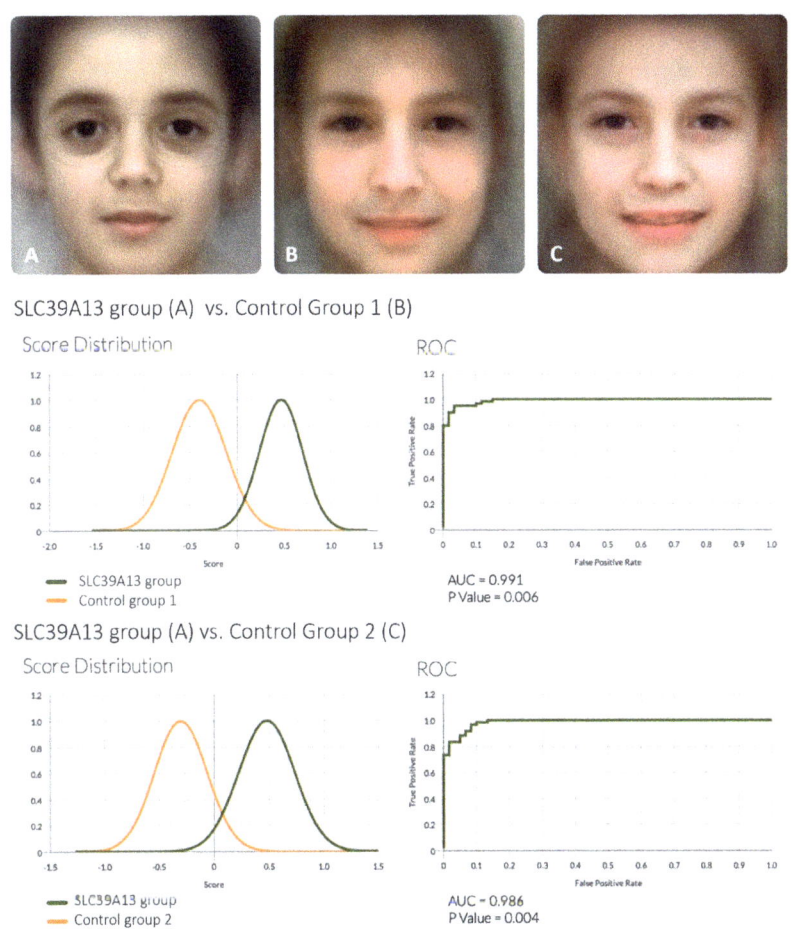

Figure 2. DeepGestalt technology analysis of facial features in SLC39A13 deficiency. The upper part shows the "averaged" artificial composite facial gestalt image of individuals with SLC39A13 deficiency (panel **A**) vs. two matched control groups (panels **B** and **C**). The lower part shows the result of the comparisons between **A** vs. **B**, and **A** vs. **C**.

Figure 3 shows the ratio of the collagen crosslink derivates, pyridinoline and deoxypyridinoline, in the urine of five patients and five roughly age-matched controls. The ratio of pyridinoline-to-deoxypyridinoline is reduced in the SLC39A13-deficient subjects, indicating a relative deficiency of the hydroxylated form, pyridinoline, relative to the non-hydroxylated form, deoxypyridinoline. The values clearly segregated in two distinct clusters with no overlap. In addition to confirming the notion of reduced lysyl hydroxylation as a pathogenic mechanism in these patients, the findings show that determination of urinary crosslink derivates can be a useful diagnostic screening method. Anecdotally, because of the role of vitamin C in collagen hydroxylation, individuals 1 and 2 took an oral dose of 1000 mg of vitamin C over four weeks and sent us urine samples for analysis before and twice during the test period. Unfortunately, there was no change in the pyridinoline-to-deoxypyridinoline ratio.

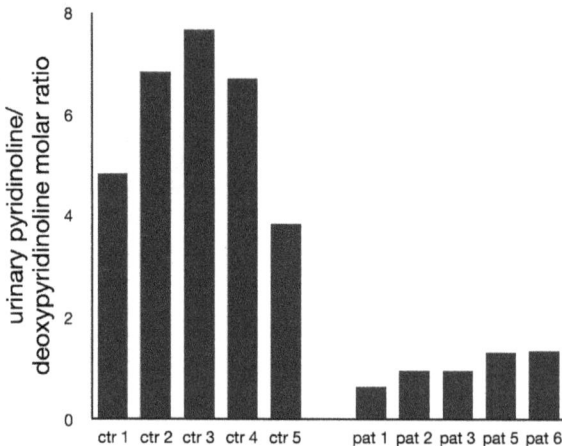

Figure 3. Ratio of the collagen crosslink derivates, pyridinoline and deoxypyridinoline, in the urine of five SLC39A3-dificient individuals and in five roughly age-matched controls. The ratio is consistently lower in SLC39A13-deficient individuals, with no overlap to the control group.

4. Discussion

4.1. Clinical Aspects of SCL39A13 Deficiency

Our observations confirm the previous findings of Giunta et al. (2008), Fukada et al. (2008), and Dusanic et al. (2018) that SLC39A13 deficiency is associated with a clinical phenotype characterized by post-natal short stature, connective tissue weakness affecting mainly the skin and peripheral joints, a characteristic facial appearance, and a moderate skeletal dysplasia. These additional observations help to delineate a more precise phenotype.

Muscular hypotonia—This has been seen in several *SLC39A13* patients in the neonatal period and in early childhood; in the patient described by Dusanic et al. (2018), muscular hypotonia was conspicuous enough to lead to investigations for myopathy with some abnormal (though non-specific) findings in a muscle biopsy at adolescence. While the prevalence of myopathy in SLC39A3 deficiency and its structural features remain to be investigated, the observations confirm the well-known phenomenon that connective tissue diseases, and particularly EDS type VI-A (lysyl hydroxylase deficiency), may present as the "floppy infant" phenomenon and/or with hypotonia with delayed motor development.

Growth and stature—postnatal reduced linear growth seems to be a salient and consistent feature of SLC39A13 deficiency; this is in contrast to other types of connective tissue disease and EDS types, where short stature is not a prominent feature. The two other forms of "spondylodysplastic EDS", namely B3GALT6 deficiency and B4GALT7 deficiency, are also associated with short stature, but in those conditions, short stature can be explained by skeletal dysplasia. In *SLC39A13*, body proportions are normal and short stature seems to be more intrinsic and represents a true growth failure and not a consequence of bone dysplasia.

Eye findings—Myopia has been reported in several *SLC39A13* patients and also seems to be a direct (though non-specific) manifestation of the primary genetic defect. This may be connected to the thin sclerae that often have a blueish or greyish color, particularly in individuals with darker pigmentation. Keratoconus, as seen in three patients, is potentially dangerous because of the possibility of perforation or rupture; ophthalmologic investigation should be recommended in all confirmed cases.

Dental features—Absence of one or several incisor teeth in the permanent dentition has been observed in a majority of SLC39A13-deficient individuals. In our case 6, the lower incisors of the primary dentition were also small and dysplastic. Hypodontia and oligodontia are consistent, if perhaps not obligate, features of the disorder.

Vascular complications—marked varicosities of the lower legs have been described in a number of *SLC39A13* patients. Clinically, more importantly, the two elder patients known (patients 1 and 2 in Fukada et al. (2008) and in this study) have both suffered cerebral hemorrhage. While this may not be statistically significant and may even be causally unrelated (e.g., hormonal contraception in patient 2), it is worth keeping in mind as a potential complication. Fortunately, other major vascular events have thus far not been recorded.

Facial features—analysis of the facial features with the DeepGestalt technology and comparison with age-matched controls confirms that there is an SLC39A13-associated facial phenotype that is significantly divergent from that of control individuals. This phenotype includes a rather flat face, mild hypertelorism, downslanting palpebral fissures, lack of periocular connective tissue giving the impression of prominent eyes, and a small mouth. The Face2Gene system evoked a few other syndromes with some resemblance to the *SLC39A13* phenotype, such as Noonan syndrome (downslanting palpebral fissures), Stickler syndrome (flat face; associated with *COL2A1* haploinsufficiency), as well as—interestingly—the vascular type of EDS (EDS IV, dominant *COL3A1*), probably because of the "hollow eyes". Thus, analogy between SLC39A13 deficiency and vascular EDS may include vascular fragility as well as a moderate resemblance of facial features.

4.2. What Are the Pathogenetic Mechanisms Leading from SLC39A13 Loss of Function to the Complex Clinical Phenotype?

The pathogenesis of SLC39A13 deficiency remains poorly understood. The observation of reduced collagen hydroxylation indicates that a partial failure of collagen crosslinking is one pathogenic mechanism [4], and the determination of pyridinoline-to-deoxypyridinoline is a useful diagnostic help. As correctly indicated by Giunta et al. (2008), the hydroxylation defect probably involves collagens other than collagen 1, such as collagen 2 and collagen 3, and this may explain the mild chondrodysplastic features (collagen 2) as well as the thin and fragile skin and the putative vascular fragility (collagen 3). However, it is difficult to ascribe the short stature, facial features, and oligodontia solely to reduced collagen hydroxylation. The hypothesis that part of the pathogenesis may involve the role of zinc as an intracellular messenger [5] may be justified, but experimental evidence is lacking; in particular, SLC39A13 individuals do not show signs of immune system dysfunction in contrast with the mouse model [5,11].

4.3. Diagnosis

From the diagnostic perspective, presentation with short stature and dysmorphic facial features may lead pediatricians and clinical geneticists to classify these patients within the group of genetic syndromes rather than connective tissue disorders; this "syndromic" presentation of SLC39A13 is not well known and this report may help raise awareness of this condition and its presenting features among clinical geneticists and dysmorphologists.

Author Contributions: Conceptualization, C.K., S.U., and A.S.-F.; methodology, A.S.-F, M.K., B.C.-X. and N.F.; software, N.F. and A.S.-F.; formal analysis, B.C.-X., M.K., N.F. and A.S.-F.; investigation, C.K., Y.H.-H., M.C. and A.S.-F.; writing—original draft preparation, C.K. and A.S.-F.; writing—review and editing, C.K., B.C.-X., C.M., M.K., S.U. and A.S.-F.; visualization, A.S.-F.; supervision, A.S.-F. All authors have read and agreed to the published version of the manuscript.

Funding: This research received no external funding.

Acknowledgments: We wish to express our gratitude to the individuals who participated in this study. Through the generous sharing of their data, they will help in improving the diagnosis and hopefully the treatment of other affected children and adults. Individual 3 originally came to our observation (YHH, SU, AS-F) through the former European Skeletal Dysplasia Network (ESDN). This study was also supported by a generous donation by the Fondation Guillaume Gentil (Lausanne) to the Division of Genetic Medicine of the Lausanne University Hospital (CHUV).

Conflicts of Interest: Superti-Furga is a member of scientific advisory board of FDNA (non-paid).

References

1. The Ehlers-Danlos Syndrome. In *Connective Tissue and Its Heritable Disorders*; Wiley-Liss, Inc.: New York, NY, USA, 2002; pp. 431–523. [CrossRef]
2. Beighton, P. Ehlers-Danlos syndrome. *Ann. Rheum. Dis.* **1970**, *29*, 332–333. [CrossRef] [PubMed]
3. Malfait, F.; Francomano, C.; Byers, P.; Belmont, J.; Berglund, B.; Black, J.; Bloom, L.; Bowen, J.M.; Brady, A.F.; Burrows, N.P.; et al. The 2017 international classification of the Ehlers-Danlos syndromes. *Am. J. Med. Genet. C Semin. Med. Genet.* **2017**, *175*, 8–26. [CrossRef] [PubMed]
4. Giunta, C.; Elcioglu, N.H.; Albrecht, B.; Eich, G.; Chambaz, C.; Janecke, A.R.; Yeowell, H.; Weis, M.; Eyre, D.R.; Kraenzlin, M.; et al. Spondylocheiro dysplastic form of the Ehlers-Danlos syndrome—An autosomal-recessive entity caused by mutations in the zinc transporter gene SLC39A13. *Am. J. Hum. Genet.* **2008**, *82*, 1290–1305. [CrossRef] [PubMed]
5. Fukada, T.; Civic, N.; Furuichi, T.; Shimoda, S.; Mishima, K.; Higashiyama, H.; Idaira, Y.; Asada, Y.; Kitamura, H.; Yamasaki, S.; et al. The zinc transporter SLC39A13/ZIP13 is required for connective tissue development; its involvement in BMP/TGF-beta signaling pathways. *PLoS ONE* **2008**, *3*, e3642. [CrossRef]
6. Mortier, G.R.; Cohn, D.H.; Cormier-Daire, V.; Hall, C.; Krakow, D.; Mundlos, S.; Nishimura, G.; Robertson, S.; Sangiorgi, L.; Savarirayan, R.; et al. Nosology and classification of genetic skeletal disorders: 2019 revision. *Am. J. Med. Genet. A* **2019**, *179*, 2393–2419. [CrossRef] [PubMed]
7. Dusanic, M.; Dekomien, G.; Lucke, T.; Vorgerd, M.; Weis, J.; Epplen, J.T.; Kohler, C.; Hoffjan, S. Novel Nonsense Mutation in SLC39A13 Initially Presenting as Myopathy: Case Report and Review of the Literature. *Mol. Syndromol.* **2018**, *9*, 100–109. [CrossRef] [PubMed]
8. Gurovich, Y.; Hanani, Y.; Bar, O.; Nadav, G.; Fleischer, N.; Gelbman, D.; Basel-Salmon, L.; Krawitz, P.M.; Kamphausen, S.B.; Zenker, M.; et al. Identifying facial phenotypes of genetic disorders using deep learning. *Nat. Med.* **2019**, *25*, 60–64. [CrossRef] [PubMed]
9. Face2Gene Research app. Available online: https://www.face2gene.com/ (accessed on 4 April 2020).
10. Amudhavalli, S.M.; Hanson, R.; Angle, B.; Bontempo, K.; Gripp, K.W. Further delineation of Ayme-Gripp syndrome and use of automated facial analysis tool. *Am. J. Med. Genet. A* **2018**, *176*, 1648–1656. [CrossRef] [PubMed]
11. Hirano, T.; Murakami, M.; Fukada, T.; Nishida, K.; Yamasaki, S.; Suzuki, T. Roles of zinc and zinc signaling in immunity: Zinc as an intracellular signaling molecule. *Adv. Immunol.* **2008**, *97*, 149–176. [CrossRef] [PubMed]

© 2020 by the authors. Licensee MDPI, Basel, Switzerland. This article is an open access article distributed under the terms and conditions of the Creative Commons Attribution (CC BY) license (http://creativecommons.org/licenses/by/4.0/).

Review

Recent Advances in the Pathophysiology of Musculocontractural Ehlers-Danlos Syndrome

Tomoki Kosho [1,2,3,*], Shuji Mizumoto [4], Takafumi Watanabe [5], Takahiro Yoshizawa [6], Noriko Miyake [7] and Shuhei Yamada [4]

1. Department of Medical Genetics, Shinshu University School of Medicine, Matsumoto 390-8621, Japan
2. Center for Medical Genetics, Shinshu University Hospital, Matsumoto 390-8621, Japan
3. Research Center for Supports to Advanced Science, Matsumoto 390-8621, Japan
4. Department of Pathobiochemistry, Faculty of Pharmacy, Meijo University, Nagoya 468-8503, Japan; mizumoto@meijo-u.ac.jp (S.M.); shuheiy@meijo-u.ac.jp (S.Y.)
5. Laboratory of Anatomy, School of Veterinary Medicine, Rakuno Gakuen University, Ebetsu 069-8501, Japan; t-watanabe@rakuno.ac.jp
6. Division of Animal Research, Research Center for Supports to Advanced Science, Shinshu University, Matsumoto 390-8621, Japan; tyoshizawa@shinshu-u.ac.jp
7. Department of Human Genetics, Yokohama City University Graduate School of Medicine, Yokohama 236-0004, Japan; nmiyake@yokohama-cu.ac.jp
* Correspondence: ktomoki@shinshu-u.ac.jp; Tel.: +81-263-37-2618; Fax: +81-263-37-2619

Received: 12 December 2019; Accepted: 23 December 2019; Published: 29 December 2019

Abstract: Musculocontractural Ehlers–Danlos Syndome (mcEDS) is a type of EDS caused by biallelic pathogenic variants in the gene for carbohydrate sulfotransferase 14/dermatan 4-O-sulfotransferase 1 (*CHST14/D4ST1*, mcEDS-*CHST14*), or in the gene for dermatan sulfate epimerase (*DSE*, mcEDS-*DSE*). Thus far, 41 patients from 28 families with mcEDS-*CHST14* and five patients from four families with mcEDS-*DSE* have been described in the literature. Clinical features comprise multisystem congenital malformations and progressive connective tissue fragility-related manifestations. This review outlines recent advances in understanding the pathophysiology of mcEDS. Pathogenic variants in *CHST14* or *DSE* lead to reduced activities of relevant enzymes, resulting in a negligible amount of dermatan sulfate (DS) and an excessive amount of chondroitin sulfate. Connective tissue fragility is presumably attributable to a compositional change in the glycosaminoglycan chains of decorin, a major DS-proteoglycan in the skin that contributes to collagen fibril assembly. Collagen fibrils in affected skin are dispersed in the papillary to reticular dermis, whereas those in normal skin are regularly and tightly assembled. Glycosaminoglycan chains are linear in affected skin, stretching from the outer surface of collagen fibrils to adjacent fibrils; glycosaminoglycan chains are curved in normal skin, maintaining close contact with attached collagen fibrils. Homozygous ($Chst14^{-/-}$) mice have been shown perinatal lethality, shorter fetal length and vessel-related placental abnormalities. Milder phenotypes in mcEDS-*DSE* might be related to a smaller fraction of decorin DS, potentially through residual DSE activity or compensation by DSE2 activity. These findings suggest critical roles of DS and DS-proteoglycans in the multisystem development and maintenance of connective tissues, and provide fundamental evidence to support future etiology-based therapies.

Keywords: musculocontractural Ehlers–Danlos Syndome; carbohydrate sulfotransferase-14 (CHST14)/dermatan 4-O-sulfotransferase-1 (D4ST1); *CHST14*; dermatan sulfate epimerase (DSE); *DSE*; dermatan sulfate (DS); decorin; collagen

---

1. Introduction

Musculocontractural Ehlers–Danlos Syndome (mcEDS) is a type of EDS, caused by biallelic pathogenic variants in the gene for carbohydrate sulfotransferase 14/dermatan 4-O-sulfotransferase 1

(*CHST14/D4ST1*, mcEDS-*CHST14*) (MIM#601776), or in the gene for dermatan sulfate epimerase (*DSE*, mcEDS-*DSE*) (MIM#615539) [1–3]. mcEDS-*CHST14* was originally described as three independent conditions: A rare type of arthrogryposis syndrome "adducted thumb-clubfoot syndrome" [4]; a specific type of EDS "EDS, Kosho type" [5,6]; and a subset of kyphoscoliosis type without lysyl hydroxylase deficiency [7]. To date, 41 patients from 28 families have been reported to have mcEDS-*CHST14* [4–22]. mcEDS-*DSE* was identified in a patient with a phenotype similar to that of patients with mcEDS-*CHST14* [23], as well as in four additional patients from three families [18,24]. These disorders were defined as subtypes of EDS, based on the International Classification of the EDSs [3]. Clinical features are highly characteristic, comprising multisystem congenital malformations such as craniofacial features (e.g., large fontanelle, hypertelorism, short and downslanting palpebral fissures, blue sclerae, short nose with hypoplastic columella, low-set and rotated ears, high palate, long philtrum, thin upper lip vermilion, small mouth and micro-retrognathia), multiple congenital contractures (e.g., adduction–flexion contractures of thumbs and talipes equinovarus), and visceral and ocular malformations. Features also include progressive connective tissue fragility-related manifestations, such as skin hyperextensibility, bruisability, and fragility with atrophic scars; recurrent dislocations; progressive talipes or spinal deformities; pneumothorax or pneumohemothorax; large subcutaneous hematomas; and/or diverticular perforation (Figure 1) [1,2]. Major diagnostic criteria of the disorder are as follows: 1) congenital multiple contractures, characteristically adduction–flexion contractures and/or talipes equinovarus (clubfoot); 2) characteristic craniofacial features, which are evident at birth or in early infancy; 3) characteristic cutaneous features including hyperextensibility, bruisability and fragility with atrophic scars, as well as increased palmer wrinkles [3]. Minor criteria as follows: 1) recurrent/chronic dislocations, 2) pectus deformities (e.g., flat or excavated), 3) spinal deformities (e.g., scoliosis or kyphoscoliosis), 4) peculiar fingers (e.g., tapered, slender, or cylindrical), 5) progressive talipes deformities (e.g., valgus, planus, or cavum), 6) large subcutaneous hematomas, 7) chronic constipation, 8) colonic diverticula, 9) pneumothorax/pneumohemothorax, 10) nephrolithiasis/cystolithiasis, 11) hydronephrosis, 12) cryptorchidism in males, 13) strabismus, 14) refractive errors (e.g., myopia or astigmatism) and/or 15) glaucoma/elevated intraocular pressure [3].

In this review, we describe the comprehensive pathophysiological findings of mcEDS, as demonstrated in previous studies including our recent reports.

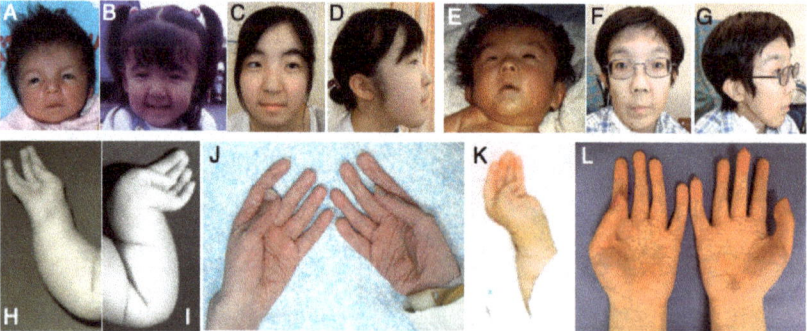

Figure 1. *Cont.*

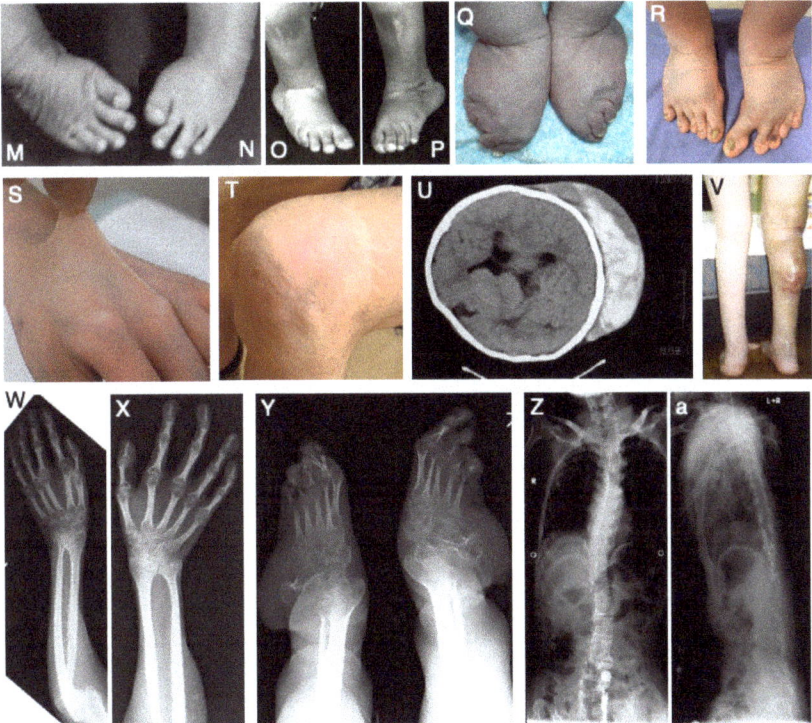

Figure 1. Clinical photographs and radiological images of patients with mcEDS-*CHST14*. Clinical photographs of a patient with heterozygous variants Pro281Leu/Try293Cys at age 23 days (**A**), 3 years (**B**), and 16 years (**C,D**); those of a patient with a homozygous variant Pro281Leu at age 2 months (**H,M,N**), 3 months (**I**), 6 years (**O,P**), and 28 years (**J,Q**); photographs of a patient with a homozygous variant "P281L" in the neonatal period (**E**) and at age 30 years (**F,G,T**); and photographs of a patient with heterozygous variants Pro281Leu /Cys289Ser at age 1 month (**K**), 16 years (**V**), and 19 years (**L,R,S**) [5]. Radiological image of a patient with heterozygous variants Pro281Leu /Try293Cys at age 6 years (**U**); images of a patient with a homozygous variant Pro281Leu at age 28 years (**W–Z,a**) [5,13]. (U, reproduced from Kosho et al. *Am. J. Med. Genet. Part A* **2005**, *138A*, 282–287, with permission from Wiley-Liss, Inc.; the other images, reproduced from Kosho et al. *Am. J. Med. Genet. Part A* **2010**, *152A*, 1333–1346, with permission from Wiley-Liss, Inc.).

2. Molecular Findings

Pathogenic variants have been detected throughout *CHST14* (NM_130468.4): 11 missense variants, five frameshift variants, and three nonsense variants in patients with mcEDS [4,6,7,14–19,22] (Figure 2). The p.(Pro281Leu) variant is most common (n = 10 families), followed by p.(Try293Cys) (n = 4), p.(Val49*) (n = 3), p.(Arg213Pro), and p.(Phe209Ser) (n = 2); p.(Arg29Glyfs*113), p.(Lys69*), p.(Gln113Argfs*14), p.(Arg135Gly), p.(Leu137Gln), p.(Cys152Leufs*10), p.(Arg218Ser), p.(Gly228Leufs*13), p.(Glu262Lys), p.(Tyr266*), p.(Arg274Pro), p.(Met280Leu), p.(Cys289Ser), p.(Trp327Cysfs*29), and p.(Glu334Glyfs*107) variants are particularly uncommon (n = 1 for all). Furthermore, p.(Gly19Trpfs*19) and p.(Lys26Alafs*16) variants have been detected in patients with features similar to those of mcEDS, among patients with hereditary connective tissue disorders and skeletal dysplasia, respectively (Figure 2A) [25,26]. Three missense variants have been detected in *DSE* (NM_013352.4): p.(Arg267Gly), p.(Ser268Leu), and p.(His588Arg); and one frameshift variant, p.(Pro384Trpfs*9), has also been detected (Figure 2B) [18,23,24]. Another frameshift variant, p.(Gly216Glufs*3), was detected in a patient with features similar to those of mcEDS, among patients with chondrodysplasia who exhibited multiple dislocations (Figure 2B) [27].

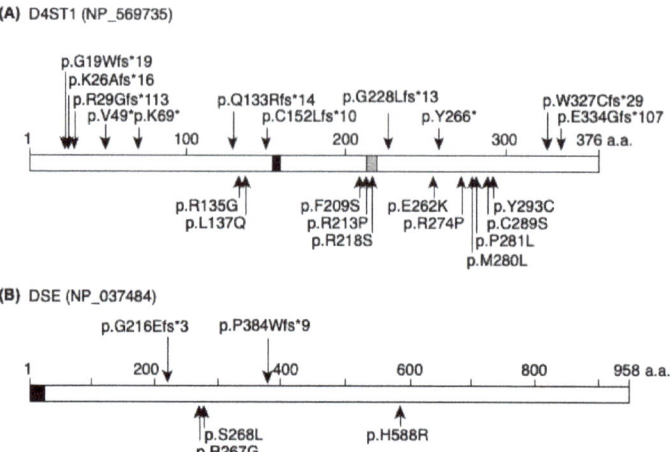

Figure 2. Published pathogenic protein changes of D4ST1 and DSE in mcEDS. (**A**) Previously published truncating and non-truncating protein alterations in D4ST1 are shown in upper and lower panels, respectively. Black box indicates 5′-phosphosulfate binding site and gray box indicates 3′-phosphate binding site. (**B**) Previously published truncating and non-truncating protein alterations in DSE are shown in upper and lower panels, respectively. Black box indicates the signal peptide.

No apparent genotype-phenotype correlations have been reported in patients with mcEDS-CHST14. Phenotypes of patients with mcEDS-DSE seem to be milder than those of patients with mcEDS-CHST14 [18,24].

3. Glycobiological Findings

Normal, biosynthetic pathways of chondroitin sulfate (CS) and dermatan sulfate (DS) are shown in Figure 3A. Reduced activity of D4ST1 in fibroblast cultures of skin from a patient with mcEDS-CHST14 caused by compound heterozygous p.(Pro281Leu)/(Tyr293Cys) substitutions in CHST14, as well as in fibroblasts from a patient with mcEDS-CHST14 caused by a homozygous p.(Pro281Leu) substitution in CHST14, showed a marked reduction in D4ST1 activity (Figure 3B); this change in activity resulted in a negligible amount of DS and an excessive amount of CS (Figure 3C) [6].

Decorin, which consists of a core protein and a single glycosaminoglycan (GAG) chain, is a major DS-proteoglycan (PG) that plays an important role in the assembly of collagen fibrils in the skin [28]; it also plays roles in the pathophysiology of mcEDS-CHST14 [1–4,6,7]. GAG chains of decorin-PG from skin fibroblasts of a patient with p.(Pro281Leu)/(Tyr293Cys) substitutions, as well as from skin fibroblasts of a patient with a homozygous p.(Pro281Leu) substitution, contained only CS and no DS; in contrast, GAG chains of decorin-PG from skin fibroblasts of healthy controls contained mainly DS (Figure 3D) [6,18]. 4-O-Sulfation in CS and DS chains functions as an inhibitor of DSE [29]. Thus, impaired 4-O-sulfation inhibition due to D4ST1 deficiency enables back-epimerization from L-iduronic acid (IdoUA) to D-glucuronic acid (GlcUA) (Figure 3E) [4,6,7]. In our laboratory, we have established a urinary disaccharide analysis of CS/DS chains through an anion-exchange chromatography after treatment with DS-specific degrading enzymes; this analysis method showed that no DS was present in the urine of eight patients with mcEDS-CHST14 [30]. This result suggested a systemic depletion of DS in patients with mcEDS-CHST14; thus we presume that our urinary disaccharide analysis method can be implemented to allow a non-invasive screening for mcEDS-CHST14 [30].

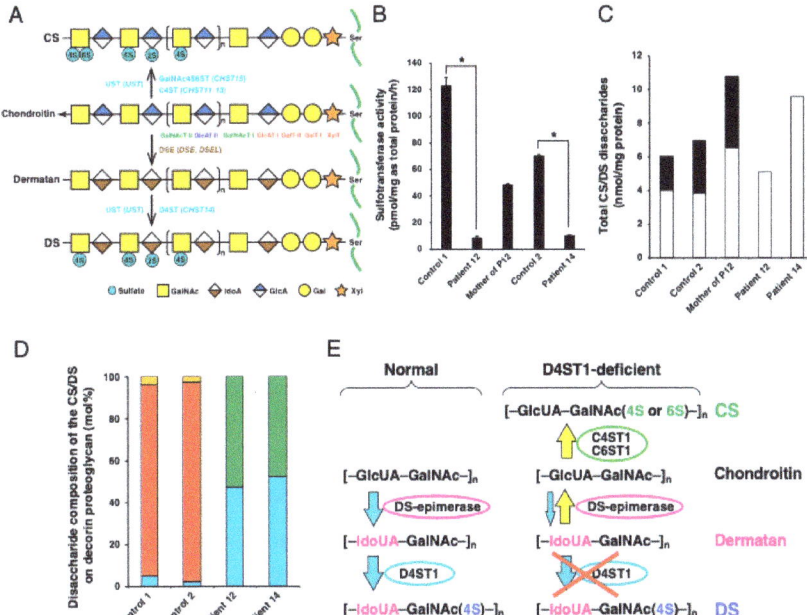

Figure 3. Biosynthesis of CS/DS and glycobiological analysis in mcEDS-*CHST14*. (**A**) Biosynthetic assembly of CS and DS chains by glycosyltransferases, epimerases, and sulfotransferases. It starts with the biosynthesis of a tetrasaccharide linker region, glucuronic acid-β1,3-galactose-β1,3-galactose-β1,4-xylose-β1-O-(GlcUA-Gal-Gal-Xyl-), onto serine residues of specific core proteins of PGs by β-xylosyltransferase (XylT), GalT-I, GalT-II and β1,3-glucuronosyltransferase-I (GlcAT-I), respectively. Subsequently, a repetitive disaccharide region [N-acetyl-D-galactosamine(GalNAc)-GlcUA]$_n$ of chondroitin is elongated by the actions of N-acetyl-D-galactosaminyltransferase-I (GalNAcT-I), N-acetyl-D-galactosaminyltransferase-II (GalNAcT-II) and glucuronyltransferase-II (GlcAT-II), which are encoded by CS N-acetylgalactosaminyltransferase-1 and -2, chondroitin synthase-1, -2, and -3 and chondroitin polymerizing factor genes. Chondroitin chains are matured to CS through modifications by chondroitin 4-*O*-sulfotransferase (C4ST), chondroitin 6-*O*-sulfotransferase (C6ST) and uronyl 2-*O*-sulfotransferase (UST). A disaccharide-repeating region of dermatan is synthesized through epimerization of a carboxyl group at C5 from GlcUA to IdoUA by DSE. A mature DS chain is synthesized through modification with sulfation by D4ST1 and UST. (**B**) Sulfotransferase activity toward dermatan in fibroblast cultures of skin from two patients with mcEDS-*CHST14* (patient 12 with heterozygous variants Pro281Leu/Tyr293Cys; patient 14 with a homozygous variant Pro281Leu), the mother of patient 12, and a sex- and age-matched healthy volunteer [6]. * $p < 0.0001$ by two-tailed unpaired *t*-test. (**C**) Total amounts of CS and DS derived from fibroblast cultures of skin [6]. Total disaccharide contents of CS (white box) and DS (black box) were calculated based on the peak area in the chromatograms of the digests with chondroitinase AC and chondroitinase B, respectively. (**D**) Proportion of the disaccharide units in the CS-DS hybrid chain in decorin-PGs secreted by the fibroblasts [6]. Cyan, green, red and orange boxes are GlcUA-GalNAc(4S), GlcUA-GalNAc(6S), IdoUA-GalNAc(4S), and IdoUA(2S)-GalNAc(4S), respectively. Abbreviations of 2S, 4S and 6S indicate 2-*O*-, 4-*O*- and 6-*O*-sulfate, respectively. (**E**) Schematic diagram of the biochemical mechanism in the replacement of DS by CS in mcEDS-*CHST14*. Defect in D4ST1 enables a back-epimerization reaction that converts IdoUA back to GlcUA to form chondroitin by DSE, followed by the 4-*O*-sulfation and/or 6-*O*-sulfation of GalNAc residues in chondroitin by C4ST1 and C6ST1, respectively. (B–D, reproduced from Miyake et al. *Hum. Mutat.* **2010**, *31*, 1233–1239, with permission from Wiley-Liss, Inc.).

Regarding patients with mcEDS-*DSE*, reduced activity of DSE in fibroblast cultures of skin from a patient with a homozygous p.(Ser268Leu) substitution resulted in marked reduction of DS disaccharides, compared with healthy controls [23]. The total amount of CS in the cell fraction from affected skin fibroblasts was increased by approximately 1.5-fold, which might reflect increased synthesis and/or reduced conversion of CS chains [23]. A minor fraction of DS from decorin-PG was present in skin fibroblasts from a patient with a homozygous p.(Ser268Leu) substitution [23]; this suggested residual DSE activity or compensation by DSE2, which might be related to milder phenotypes in patients with mcEDS-*DSE* than in patients with mcEDS-*CHST14* [18].

4. Pathological Findings

The pathology of mcEDS-*CHST14*, as the simplest model for complete depletion of DS, has been extensively investigated using affected skin specimens. Light microscopy of skin specimens (hematoxylin and eosin staining) from patients with compound heterozygous p.(Pro281Leu)/(Cys289Ser) or p.(Pro281Leu)/(Tyr293Cys) substitutions showed that fine collagen fibers were predominantly present in the reticular to papillary dermis; marked reduction of normally thick collagen bundles were also observed (Figure 4A) [6]. Immunohistochemistry staining of decorin core protein in skin specimens from patients with compound heterozygous p.(Pro281Leu)/(CYs289Ser) or p.(Pro281Leu)/(Tyr293Cys) substitutions showed that decorin core protein was present on collagen fibers that were thin and filamentous without clear boundaries; in contrast, skin specimens from healthy controls showed decorin core protein on collagen fibers that were thick bundles with clear boundaries [31] (Figure 4B). Transmission electron microscopy of skin specimens from five patients with compound heterozygous p.(Pro281Leu)/(Cys289Ser), p.(Pro281Leu)/(Tyr293Cys), or p.(Phe209Ser)/(Pro281Leu) substitutions showed that collagen fibrils were dispersed in the papillary to reticular dermis, whereas skin specimens from healthy controls exhibited collagen fibrils that were regularly and tightly assembled (Figure 4C) [6,31]. Transmission electron microscopy-based cupromeronic blue staining to visualize GAG chains on affected skin samples showed that GAG chains were linear, stretching from the outer surface of collagen fibrils to adjacent fibrils, whereas skin samples from healthy controls exhibited curved GAG chains that maintained close contact with attached collagen fibrils (Figure 4D) [31]. This structural alteration of GAG chains of decorin is presumably related to the biochemical alteration from DS to CS: the structure of DS-GAG chains is flexible because L-IdoUA residues in DS can easily adopt any of the nearly equi-energetic $^1C_4$, $^2S_0$, and $^4C_1$ conformations, whereas the structure of CS-GAG chains is rigid because D-GlcUA in CS only adopts the $^4C_1$ conformation [6,32,33].

Furthermore, focused ion beam scanning electron microscopy using cupromeronic blue staining uncovered the structure of collagen fibrils in association with GAG chains (likely comprising decorin): GAG chains form a ring mesh-like structure with each ring surrounding a collagen fibril at its D band, fusing with adjacent rings to form a planar network [34]. Abnormally stretching CS-GAG chains of decorin in the affected skin would disrupt the ring-mesh structure of collagen fibrils (Figure 4E), which could result in substantial fragility.

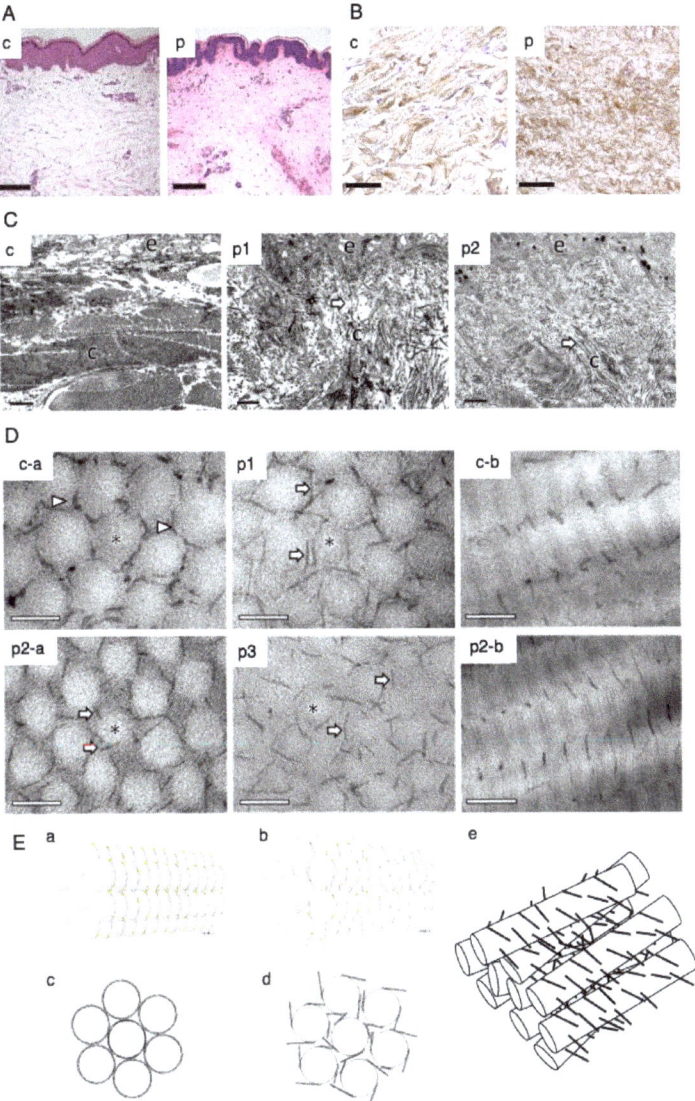

Figure 4. Skin pathology of mcEDS-*CHST14*. (**A**) Light microscopy (hematoxylin and eosin staining). In the skin specimen from a patient with heterozygous variants Pro281Leu/ Cys289Ser (panel p), fine collagen fibers are present predominantly in the reticular to papillary dermis with marked reduction of thick collagen bundles; thick collagen bundles are observed in a skin specimen from a healthy control volunteer (panel c) [6]. (**B**) Immunohistochemical staining for decorin core protein. Decorin core protein is present on collagen fibers in thick bundles in a skin specimen from a healthy control volunteer (panel c), but on thin and filamentous collagen fibers without clear boundaries in a skin specimen from a patient with heterozygous variants Pro281Leu/ Cys289Ser (panel p) [31] (**C**) Transmission electron microscopy. Collagen fibrils are regularly and tightly assembled in a skin specimen from a healthy control volunteer (panel c), but are dispersed in the papillary to reticular dermis in skin specimens from a patient with heterozygous variants Pro281Leu/ Tyr293Cys (panel p1)

and a patient with a novel homozygous variant (panel p2) [31]. (**D**) Transmission electron microscopy-based cupromeronic blue staining. GAG chains are curved and maintain close contact with attached collagen fibrils in the skin specimens from a healthy control volunteer (panels c-a, c-b); conversely, they are linear and stretch from the outer surface of collagen fibrils to adjacent fibrils in skin specimens from a patient with heterozygous variants Pro281Leu/ Tyr293Cys (panel p1), a patient with a novel homozygous variant (panels p2a and p2b) and another patient with heterozygous variants Pro281Leu/ Tyr293Cys (panel p3) [31]. (**E**) Schematic representations of collagen fibrils and GAG chains. Decorin core protein binds to D bands of collagen fibrils both in normal skin and in affected skin. GAG chains composed of DS adhere to collagen fibrils along D bands, beginning from the core protein (panels a and c), whereas GAG chains composed of CS extend linearly and perpendicularly to collagen fibrils from the core protein (panels b and d) [31]. Putative spatial disorganization of collagen fibril networks in the skin of patients (panel e) (A, reproduced from Miyake et al. *Hum. Mutat.* **2010**, *31*, 1233–1239, with permission from Wiley-Liss, Inc.; B–E, reproduced from Hirose et al. *Biochim. Biophys. Acta Gen. Subj.* **2019**, *1863*, 623–631, with permission from Elsevier, Inc.).

5. Animal Model-Based Findings

Knockout ($Chst14^{-/-}$) mice were generated through homologous recombination that targeted the only coding exon 1 (i.e., exon 1) of *Chst14* [35,36]. F2 mice showed reduced weight and/or length and reduced bone volume/thickness/density of their lumbar vertebrae [35]. $Chst14^{-/-}$ mice showed reduced neurogenesis and diminished proliferation of neural stem cells, accompanied by increased expression of glutamate aspartate transporter and epidermal growth factor, compared with findings in both wild-type ($Chst14^{+/+}$) mice and other knockout ($Chst11^{-/-}$) mice, a model for chondroitin 4-O-sulfotransferase 1 (C4ST1) deficiency [36]. $Chst14^{-/-}$ mice also had a smaller body mass (Figure 5A, B), reduced fertility, a kinked tail, and increased skin fragility compared with their wild-type ($Chst14^{+/+}$) littermates; however, brain weight and gross anatomy were not affected [36,37]. Schwann cells from $Chst14^{-/-}$ mice formed longer processes in vitro and exhibited greater proliferation than those from $Chst14^{+/+}$ mice. Functional recovery and axonal regrowth in $Chst14^{-/-}$ mice were initially accelerated after femoral nerve transection and suture; after 3 months, these characteristics were similar to those in $Chst14^{+/+}$ littermates. These findings suggested that DS, synthesized by Chst14/D4st1, might be of limited importance for neural development; moreover, it might contribute to the regeneration-restricting environment in the adult mammalian nervous system [38]. Only a few adult $Chst14^{-/-}$ mice were generated because of perinatal lethality in most of the homozygous mice; these adult $Chst14^{-/-}$ mice showed significantly shorter crown-rump length, compared with wild-type or heterozygous mice (Figure 5A–C) [37,38]. The placentas of $Chst14^{-/-}$ fetuses showed a reduced weight, alterations in the vascular structure and ischemic and/or necrotic-like changes (Figure 5D–G). Transmission electron microscopy of homozygous placentas demonstrated an abnormal capillary basement membrane structure in the placental villus, compared with wild-type or heterozygous placentas (Figure 5H–J). These findings showed that DS was essential for placental vascular development and perinatal fetal survival. In addition, DS was suggested to be related to the structure and/or function of capillary basement membranes, which constitutes an extracellular matrix [37].

C4st1, encoded by *Chst11*, transfers a sulfate group from 3'-phosphoadenosine 5'-phosphosulfates to the C4 position of GalNAc residues in chondroitin (Figure 3A). *Chst11*-mutant mice died within 6 hours of birth due to respiratory failure, severe dwarfism and chondrodysplasia (i.e., abnormalities in the cartilage growth plate and chondrocyte columns) [39]. Furthermore, marked reductions in the content and 4-O-sulfation of CS, as well as the downregulation of bone morphogenetic protein signaling and upregulation of transforming growth factor-β, have also been observed in *Chst11*-mutant mice. These findings suggest that Chst11/C4st1 and the 4-O-sulfation of CS chains are essential for early development and for bone morphogenetic protein and transforming growth factor-β in signaling pathways in cartilage. When considered in the context of $Chst14^{-/-}$ mouse phenotypes, the above findings suggest that 4-O-sulfation is required for the maturation of both CS and DS; moreover, CS and

DS exert distinct functions in the development of cartilage and skin, respectively. However, the DS content in *Chst11*-mutant mice remains to be elucidated [39].

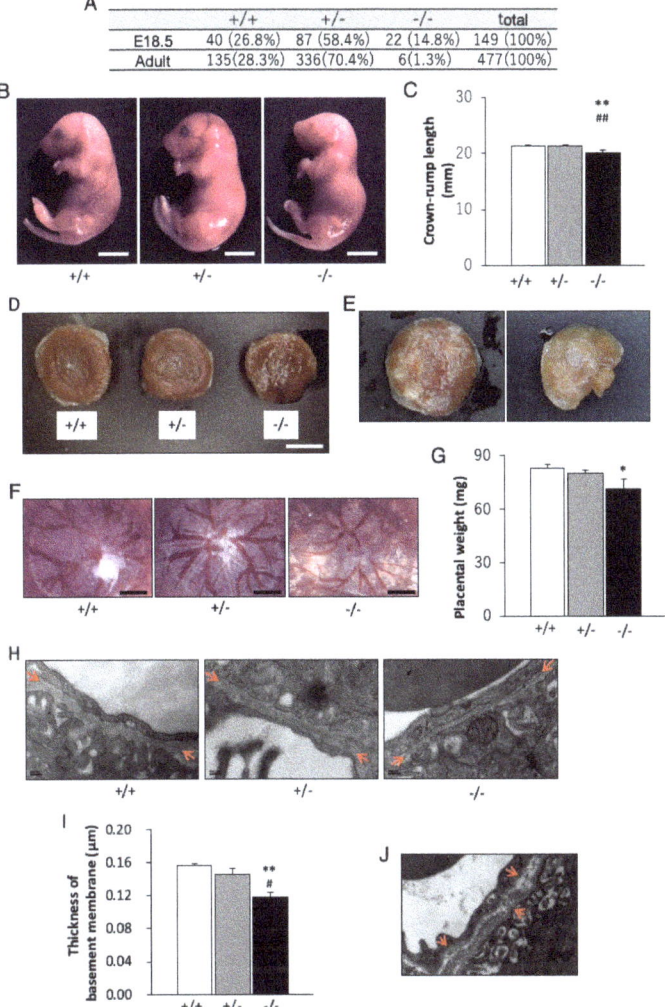

Figure 5. Mouse model for mcEDS-CHST14. (**A**) Numbers (percentages) of each type at embryonic (E18.5) and postnatal (adult) ages. Markedly greater numbers of homozygous fetuses are observed, compared with homozygous adults. (**B**) General appearances of wild-type (*Chst14*$^{+/+}$), heterozygous (*Chst14*$^{+/-}$) and homozygous (*Chst14*$^{-/-}$) fetuses [37]. (**C**) Crown-rump length of each type of fetus (bar: 5 mm). Homozygous mice demonstrate significantly shorter crown-rump length, compared with wild-type or heterozygous mice [37]. (**D**) Appearance of each type of placenta (bar: 5 mm) [37]. (**E**) Homozygous placentas exhibit appearances indicative of hypoxia (left) and necrosis (right) [37]. (**F**) Microphotographs of the chorionic plate side of the placentas (bar: 1 mm). Homozygous placentas demonstrate smaller vascular diameters, compared with wild-type or heterozygous placentas [37]. (**G**) Weight of each type of placenta (mean + standard error of the mean). (**H**) Transmission electron microscopy of capillary basement membrane in the labyrinth zone of each type of placenta (bar: 200 nm). Arrows indicate capillary basement membrane [37]. (**I**) Homozygous placentas show a significantly

thinner capillary basement membrane, compared with wild-type or heterozygous placentas [37]. (J) Homozygous placentas show structural abnormalities (arrows) in capillary basement membrane (bar: 0.2 µm) [37]. * $p < 0.05$, ** $p < 0.01$, compared with wild-type; # $p < 0.05$, ## $p < 0.01$, compared with heterozygous; one-way analysis of variance (ANOVA) followed by the Tukey–Kramer post hoc test. (A–J, reproduced from Yoshizawa et al. *Glycobiology* **2018**, *28*, 80–89, with permission from Oxford University Press, Inc.).

Biosynthesis of DS requires D4st1, dermatan sulfate epimerase 1 (DS-epi1) and dermatan sulfate epimerase 2 (DS-epi2), which are encoded by *Chst14*, *Dse* and *Dsel*, respectively [40–43]. Furthermore, D4ST1 interacts directly with DS-epi1, but not with DS-epi2, to form a hetero-complex that is required for the formation of IdoUA blocks in DS chains [44]. *DS-epi1*$^{-/-}$ mice showed greater skin fragility compared with wild-type littermates, due to altered collagen fibril morphology [45]; this phenotype was similar to that of *Chst14*$^{-/-}$ mice [38]. The numbers of IdoUA blocks were dramatically reduced in DS side chains of decorin-PG and biglycan-PG from the skin of *DS-epi1*$^{-/-}$ mice [45], which suggests that DS-epi1 mainly synthesizes IdoUA blocks in vivo. *DS-epi1*$^{-/-}$ embryos and newborn mice showed kinked tails, which is a common feature in *Chst14*$^{-/-}$ mice [38]; the *DS-epi1*$^{-/-}$ embryos and newborn mice also showed significantly thicker epidermal layers through histological staining, compared with heterozygous or wild-type littermates [45,46]. Immunohistochemical staining of epidermal layers in *DS-epi1*$^{-/-}$ newborns showed increased expression of keratin 5 in the basal layer and keratin 1 in the spinous layer [46]. Furthermore, a small portion of *DS-epi1*$^{-/-}$ embryos showed an abdominal wall defect with herniated intestines, exencephaly and spina bifida. Defective collagen structure in the dermis and imbalanced keratocyte maturation were presumed to cause developmental defects in *DS-epi1*$^{-/-}$ mice [46]. These observations indicate that *DS-epi1*$^{-/-}$ mice may constitute a useful model of mcEDS-DSE. *DS-epi2*$^{-/-}$ mice displayed no significant defects and DS-epi1 compensated for the absence of DS-epi2 in most tissue, which indicated that DS-epi1 is the major contributor of epimerase activity [47]. DS-epi2 exhibits higher expression than DS-epi1 in developing the mouse brain [48]. Although CS/DS chains in the brains of *DS-epi2*$^{-/-}$ newborn mice demonstrated a 38% reduction in IdoUA content, compared with wild-type littermates, the brains of adult knockout mice showed normal extracellular matrix features [47]. *DS-epi1*$^{-/-}$/*DS-epi2*$^{-/-}$ mice experienced perinatal death with variable phenotypes at late embryological stages and birth; these phenotypes included umbilical hernia, exencephaly, and a kinked tail, as well as complete loss of IdoUA residues in CS/DS chains [49]. However, a minority of embryos exhibited normal lung, bone, and cartilage features.

These findings indicate that DS, DS-PGs, DS-epi1 and/or DS-epi2 are important in early embryonic development and perinatal survival [49].

6. Ongoing Projects and Future Perspectives

Our group established a multicenter collaboration network to promote comprehensive basic research on mcEDS. In this network, genetic testing is provided as part of routine medical care covered by national health insurance, using a custom next generation sequencing-based panel that includes all EDS-related genes and genes for hereditary connective tissue disorders [50]. Whole exome sequencing is performed to identify other potential causative genes for mcEDS in patients with similar features who do not exhibit pathogenic variants in *CHST14* or *DSE*. Biochemical abnormalities of the extracellular matrix are investigated (e.g., various types of collagen and DS-PGs, including decorin and biglycan), using patient skin fibroblasts. Furthermore, mass spectrometry is performed to identify appropriate serum/plasma biomarker(s) that may be useful in the diagnosis or surveillance of disease progression. A subsequent pathological approach includes an elucidation of the structural alterations of collagen fibril networks and GAG chains of decorin in skin specimens from patients with mcEDS-DSE. Determination of the crystal structure of D4ST1 would be useful for understanding the effects of common missense variants (e.g., p.(Pro281Leu)) in *CHST14*. A technical breakthrough in the efficient generation of knockout mice for *Chst14* (e.g., CRISPR/cas9) is needed to continue mouse-based phenotypic and pathophysiological investigations. Considering the differences in the phenotypes between patients with mcEDS and $Chst14^{-/-}$ or $DS\text{-}epi1^{-/-}$ mice, experiments using other models that could reflect phenotypic and pathophysiological abnormalities in patients with mcEDS, such as induced pluripotent stem cells, would be particularly useful. All of these approaches will be valuable for further elucidating the critical roles of DS and DS-PGs, including decorin and biglycan, in the multisystem development and maintenance of connective tissues; in addition, they provide fundamental evidence for future etiology-based therapies, such as adeno-associated virus-based gene therapy.

Acknowledgments: We thank all the collaborators engaged in this project. We are also grateful to the patients and their families for participating in the studies described here. We thank Ryan Chastain-Gross, from Edanz Group (www.edanzediting.com/ac) for editing a draft of this manuscript. The drafting and revision of this review was supported by the Practical Research Project for Rare/Intractable Diseases #105 from the Japan Agency for Medical Research Development (AMED) (T.K., S.M., T.Y., N.M., and S.Y.); Grant-in-Aid for Young Scientists (B) (16K19396) (T.Y.); Grant-in-Aid for Scientific Research (C) from the Ministry of Education, Culture, Sports, Science and Technology of Japan (#16K08251, #19K07054) (S.M.) (#19K08745) (T.Y.); Grant-in-Aid for Scientific Research (B) from the Ministry of Education, Culture, Sports, Science and Technology of Japan (#19H03616) (T.K., S.M., T.W., T.Y., N.M., and S.Y.); Grant-in Aid for Research Center for Pathogenesis of Intractable Diseases from the Research Institute of Meijo University (S.M. and S.Y.).

Conflicts of Interest: The authors declare no conflict of interest.

References

1. Kosho, T. *CHST14/D4ST1* deficiency: New form of Ehlers-Danlos syndrome. *Pediatr. Int.* **2016**, *58*, 88–99. [CrossRef]
2. Brady, A.F.; Demirdas, S.; Fournel-Gigleux, S.; Ghali, N.; Giunta, C.; Kapferer-Seebacher, I.; Kosho, T.; Mendoza-Londono, R.; Pope, M.F.; Rohrbach, M.; et al. The Ehlers-Danlos syndromes, rare types. *Am. J. Med. Genet. C Semin. Med. Genet.* **2017**, *175*, 70–115. [CrossRef] [PubMed]
3. Malfait, F.; Francomano, C.; Byers, P.; Belmont, J.; Berglund, B.; Black, J.; Bloom, L.; Bowen, J.M.; Brady, A.F.; Burrows, N.P.; et al. The 2017 international classification of the Ehlers-Danlos syndromes. *Am. J. Med. Genet. C Semin. Med. Genet.* **2017**, *175*, 8–26. [CrossRef] [PubMed]
4. Dündar, M.; Müller, T.; Zhang, Q.; Pan, J.; Steinmann, B.; Vodopiutz, J.; Gruber, R.; Sonoda, T.; Krabichler, B.; Utermann, G.; et al. Loss of dermatan-4-sulfotransferase 1 function results in adducted thumb-clubfoot syndrome. *Am. J. Hum. Genet.* **2009**, *85*, 873–882. [CrossRef] [PubMed]
5. Kosho, T.; Miyake, N.; Hatamochi, A.; Takahashi, J.; Kato, H.; Miyahara, T.; Igawa, Y.; Yasui, H.; Ishida, T.; Ono, K.; et al. A new Ehlers-Danlos syndrome with craniofacial characteristics, multiple congenital contractures, progressive joint and skin laxity, and multisystem fragility-related manifestations. *Am. J. Med. Genet. Part A* **2010**, *152A*, 1333–1346. [CrossRef] [PubMed]
6. Miyake, N.; Kosho, T.; Mizumoto, S.; Furuichi, T.; Hatamochi, A.; Nagashima, Y.; Arai, E.; Takahashi, K.; Kawamura, R.; Wakui, K.; et al. Loss-of-function mutations of *CHST14* in a new type of Ehlers-Danlos syndrome. *Hum. Mutat.* **2010**, *31*, 966–974. [CrossRef] [PubMed]
7. Malfait, F.; Syx, D.; Vlummens, P.; Symoens, S.; Nampoothiri, S.; Hermanns-Lê, T.; Van Laer, L.; De Paepe, A. Musculocontractural Ehlers-Danlos syndrome (former EDS type VIB) and adducted thumb clubfoot syndrome (ATCS) represent a single clinical entity caused by mutations in the dermatan-4-sulfotransferase 1 encoding *CHST14* gene. *Hum. Mutat.* **2010**, *31*, 1233–1239. [CrossRef] [PubMed]
8. Dündar, M.; Demiryilmaz, F.; Demiryilmaz, I.; Kumandas, S.; Erkilic, K.; Kendirch, M.; Tuncel, M.; Ozyazgan, I.; Tolmie, J.L. An autosomal recessive adducted thumb-club foot syndrome observed in Turkish cousins. *Clin. Genet.* **1997**, *51*, 61–64. [CrossRef]
9. Sonoda, T.; Kouno, K. Two brothers with distal arthrogryposis, peculiar facial appearance, cleft palate, short stature, hydronephrosis, retentio testis, and normal intelligence: A new type of distal arthrogryposis? *Am. J. Med. Genet.* **2000**, *91*, 280–285. [CrossRef]
10. Dündar, M.; Kurtoglu, S.; Elmas, B.; Demiryilmaz, F.; Candemir, Z.; Ozkul, Y.; Durak, A.C. A case with adducted thumb and club foot syndrome. *Clin. Dysmorphol.* **2001**, *10*, 291–293. [CrossRef]
11. Janecke, A.R.; Unsinn, K.; Kreczy, A.; Baldissera, I.; Gassner, I.; Neu, N.; Utermann, G.; Müller, T. Adducted thumb-club foot syndrome in sibs of a consanguineous Austrian family. *J. Med. Genet.* **2001**, *38*, 265–269. [CrossRef] [PubMed]

12. Yasui, H.; Adachi, Y.; Minami, T.; Ishida, T.; Kato, Y.; Imai, K. Combination therapy of DDAVP and conjugated estrogens for a recurrent large subcutaneous hematoma in Ehlers-Danlos syndrome. *Am. J. Hematol.* **2003**, *72*, 71–72. [CrossRef] [PubMed]
13. Kosho, T.; Takahashi, J.; Ohashi, H.; Nishimura, G.; Kato, H.; Fukushima, Y. Ehlers-Danlos syndrome type VIB with characteristic facies, decreased curvatures of the spinal column, and joint contractures in two unrelated girls. *Am. J. Med. Genet. Part A* **2005**, *138A*, 282–287. [CrossRef] [PubMed]
14. Shimizu, K.; Okamoto, N.; Miyake, N.; Taira, K.; Sato, Y.; Matsuda, K.; Akimaru, N.; Ohashi, H.; Wakui, K.; Fukushima, Y.; et al. Delineation of dermatan 4-O-sulfotransferase 1 deficient Ehlers-Danlos syndrome: Observation of two additional patients and comprehensive review of 20 reported patients. *Am. J. Med. Genet. A* **2011**, *155A*, 1949–1958. [CrossRef]
15. Mendoza-Londono, R.; Chitayat, D.; Kahr, W.H.; Hinek, A.; Blaser, S.; Dupuis, L.; Goh, E.; Badilla-Porras, R.; Howard, A.; Mittaz, L.; et al. Extracellular matrix and platelet function in patients with musculocontractural Ehlers-Danlos syndrome caused by mutations in the *CHST14* gene. *Am. J. Med. Genet. A* **2012**, *158A*, 1344–1354. [CrossRef]
16. Voermans, N.C.; Kempers, M.; Lammens, M.; van Alfen, N.; Janssen, M.C.; Bönnemann, C.; van Engelen, B.G.; Hamel, B.C. Myopathy in a 20-year-old female patient with D4ST-1 deficient Ehlers-Danlos syndrome due to a homozygous *CHST14* mutation. *Am. J. Med. Genet. A* **2012**, *158A*, 850–855. [CrossRef]
17. Winters, K.A.; Jiang, Z.; Xu, W.; Li, S.; Ammous, Z.; Jayakar, P.; Wierenga, K.J. Re-assigned diagnosis of D4ST1-deficient Ehlers-Danlos syndrome (adducted thumb-clubfoot syndrome) after initial diagnosis of Marden-Walker syndrome. *Am. J. Med. Genet. A* **2012**, *158A*, 2935–2940. [CrossRef]
18. Syx, D.; Van Damme, T.; Symoens, S.; Maiburg, M.C.; van de Laar, I.; Morton, J.; Suri, M.; Del Campo, M.; Hausser, I.; Hermanns-Lê, T.; et al. Genetic heterogeneity and clinical variability in musculocontractural Ehlers-Danlos syndrome caused by impaired dermatan sulfate biosynthesis. *Hum. Mutat.* **2015**, *36*, 535–547. [CrossRef]
19. Janecke, A.R.; Li, B.; Boehm, M.; Krabichler, B.; Rohrbach, M.; Müller, T.; Fuchs, I.; Golas, G.; Katagiri, Y.; Ziegler, S.G.; et al. The phenotype of the musculocontractural type of Ehlers-Danlos syndrome due to *CHST14* mutations. *Am. J. Med. Genet. A* **2016**, *170A*, 103–115. [CrossRef]
20. Kono, M.; Hasegawa-Murakami, Y.; Sugiura, K.; Ono, M.; Toriyama, K.; Miyake, N.; Hatamochi, A.; Kamei, Y.; Kosho, T.; Akiyama, M. A 45-year-old woman with Ehlers-Danlos syndrome caused by dermatan 4-O-sulfotransferase-1 deficiency: Implications for early ageing. *Acta Derm. Venereol.* **2016**, *96*, 830–831. [CrossRef]
21. Mochida, K.; Amano, M.; Miyake, N.; Matsumoto, N.; Hatamochi, A.; Kosho, T. Dermatan 4-O-sulfotransferase 1-deficient Ehlers-Danlos syndrome complicated by a large subcutaneous hematoma on the back. *J. Dermatol.* **2016**, *43*, 832–833. [CrossRef] [PubMed]
22. Sandal, S.; Kaur, A.; Panigrahi, I. Novel mutation in the *CHST14* gene causing musculocontractural type of Ehlers-Danlos syndrome. *BMJ Case Rep.* **2018**, *pii*, bcr-2018-226165. [CrossRef] [PubMed]
23. Müller, T.; Mizumoto, S.; Suresh, I.; Komatsu, Y.; Vodopiutz, J.; Dundar, M.; Straub, V.; Lingenhel, A.; Melmer, A.; Lechner, S.; et al. Loss of dermatan sulfate epimerase (DSE) function results in musculocontractural Ehlers-Danlos syndrome. *Hum. Mol. Genet.* **2013**, *22*, 3761–3772. [CrossRef] [PubMed]
24. Schirwani, S.; Metcalfe, K.; Wagner, B.; Berry, I.; Sobey, G.; Jewell, R. DSE associated musculocontractural EDS, a milder phenotype or phenotypic variability. *Eur. J. Med. Genet.* **2019**. [CrossRef] [PubMed]
25. Alazami, A.M.; Al-Qattan, S.M.; Faqeih, E.; Alhashem, A.; Alshammari, M.; Alzahrani, F.; Al-Dosari, M.S.; Patel, N.; Alsagheir, A.; Binabbas, B.; et al. Expanding the clinical and genetic heterogeneity of hereditary disorders of connective tissue. *Hum. Genet.* **2016**, *135*, 525–540. [CrossRef] [PubMed]
26. Maddirevula, S.; Alsahli, S.; Alhabeeb, L.; Patel, N.; Alzahrani, F.; Shamseldin, H.E.; Anazi, S.; Ewida, N.; Alsaif, H.S.; Mohamed, J.Y.; et al. Expanding the phenome and variome of skeletal dysplasia. *Genet. Med.* **2018**, *20*, 1609–1616. [CrossRef]
27. Ranza, E.; Huber, C.; Levin, N.; Baujat, G.; Bole-Feysot, C.; Nitschke, P.; Masson, C.; Alanay, Y.; Al-Gazali, L.; Bitoun, P.; et al. Chondrodysplasia with multiple dislocations: Comprehensive study of a series of 30 cases. *Clin. Genet.* **2017**, *91*, 868–880. [CrossRef]
28. Nomura, Y. Structural change in decorin with skin aging. *Connect Tissue Res.* **2006**, *47*, 249–255. [CrossRef]

29. Malmström, A. Biosynthesis of dermatan sulfate. II. Substrate specificity of the C-5 uronosyl epimerase. *J. Biol. Chem.* **1984**, *259*, 161–165.
30. Mizumoto, S.; Kosho, T.; Hatamochi, A.; Honda, T.; Yamaguchi, T.; Okamoto, N.; Miyake, N.; Yamada, S.; Sugahara, K. Defect in dermatan sulfate in urine of patients with Ehlers-Danlos syndrome caused by a *CHST14/D4ST1* deficiency. *Clin. Biochem.* **2017**, *50*, 670–677. [CrossRef]
31. Hirose, T.; Takahashi, N.; Tangkawattana, P.; Minaguchi, J.; Mizumoto, S.; Yamada, S.; Miyake, N.; Hayashi, S.; Hatamochi, A.; Nakayama, J.; et al. Structural alteration of glycosaminoglycan side chains and spatial disorganization of collagen networks in the skin of patients with mcEDS-*CHST14*. *Biochim. Biophys. Acta Gen. Subj.* **2019**, *1863*, 623–631. [CrossRef] [PubMed]
32. Casu, B.; Petitou, M.; Provasoli, M.; Sinaÿ, P. Conformational flexibility: A new concept for explaining binding and biological properties of iduronic acid-containing glycosaminoglycans. *Trends Biochem. Sci.* **1988**, *13*, 221–225. [CrossRef]
33. Catlow, K.R.; Deakin, J.A.; Wei, Z.; Delehedde, M.; Femig, D.G.; Gherardi, E.; Gallagher, J.T.; Pavão, M.S.G.; Lyon, M. Interactions of hepatocyte growth factor/scatter factor with various glycosaminoglycans reveal an important interplay between the presence of iduronate and sulfate density. *J. Biol. Chem.* **2008**, *283*, 5235–5248. [CrossRef] [PubMed]
34. Watanabe, T.; Kametani, K.; Koyama, Y.I.; Suzuki, D.; Imamura, Y.; Takehana, K.; Hiramatsu, K. Ring-mesh model of proteoglycan glycosaminoglycan chains in tendon based on three-dimensional reconstruction by focused ion beam scanning electron microscopy. *J. Biol. Chem.* **2016**, *291*, 23704–23708. [CrossRef] [PubMed]
35. Tang, T.; Li, L.; Tang, J.; Li, Y.; Lin, W.Y.; Martin, F.; Grant, D.; Solloway, M.; Parker, L.; Ye, W.; et al. A mouse knockout library for secreted and transmembrane proteins. *Nat. Biotechnol.* **2010**, *28*, 749–755. [CrossRef] [PubMed]
36. Bian, S.; Akyüz, N.; Bernreuther, C.; Loers, G.; Laczynska, E.; Jakovcevski, I.; Schachner, M. Dermatan sulfotransferase Chst14/D4st1, but not chondroitin sulfotransferase Chst11/C4st1, regulates proliferation and neurogenesis of neural progenitor cells. *J. Cell Sci.* **2011**, *124*, 4051–4063. [CrossRef]
37. Yoshizawa, T.; Mizumoto, S.; Takahashi, Y.; Shimada, S.; Sugahara, K.; Nakayama, J.; Takeda, S.; Nomura, Y.; Nitahara-Kasahara, Y.; Okada, T.; et al. Vascular abnormalities in the placenta of *Chst14*[-/-] fetuses: Implications in the pathophysiology of perinatal lethality of the murine model and vascular lesions in human *CHST14/D4ST1* deficiency. *Glycobiology* **2018**, *28*, 80–89. [CrossRef]
38. Akyüz, N.; Rost, S.; Mehanna, A.; Bian, S.; Loers, G.; Oezen, I.; Mishra, B.; Hoffmann, K.; Guseva, D.; Laczynska, E.; et al. Dermatan 4-*O*-sulfotransferase1 ablation accelerates peripheral nerve regeneration. *Exp. Neurol.* **2013**, *247*, 517–530. [CrossRef]
39. Klüppel, M.; Wight, T.N.; Chan, C.; Hinek, A.; Wrana, J.L. Maintenance of chondroitin sulfation balance by chondroitin-4-sulfotransferase 1 is required for chondrocyte development and growth factor signaling during cartilage morphogenesis. *Development* **2005**, *132*, 3989–4003. [CrossRef]
40. Evers, M.R.; Xia, G.; Kang, H.G.; Schachner, M.; Baenziger, J.U. Molecular cloning and characterization of a dermatan-specific *N*-acetylgalactosamine 4-*O*-sulfotransferase. *J. Biol. Chem.* **2001**, *276*, 36344–36353. [CrossRef]
41. Mikami, T.; Mizumoto, S.; Kago, N.; Kitagawa, H.; Sugahara, K. Specificities of three distinct human chondroitin/dermatan *N*-acetylgalactosamine 4-*O*-sulfotransferases demonstrated using partially desulfated dermatan sulfate as an acceptor: Implication of differential roles in dermatan sulfate biosynthesis. *J. Biol. Chem.* **2003**, *278*, 36115–36127. [CrossRef] [PubMed]
42. Maccarana, M.; Olander, B.; Malmström, J.; Tiedemann, K.; Aebersold, R.; Lindahl, U.; Li, J.P.; Malmström, A. Biosynthesis of dermatan sulfate: Chondroitin-glucuronate C5-epimerase is identical to SART2. *J. Biol. Chem.* **2006**, *281*, 11560–11568. [CrossRef] [PubMed]
43. Pacheco, B.; Malmström, A.; Maccarana, M. Two dermatan sulfate epimerases form iduronic acid domains in dermatan sulfate. *J. Biol. Chem.* **2009**, *284*, 9788–9795. [CrossRef] [PubMed]
44. Tykesson, E.; Hassinen, A.; Zielinska, K.; Thelin, M.A.; Frati, G.; Ellervik, U.; Westergren-Thorsson, G.; Malmström, A.; Kellokumpu, S.; Maccarana, M. Dermatan sulfate epimerase 1 and dermatan 4-*O*-sulfotransferase 1 form complexes that generate long epimerized 4-*O*-sulfated blocks. *J. Biol. Chem.* **2018**, *293*, 13725–13735. [CrossRef]

45. Maccarana, M.; Kalamajski, S.; Kongsgaard, M.; Magnusson, S.P.; Oldberg, A.; Malmström, A. Dermatan sulfate epimerase 1-deficient mice have reduced content and changed distribution of iduronic acids in dermatan sulfate and an altered collagen structure in skin. *Mol. Cell Biol.* **2009**, *29*, 5517–5528. [CrossRef]
46. Gustafsson, R.; Stachtea, X.; Maccarana, M.; Grottling, E.; Eklund, E.; Malmström, A.; Oldberg, A. Dermatan sulfate epimerase 1 deficient mice as a model for human abdominal wall defects. *Birth Defects Res. A Clin. Mol. Teratol.* **2014**, *100*, 712–720. [CrossRef]
47. Bartolini, B.; Thelin, M.A.; Rauch, U.; Feinstein, R.; Oldberg, A.; Malmström, A.; Maccarana, M. Mouse development is not obviously affected by the absence of dermatan sulfate epimerase 2 in spite of a modified brain dermatan sulfate composition. *Glycobiology* **2012**, *22*, 1007–1016. [CrossRef]
48. Akatsu, C.; Mizumoto, S.; Kaneiwa, T.; Maccarana, M.; Malmström, A.; Yamada, S.; Sugahara, K. Dermatan sulfate epimerase 2 is the predominant isozyme in the formation of the chondroitin sulfate/dermatan sulfate hybrid structure in postnatal developing mouse brain. *Glycobiology* **2011**, *21*, 565–574. [CrossRef]
49. Stachtea, X.N.; Tykesson, E.; van Kuppevelt, T.H.; Feinstein, R.; Malmström, A.; Reijmers, R.M.; Maccarana, M. Dermatan sulfate-free mice display embryological defects and are neonatal lethal despite normal lymphoid and non-lymphoid organogenesis. *PLoS ONE* **2015**, *10*, e0140279. [CrossRef]
50. Koitabashi, N.; Yamaguchi, T.; Fukui, D.; Nakano, T.; Umeyama, A.; Toda, K.; Funada, R.; Ishikawa, M.; Kawamura, R.; Okada, K.; et al. Peripartum iliac arterial aneurysm and rupture in a patient with vascular Ehlers-Danlos syndrome diagnosed by next-generation sequencing. *Int. Heart J.* **2018**, *59*, 1180–1185. [CrossRef]

© 2019 by the authors. Licensee MDPI, Basel, Switzerland. This article is an open access article distributed under the terms and conditions of the Creative Commons Attribution (CC BY) license (http://creativecommons.org/licenses/by/4.0/).

Article

Defining the Clinical, Molecular and Ultrastructural Characteristics in Occipital Horn Syndrome: Two New Cases and Review of the Literature

Aude Beyens [1,2], Kyaran Van Meensel [1], Lore Pottie [1], Riet De Rycke [3,4,5], Michiel De Bruyne [3,4,5], Femke Baeke [3,4,5], Piet Hoebeke [6], Frank Plasschaert [7], Bart Loeys [8], Sofie De Schepper [2], Sofie Symoens [1] and Bert Callewaert [1,*]

1. Center for Medical Genetics Ghent, Ghent University Hospital, 9000 Ghent, Belgium
2. Department of Dermatology, Ghent University Hospital, 9000 Ghent, Belgium
3. Department for Biomedical Molecular Biology, Ghent University, 9000 Ghent, Belgium
4. VIB Center for Inflammation Research, 9000 Ghent, Belgium
5. Ghent University Expertise Centre for Transmission Electron Microscopy and VIB BioImaging Core, 9000 Ghent, Belgium
6. Department of Urology, Ghent University Hospital, 9000 Ghent, Belgium
7. Department of Orthopedic Surgery, Ghent University Hospital, 9000 Ghent, Belgium
8. Center for Medical Genetics, University of Antwerp/Antwerp University Hospital, Antwerp, Belgium
* Correspondence: bert.callewaert@ugent.be; Tel.: +32-9-332-5026

Received: 28 June 2019; Accepted: 11 July 2019; Published: 12 July 2019

Abstract: Occipital horn syndrome (OHS) is a rare connective tissue disorder caused by pathogenic variants in ATP7A, encoding a copper transporter. The main clinical features, including cutis laxa, bony exostoses, and bladder diverticula are attributed to a decreased activity of lysyl oxidase (LOX), a cupro-enzyme involved in collagen crosslinking. The absence of large case series and natural history studies precludes efficient diagnosis and management of OHS patients. This study describes the clinical and molecular characteristics of two new patients and 32 patients previously reported in the literature. We report on the need for long-term specialized care and follow-up, in which MR angiography, echocardiography and spirometry should be incorporated into standard follow-up guidelines for OHS patients, next to neurodevelopmental, orthopedic and urological follow-up. Furthermore, we report on ultrastructural abnormalities including increased collagen diameter, mild elastic fiber abnormalities and multiple autophagolysosomes reflecting the role of lysyl oxidase and defective ATP7A trafficking as pathomechanisms of OHS.

Keywords: occipital horn syndrome; Ehlers–Danlos syndrome type IX; Menkes syndrome; ATP7A; review; copper transport; elastic fiber; collagen

1. Introduction

Occipital horn syndrome (OHS, OMIM#304150), previously known as Ehlers–Danlos syndrome type IX or X-linked cutis laxa, is a rare disorder characterized by prominent connective tissue abnormalities including cutis laxa, hernias, joint laxity and bladder diverticula, and pathognomonic exostoses with "occipital horns" or downward pointing exostoses situated in the tendinous insertions of the sternocleidomastoid and trapezius muscles. In addition, patients may show mild to moderate intellectual disability [1–6]. Being allelic with Menkes disease (MD, OMIM#30011), OHS is considered the milder end of the phenotypic spectrum due to pathogenic variants in *ATP7A* [4] that encodes a copper transporter. Several clinical features in both disorders are related to the malfunctioning of cupro-enzymes, including lysyl oxidase, dopamine ß-hydroxylase, tyrosinase and cytochrome C oxidase [7–10].

ATP7A mediates (1) copper transport from the gastrointestinal tract into the bloodstream, (2) intracellular delivery of copper to cupro-enzymes in the secretory pathway and (3) efflux of excess copper from the cell [11,12]. Intracellular ATP7A trafficking is essential for copper homeostasis. Under basal low copper conditions, ATP7A is found in the trans-Golgi network, where it is essential for cupro-enzyme biogenesis. At higher copper concentrations, it reversibly localizes to the plasma membrane and post-Golgi compartments, where it is responsible for the extrusion of excess copper from the cell [13–15]. The *ATP7A* gene maps to Xq13.3 and spans a genomic region of ~140 kb. Its predominant transcription product contains 23 exons [16–20]. ATP7A belongs to the large P-type ATPase family, which are ATP-driven membrane pumps essential in the maintenance of electrochemical gradients (such as Na^+/K^+-, H^+/K^+- and Ca^{2+}-pumps), as well as lipid and cationic homeostasis [21], including ATP7A and ATP7B, both important for copper homeostasis [13,22]. P-type ATPases share a structural core containing a transmembrane (TM) domain responsible for transport and an N-, P- and A- soluble catalytic domain, respectively involved in nucleotide binding, phosphorylation and dephosphorylation [23]. The transmembrane domain of ATP7A contains eight transmembrane helices (TM1–TM8). Six amino terminal domains contain metal-binding CXXC motifs (MBD1–MBD6) [24,25].

MD is mostly caused by truncating variants, including nonsense variants, frameshift variants and splice-site variants that lead to out of frame transcripts and large deletions, which result in very low to absent ATP7A levels [12,26]. On the contrary, the milder symptoms in OHS are the consequence of pathogenic variants in *ATP7A*, mostly "leaky" splice-site variants resulting in exon-skipping, which permit the production of a small amount of normal ATP7A mRNA [6,13,27–30]. This low amount of normal mRNA can result in residual protein activity and allow effective cation transfer to some copper-dependent enzymes. ATP7A levels as low as 2–5% have been shown to be sufficient to result in the milder OHS phenotype [31]. Connective tissue abnormalities in OHS have been attributed to a decreased activity of lysyl oxidase (LOX), a cupro-enzyme that normally deaminates lysine and hydrolysine residues in the first step of collagen crosslinking [7,32]. Patients may also encounter mild neurological signs and dysautonomia due to partial defects in dopamine-ß-hydroxylase enzyme activity, that converts dopamine to norepinephrine, a crucial neurotransmitter in norepinephrinergic neurons. As such, patients with OHS typically have low to normal levels of serum copper and ceruloplasmin together with abnormal plasma and cerebrospinal fluid (CSF) catecholamine levels [3,33,34].

Due to the rarity of OHS and the limited number of cases scattered in literature, the natural history and clinical phenotype remain incompletely studied. This study reports the clinical, molecular and ultrastructural data for two new cases of OHS. In addition, we performed a clinical review of 32 previously described patients [6,7,11,27–31,35–51].

2. Materials and Methods

2.1. Patients

Informed consent was obtained from subject 1 and his parents, and from subject 2. Additional consent was given to publish the clinical pictures in Figure 1. Clinical information was obtained by evaluating both patients at our center with the use of a detailed clinical checklist. We performed a literature review of 32 OHS patients with confirmed *ATP7A* pathogenic variants in the Pubmed and Embase databases, only including studies with ample clinical data. Selection of the reports was undertaken independently by two different authors (A.B and K.V.M). Clinical data from the literature were evaluated using the same methodology (Supplementary Table S1). This study was conducted in accordance with the Declaration of Helsinki and approved by the Ghent University Hospital ethics committee (registration number B670201319316).

2.1.1. Clinical Reports

Case 1

Subject 1 (F1:II-1) was born at term after an uncomplicated pregnancy. At birth, his weight was 2400 g, length 45.5 cm and head circumference 32.4 cm. Developmental milestones were achieved normally. At three years of age, he presented with micturition problems evolving to urinary retention, which led to the discovery of giant bladder diverticula. Over subsequent years, several diverticulectomies were performed, followed by continent diversion and intermittent catheterization.

Clinical examination at 11 years of age showed mild facial dysmorphism (Figure 1) with a flat face, deep-set eyes, a long narrow nose, large ears and mild webbing of the neck. His weight was 24.9 kg (P1, −2.37 SD), height 139.2 cm (P26, −0.649 SD) and head circumference 56.2 cm (P98, +2.1 SD). Skeletal abnormalities included joint hyperlaxity of the wrists, metacarpophalangeal joints and fingers, a flat back, marked bony thickening of the proximal radius and tibia, genua valga and pedes plani. There was skin webbing between the second and third toe. His skin was soft and mildly hyperextensible with fine wrinkles over the abdomen and dorsum of hands and feet. He had no hair abnormalities. Neurological examination was normal. He had recurrent inguinal hernias. He reported bruising easily, but no impaired wound healing. He suffered from chronic joint pain. The patient attended normal education.

Additional examinations included a total skeletal radiography and MRI/MRA (Figure 2). X-rays show severe skeletal dysplasia, occipital horns, undertubulation of bony structures and broadening of the ventral end of the first left rib, the distal end of both claviculae and the scapular neck. The long bones showed bowing with mid-diaphyseal broadening. There was bilateral luxation of the radial heads, bilateral overgrowth of the ulnar coronoid processes and prominent trochanter minor of both femurs. We observed a short fibula and bilateral coxa valga with rounded iliac wings. The metatarsals were broad. The vertebrae showed platyspondyly. Brain MRI and angiography showed tortuous intracranial arteries. Echocardiography, lung radiography and spirometry were normal. Serum copper and ceruloplasmin levels were within normal ranges (89.6µg/dL (70–140) and 0.287 g/L (0.2–0.6), respectively).

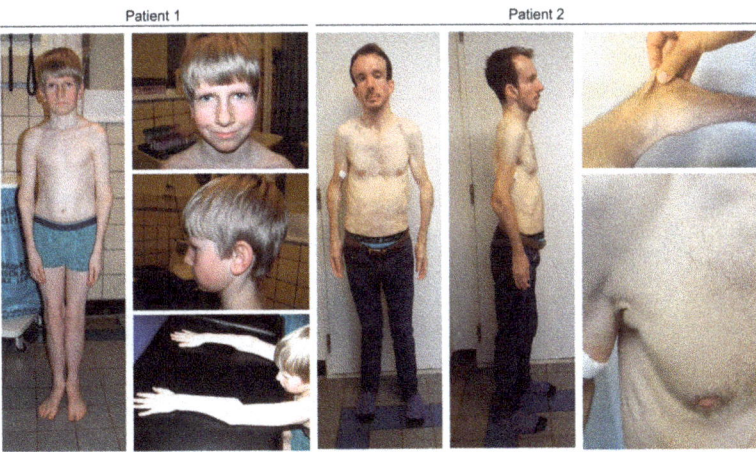

Figure 1. Clinical characteristics in subjects F1:II-1 and F2:II-1. Subject F1:II-1 displays mild facial dysmorphism with a flat face, deep-set eyes, a long and narrow nose, large ears and mild webbing of the neck. Craniofacial features in patient F2:II-1 are more pronounced and include dolichocephaly, a high forehead, biparietal narrowing, downslanting eyes, a convex nasal ridge, midfacial hypoplasia, micrognathia and low-set ears. No hair abnormalities were observed. Both have lax and hyperextensible skin, marked joint hypermobility and deformities of the proximal radius.

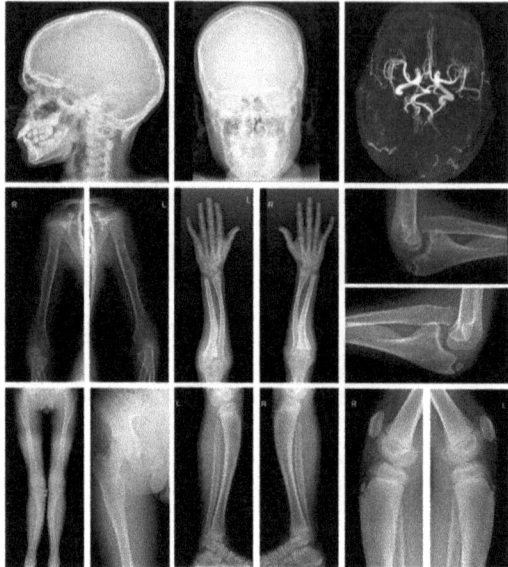

Figure 2. Total skeletal radiography and MRI/MRA studies in subject F1:II-1 showing severe skeletal dysplasia with occipital horns, undertubulation of bony structures and broadening of the ventral end of the first left rib, both claviculae and scapular neck, bowing of the long bones with mid-diaphyseal broadening, bilateral luxation of the radial heads, bilateral overgrowth of the ulnar coronoid processes, prominent trochanter minor of both femurs, short fibula, bilateral coxa valga with rounded iliac wings and broad metatarsals. Brain MRI and angiography showed tortuous intracranial arteries and normal brain structure.

Case 2

Subject 2 (F2:II-1) was evaluated at the age of 30. He was born at term after an uncomplicated pregnancy. He developed an inguinal hernia at six weeks of age. Since then, he had multiple inguinal and femoral hernias requiring repeated mesh repairs. Starting from two years of age, he had severe bladder problems, including recurrent urinary tract infections and urinary retention, due to multiple bladder diverticula. After more than 20 surgical bladder repairs, he started self-catheterization at 8 years of age. Aged 20 years, he underwent left nephrectomy for a non-functioning kidney. He reported a mild motor delay and walked at two years of age. Joint hypermobility resulted in repeated shoulder and elbow subluxations. He was of normal intelligence and pursued higher education. He suffered from migraine, worsening musculoskeletal pain, fatigue, and depression for which he was treated with gabapentin and amitriptyline. He had a history of mitral valve prolapse, orthostatic hypotension, palpitations and a prolonged QT interval on electrocardiogram. Finally, he had a longstanding history of upper and lower gastrointestinal symptoms including nausea, post-prandial vomiting and chronic diarrhea, possibly in the context of dysautonomia. He was diagnosed with asthma.

His height was 169 cm, his weight was 50.6 kg and his head circumference was 58.5 cm. Craniofacial features consisted of dolichocephaly, a high forehead, biparietal narrowing, downslanting eyes, a convex nasal ridge, midfacial hypoplasia, micrognathia and low-set ears. He had lax and hyperextensible skin (Figure 1). He had no hair abnormalities. Skeletal abnormalities included occipital horns, a loss of sagittal spine curvature, scoliosis, pectus excavatum, and restricted elbow extension. He had bowing of the ulna and radius, bilateral exostoses of the radii and tibiae, bilateral rounding of the iliac wings, cranial displacement of the right humeral head and genua vara. Neurological examination and brain

MRI were normal. Serum copper and ceruloplasmin levels were 8.3 µmol/L (11–20) and 0.13 g/L (0.20–0.50), respectively.

2.2. Molecular Analysis

For individual 1, genomic DNA was extracted from blood leukocytes using standard procedures. Array comparative genomic hybridization (180 k arrayCGH, Agilent Technologies, Santa Clara, CA, USA) showed a 300 kb deletion of chromosome band 1q31.1 (1q31.1.1q31.1(188407205-188707983)x1)(hg19/GRCh37), which does not contain any known genes. *FBLN5* (OMIM#219100, GenBank RefSeq accession number NM_006329.3), *LTBP4* (OMIM#613177, NM_003573.2) and *ATP7A* (OMIM#304150, NM_000052.7) gene sequencing was performed without the identification of a pathogenic mutation. Total RNA was extracted from cultured fibroblasts using the RNeasy Kit (Qiagen, Hilden, Germany) and cDNA was synthesized with the iScript cDNA Synthesis Kit (Bio-Rad Laboratories, Hercules, CA, USA). *ATP7A* cDNA analysis was performed followed by confirmation with long-range RT-PCR and next-generation sequencing (Miseq, Illumina, San Diego, CA, USA). For individual 2, array comparative genome hybridization was normal after which *ATP7A* gene sequencing (GenBank RefSeq accession number NM_000052.7) was performed.

2.3. Transmission Electron Microscopy (TEM)

For transmission electron microscopy, 3 mm skin fragments from both individuals and an age- and sex-matched control were immersed in a fixative solution of 2.5% glutaraldehyde and 4% formaldehyde in Na-Cacodylate buffer 0.1 M, placed in a vacuum oven for 30 min and left rotating for 3 hours at room temperature. This solution was later replaced with fresh fixative and samples were left rotating over night at 4 °C. After washing, samples were post-fixed in 1% OsO_4 with $K_3Fe(CN)6$ in 0.1 M NaCacodylate buffer, pH 7.2. After washing in double-distilled H_2O, samples were subsequently dehydrated through a graded ethanol series, including bulk staining with 2% uranyl acetate at the 50% ethanol step, followed by embedding in Spurr's resin. To select the area of interest on the block and in order to have an overview of the phenotype, semi-thin sections were first cut at 0.5 µm and stained with toluidine blue. Ultrathin sections of a gold interference color were cut using an ultramicrotome (Leica EM UC6, Wetzlar, Germany), followed by post-staining in a Leica EM AC20 for 40 min in uranyl acetate at 20 °C and for 10 min in lead stain at 20 °C. Sections were collected on formvar-coated copper slot grids. Grids were viewed with a JEM 1400plus transmission electron microscope (JEOL, Tokyo, Japan) operating at 80 kV.

2.4. Collagen Biochemical Analysis

Fibroblast cultures were started from skin biopsies from individuals 1 and 2, and a matched control. Biochemical (pro)collagen sodium dodecyl sulfate polyacrylamide gel electrophoresis (SDS-PAGE analysis) was performed on the medium and cellular fractions of the cultured skin fibroblasts. When confluent, cells were labeled with 14C-proline [52]. After SDS-PAGE separation, further processing of the gels was performed as described previously by our group [53].

3. Results

3.1. Clinical Characteristics

We describe the clinical characteristics of two new individuals with OHS and 32 patients described in the literature (Table 1, Supplementary Table S1). All patients were male, except for patient F17:II-2. The mean reported age of patients was 19.6 years (mean 17, range 1–50). Disease-related mortality was observed in 5 out of 34 patients. The mean age at death was 21.2 years and causes included respiratory failure (post-surgery apnea), gastric perforation and diffuse intravascular coagulation due to a large cephalhematoma at birth.

Table 1. Overview of the phenotypic presentation in occipital horn syndrome.

		Total (Percentage)
Craniofacial		
	Long face	6/13 (46%)
	Large ears	5/13 (38%)
	Sagging/ full cheeks	5/11 (45%)
	Hair abnormalities	14/19 (74%)
	Pili torti on trichoscopy	7/10 (70%)
Connective tissue		
	Cutis laxa	25/27 (93%)
	Inguinal hernia	15/24 (63%)
	Umbilical hernia	1/22 (5%)
Osteoarticular		
	Occipital horns	25/26 (96%)
	Radial/tibial exostoses	6/16 (38%)
	Hammer-shaped clavicula	9/19 (47%)
	Bowing of long bones	4/17 (23%)
	Mid-diaphyseal broadening	2/17 (11%)
	Rounding of the iliac wings	3/17 (18%)
	Coxa valga	6/16 (38%)
	Genu valgum	6/17 (35%)
	Metaphyseal spurring	2/16 (13%)
	Scoliosis	8/21 (38%)
	Pectus deformity	13/23 (57%)
	Dislocations	12/20 (60%)
	Contractures of large joints	3/19 (16%)
	Joint hyperlaxity	16/26 (62%)
	Fractures	2/21 (10%)
Neurological		
	Intellectual disability	17/33 (52%)
	Seizures	5/25 (20%)
	Muscle hypotonia	10/25 (40%)
	Stroke	1/23 (4%)
Cardiovascular		
	Aneurysm formation	5/7 (71%)
	Dilatation of the large veins	2/6 (33%)
	Intracranial tortuosity	7/11 (64%)
	Extracranial tortuosity	3/4 (75%)
	Dysautonomia	13/15 (87%)
Urogenital		
	Bladder diverticula	25/30 (83%)
	Renal abnormalities	6/20 (30%)
	Urinary tract infections	17/23 (74%)
	Vesicourethral reflux	7/23 (30%)
Laboratory findings		
	Serum copper	20/28 (71%)
	Serum ceruloplasmin	20/27 (74%)

Initial presentations leading to the diagnosis are presented in Table 2. One-third of patients presented with neurological features including hypotonia, developmental delay, and seizures. Nine individuals presented with connective tissue problems such as cephalhematoma and inguinal hernia. Other presenting symptoms included bladder problems such as bladder diverticula, urinary tract infections and skeletal problems.

Table 2. Initial presentations leading to the diagnosis in occipital horn syndrome.

Initial Presentation	Number of Patients
Neurological	11
Seizures	1
Developmental delay	4
Hypotonia	6
Connective tissue	9
Cephalhematoma	4
Generalized CTD	3
Inguinal hernia	2
Urogenital	5
Bladder diverticula	3
Urinary infections	2
Skeletal	3
Pectus deformity	1
Skeletal dysplasia	1
Joint pain	1
Other	3
Vomiting and diarrhea	1
Dysautonomia	1
Apnea	1
Segregation analysis	1
Unknown	2

"CTD": connective tissue disease.

Patients displayed variable craniofacial features, including a long face (6/13, 46%), large ears (5/13, 38%), sagging cheeks (5/11, 45%) and hair abnormalities (14/19, 74%). Trichoscopy confirmed pili torti in 7 out of 10 examined individuals. Connective tissue abnormalities were reported in all subjects. Cutis laxa was present in the majority of patients (25/26, 94%), as well as inguinal hernias (15/25, 63%). An umbilical hernia and hiatal hernia were observed in only one patient each (F25:II-1 and F15:II-1, respectively). Wound healing was reported to be normal after trauma or surgery. Occipital horns were reported in nearly all cases (25/26, 96%). Other recurrent skeletal manifestations included hammer-shaped claviculae (9/19, 47%), scoliosis (8/21, 38%), pectus deformities (13/23, 57%), radial/tibial exostoses (6/16, 38%), coxa valga (6/16, 38%) and genua valga (6/17, 35%). Dislocations, mainly of the radial heads and joint hypermobility were present in about two-thirds of individuals, 12/20 (60%) and 16/26 (62%), respectively. Less frequently reported skeletal manifestations include bowing of the long bones, mid-diaphyseal broadening, metaphyseal spurring and rounding of the iliac wings. Osteopenia or osteoporosis was reported in four patients and fractures were observed in two subjects.

Mild, mainly motor, developmental delay, was observed in 17 subjects (17/30, 57%). Intellectual disability was present in an equal number of patients (17/33, 52%) of whom seven had mild, four had moderate and three had severe intellectual disability. In three patients, the severity of intellectual disability was not specified. Several subjects displayed muscle hypotonia, especially at a young age (before the age of two). Seizures were documented in five patients (5/25, 20%). One individual was diagnosed with a spontaneous intracerebral hemorrhage at the age of 15, resulting in aphasia and spastic quadriparesis. MRI brain studies were performed in 11 patients and documented to be abnormal in 7 subjects, showing mild atrophy and/or delayed myelination. Intracranial tortuosity was present in two-thirds of patients (7/11, 64%). Tortuosity of extracranial arteries was found in three cases and was located in the cervical, splenic and hepatic arteries. Aneurysm formation was found in five patients, mainly affecting the same arteries. One individual was diagnosed with a type A aortic dissection at the age of 39, which was successfully surgically manage. Two patients showed dilatation of the large veins. Orthostatic hypotension, temperature instability, chronic diarrhea, and other symptoms of dysautonomia were seen in a high number of patients (13/15, 87%). Severe urological complications are an important concern in OHS, with bladder diverticula in 83% (25/30) of reported patients, of which three were eventually diagnosed with a bladder rupture. Recurrent urinary tract infections based on incomplete emptying of the bladder were seen in 74% (17/23) of patients. One-third of patients had vesicourethral reflux (VUR), sometimes complicated by secondary renal problems. Asthma was reported in four cases. Low levels of serum copper and ceruloplasmin were found in two-thirds of cases, 20/28 (71%) and 20/27 (74%), respectively.

3.2. Molecular Characteristics

After normal *ATP7A* gene sequencing, cDNA analysis showed a deletion of exon 10 in subject F1:II-1. Subsequent confirmation with long-range RT-PCR revealed a deep intronic c.2407-516G>T variant, underlying the exon 10 skipping, likely due to the activation of a cryptic splice acceptor. In patient F2:II-1, *ATP7A* gene sequencing identified a hemizygous c.2917-4A>G splice-site variant in *ATP7A*.

A total of 16 different *ATP7A* pathogenic variants were reported (Figure 3); these included seven splice-site, four missense and two frameshift variants, two deep intronic variants and one deletion. In eight patients, the specific variant was not reported, including a 98-bp deletion in the regulatory region (F9:II-1), a deep intronic intron six splice donor site variant (F12:II-1) and a duplication in exon 11 and 12 (F23:II-1). Nonsense ATP7A variants were not observed. Pathogenicity information and CADD scores can be found in Supplementary Table S2. We did not observe any clear genotype-phenotype correlation and there was no correlation with serum copper and ceruloplasmin levels (Supplementary Table S3).

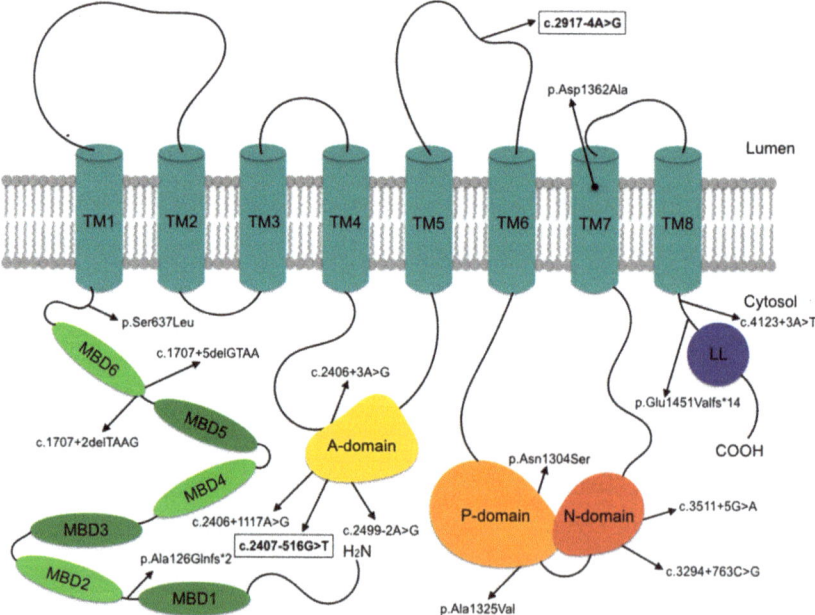

Figure 3. ATP7A pathogenic variants reported in this study. ATP7A consists of six amino terminal metal binding domains (MBD1-MBD6), a transmembrane domain containing eight transmembrane helices (TM1-TM8) and an N-, P- and A- soluble catalytic domain. The pathogenic variants identified in subjects F1:II-1 and F2:II-1 are indicated in bold and outlined.

3.3. Transmission Electron Microscopy

Transmission electron microscopy (TEM) findings in skin samples of subjects F1:II-1, F2:II-1 and a matched control are shown in Figure 4. Collagen fibrils in the dermis of the control sample are regularly organized in curvilinear bundles with uniform diameters. In the dermis of both OHS patients, the collagen fiber diameter appears to be larger and more variable. The fibrils are more loosely packed (especially in F2:II-1) but retain a normal shape. TEM of a control elastic fiber demonstrates a solid and dense core of elastin surrounded by a sparse mantle of microfibrils. In both individuals, the elastin core is surrounded by a debris extending outside the elastic fiber with loss of the microfibrillar organization. In addition, the elastin core in subject F2:II-1 is almost completely disrupted and difficult to discriminate. Fibroblast abnormalities are observed in both patients, with multiple autophagolysomes containing sickle-shaped electron dense deposits (Supplementary Figure S1). The mitochondria, Golgi complex, and endoplasmatic reticulum appear normal.

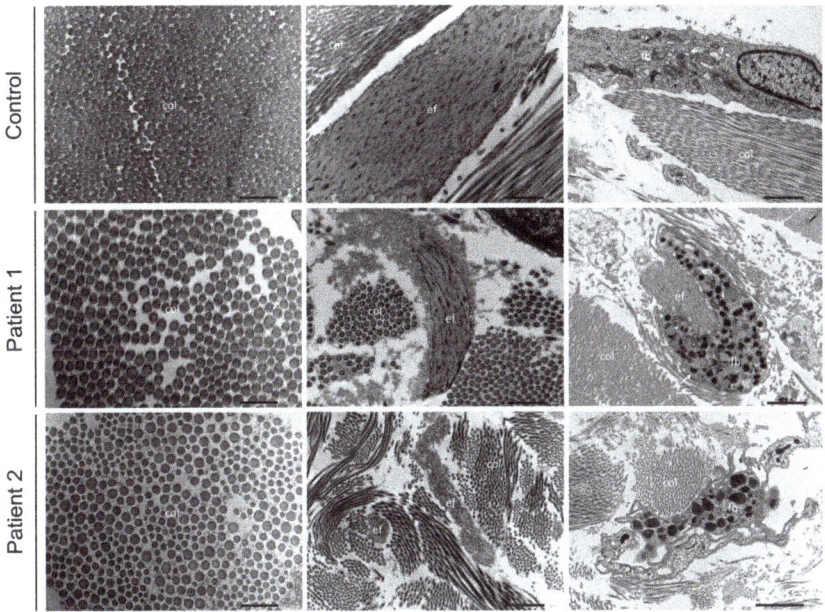

Figure 4. Transmission electron microscopy findings in occipital horn syndrome. Transmission electron microscopy (TEM) findings in skin samples of patients F1:II-1 and F2:II-1 and a matched control showing collagen (column 1), elastic fibers (column 2) and a fibroblast (column 3). col = collagen; ef = elastic fiber; fb = fibroblast. Magnification for column 1: ×15,000, column 2: ×8000 and column 3: ×4000.

3.4. Collagen Biochemistry

Considering the collagen abnormalities observed in both individuals with transmission electron microscopy, SDS-PAGE analysis of the intracellular and secreted (pro)collagen proteins produced by skin fibroblasts was performed for subjects 1 and 2 and a matched control. In both cases, we observed normal SDS-PAGE migration patterns (Supplementary Figure S2).

4. Discussion and Conclusions

The absence of large case series and natural history studies precludes efficient diagnosis and management of OHS patients. This study describes the clinical and molecular characteristics of two new patients and 32 patients previously reported in the literature. OHS is considered a mild but rare allelic variant of Menkes disease. OHS patients show prominent connective tissue abnormalities and have a better prognosis and survival rate than patients with MD [3], with low childhood mortality (1/34) in OHS. However, there is still significant mortality in (young) adults, often related to gastrointestinal, respiratory, or bleeding complications, such as an intestinal perforation due to a gastric ulcer and (post-surgery) apnea.

The clinical characteristics of OHS are listed in Table 2. Craniofacial features include mild facial dysmorphism with a long face, large ears and sagging cheeks in about half of reported patients. Hair abnormalities, including coarse hair and pili torti, are frequently present. Ubiquitous connective tissue manifestations in OHS include increased skin laxity and inguinal hernia.

Concerning the skeletal features, the pathognomonic occipital horns are present in all patients but one, which is probably an age-dependent discrepancy [41,54,55]. Other less specific skeletal abnormalities include hammer-shaped claviculae and radial/tibial exostoses. Scoliosis, pectus deformities, coxa and genua valga, joint luxations, mainly of the radial heads and general joint hypermobility are frequent but non-specific. These skeletal problems may limit daily activities and often require multiple surgical interventions.

Urological complications, mostly giant bladder diverticula and vesicourethal reflux, are found in over 80% of subjects. Secondary problems such urinary tract infections, urinary retention, bladder rupture, or VUR-mediated renal failure, which are all related to the bladder diverticula, are of great concern. Surgical interventions removing the diverticula often fail, with frequent relapse of the diverticula. Self-catheterization, vesicostomy, or continent diversion are often required for proper emptying of the bladder.

Previous reports describe a mild neurological phenotype limited to slight generalized muscle weakness and symptoms of dysautonomia [3,27], however, more detailed case studies reveal some conspicuous observations including developmental delay and intellectual disability in two-thirds of subjects and documented seizures in five individuals. Nevertheless, neurological affliction remains rather mild and normal intelligence is possible. Symptoms of dysautonomia, including chronic diarrhea, temperature instability, and orthostatic hypotension are present in almost 90% of patients and can be disabling.

Brain imaging studies are rarely performed in OHS patients and are often reported as normal. However, intracranial arterial tortuosity are noted in approximately two thirds of patients. Extracranial arterial tortuosity is less frequently observed, but has been reported in the cervical, splenic and hepatic vasculature. Aneurysm formation may complicate pre-existing tortuosity and may affect the arterial and venous circulation.

OHS is often diagnosed upon the identification of connective tissue anomalies, including cutis laxa, joint hypermobility with hypotonia, mild intellectual impairment and bladder diverticula. Later on, the pathognomonic occipital horns become more evident. The initial differential diagnosis includes other forms of cutis laxa, including *FBLN5*- and *LTBP4*-related cutis laxa (OMIM#219100 and #613177) [56] and the dermatosparaxis type of Ehlers–Danlos syndrome (EDS type VIIC, OMIM#225410) [57], which may present with cutis laxa, hypotonia and bladder diverticula, but with normal mental development. However, in OHS there is no early onset emphysema as in *FBLN5*- and *LTBP4*- related cutis laxa, nor easy bruising or severe skin fragility as in the dermatosparaxis type of EDS. *ATP6V0A2*-related cutis laxa (OMIM#219200) may also present with mild intellectual impairment but does not commonly show bladder diverticula or exostoses. Later on the exostoses can be mistaken for hereditary multiple exostoses due to defects in *EXT1* or *EXT2* (OMIM#133700 and #133701) [58].

Concerning follow-up, routine evaluation of neurodevelopment and early intervention with physiotherapy is recommended, as well as regular urological evaluation, including video urodynamic studies of bladder function. Any pain, or vascular or neurological symptoms that may relate to exostoses should be promptly investigated. To date, there are no data on a possible risk for chondrosarcomas in exostoses although these have been described in hereditary multiple exostoses [59]. It could be recommended to perform MR angiography and echocardiography from puberty onwards. Indeed, aortic root dilatation and dissection, and an increased risk for intracranial bleeds and ischemic complications have been described in other related connective tissue disorders such as Ehlers–Danlos, arterial tortuosity, Marfan and some cutis laxa syndromes. Furthermore, spirometry might detect asthma at an early stage as it was reported in four individuals. Prolonged vigilance for apnea is necessary following surgery.

The molecular data confirmed that pathogenic variants in *ATP7A* causing OHS mainly included splice-site variants, while missense and frameshift variants were less frequently observed. This observation may fit the hypothesis that a low amount of normal *ATP7A* mRNA and residual ATP7A functioning is sufficient to prevent the more severe MD [13,27,29,31,60]. No mutational hot spots were observed (Figure 3). Importantly, nearly all pathogenic variants are located in the endofacial domains and only two affect a transmembrane domain or cytosolic loop, respectively. This highlights the critical role of the transmembrane domains forming the channel for copper translocation [13]. Variants in OHS mainly affect residues implicated in ATPase-coupled copper transport. Some of these variants are predicted to inhibit the catalytic dephosphorylation leading to permanent distribution of the protein to the plasma membrane and cytosolic vesicles [14,15,21,61,62].

Our transmission electron microscopy data in skin samples of OHS individuals show larger diameters of collagen fibrils that were more loosely packed compared to those of the control sample. Elastic fibers were affected to a lesser extent but are fragmented and/or surrounded by extracellular matrix debris. OHS is associated with decreased activity of the copper dependent lysyl oxidase (LOX), that oxidates lysyl and hydroxylysyl residues in collagen and lysyl residues in elastin to aldehydes [7]. These aldehyde residues are than condensed to form cross-links, essential for collagen fibril stabilization and the integrity and elasticity of mature elastin [32,63]. Overall, this would not affect the amount or migration of soluble collagens from the medium or cellular fraction upon electrophoresis but will affect the assembly in the extracellular matrix. Also, lysyl oxidase may form cross-links with the N-propeptide domain of collagen type V, essential in collagen diameter regulation [64,65]. Fibroblasts synthesize only low amounts of type V collagen which is difficult to evaluate by electrophoresis [66,67]. Indeed, irregular fibril diameters are reminiscent of classic Ehlers–Danlos syndrome, despite the absence of cauliflower fibrils [68].

The observation of major fibroblast abnormalities, with multiple autophagolysosomes containing sickle-shaped electron dense deposits, is novel and might point to defective trafficking and trapping of proteins needing cuproenzyme-dependent processing in the post-Golgi network [14,15,21].

In conclusion, this study sheds light on the natural history of OHS by describing two new patients and 32 patients reported in the literature. Mortality in adulthood warrants long-term specialized care and follow-up, incorporating MR angiography, echocardiography and spirometry into standard follow-up guidelines for OHS patients, next to orthopedic and urological follow-up. Ultrastructural abnormalities including increased collagen diameter, mild elastic fiber abnormalities, and multiple autophagolysosomes reflect the role of LOX and defective ATP7A trafficking in the pathomechanism of OHS.

Supplementary Materials: The following are available online at http://www.mdpi.com/2073-4425/10/7/528/s1; Supplemental Figure S1: Fibroblast abnormalities in OHS. Magnification: x4000, x8000 and x15000; Supplemental Figure S2: SDS-PAGE collagen analysis in patients F1:II-1 and F2:II-1 and a matched control. Panel A: medium procollagen fraction on fibroblasts in culture for three days. Panel B: cellular collagen fraction; Supplemental Table S1. Clinical characteristics of all patients. '"+": present, "-": absent, "M": male, "F": female, "y": year(s), "mo": month(s), "d":day(s), "CVA": cerebrovascular accident, "DIC": diffuse intravascular coagulation, "FTT": failure to thrive, "GER": gastro-esophageal reflux, "SC": subcutaneous, "UTI": urinary tract infection; Supplemental Table S2. Pathogenicity information for the ATP7A variants in this study. The pathogenicity classification used in this table is based on the Alamut®mutation analysis software, which uses the following prediction tools: Align GVGD (Huntsman Cancer Institute, University of Utah) combines the biophysical characteristics of amino acids and protein multiple sequence alignments to where missense substitutions in genes of interest fall from enriched deleterious to enriched neutral. SIFT predicts whether amino acid substitution affects protein function. The MutationTaster prediction program integrates information from different biomedical databases, evolutionary conservation, splice-site changes and loss of protein features. Protein conservation is based on MARRVEL®multiple protein alignment. The splice prediction predicts the chance of variants having an impact on alternative splicing. Gnomad allele frequencies and Clinvar data can also be found. In the last column, CADD scores are depicted (Kircher et al., 2014). "NS": not specified, "NA": not applicable; Supplemental Table S3. Copper and ceruloplasmin levels in all patients.

Author Contributions: Conceptualization, A.B. and B.C.; Data curation, A.B., K.V.M., P.H., F.P., B.L., S.S. and B.C.; Formal analysis, K.V.M.; Funding acquisition, B.C.; Investigation, A.B., K.V.M., L.P., R.D.R., M.D.B., F.B., S.S. and B.C.; Methodology, A.B., L.P., R.D.R., S.S. and B.C.; Project administration, B.C.; Resources, K.V.M., R.D.R., P.H., F.P., B.L., S.S. and B.C.; Supervision, S.D.S., S.S. and B.C.; Validation, A.B., K.V.M. and B.C.; Visualization, A.B., L.P. and R.D.R.; Writing—original draft, A.B.; Writing—review and editing, A.B., K.V.M., L.P., R.D.R., M.D.B., F.B., P.H., F.P., B.L., S.D.S., S.S. and B.C.

Funding: This work is supported by the Special Research Fund, Flanders of Ghent University (grant 01N04516C to B.C.) and the European Academy of Dermatovenereology (EADV project PPRC-2018-50 to B.C.). B.C. and B.L. are senior clinical investigators of the Research Foundation Flanders.

Acknowledgments: We are thankful to all the families and patients for their interest and participation. We want to thank Zoë Malfait and Lisa Caboor for their technical support. The Center for Medical Genetics and Department of Dermatology of the Ghent University Hospital are members of the European Reference Network (ERN)-Skin. The Centers for Medical Genetics of both Ghent and Antwerp are members of VASCERN.

Conflicts of Interest: The authors declare no conflicts of interest.

References

1. Tumer, Z.; Horn, N. Menkes disease: Recent advances and new aspects. *J. Med. Genet.* **1997**, *344*, 265–274. [CrossRef] [PubMed]
2. Tumer, Z.; Moller, L.B. Menkes disease. *Eur. J. Hum. Genet.* **2010**, *185*, 511–518. [CrossRef] [PubMed]
3. Kaler, S.G. ATP7A-related copper transport diseases-emerging concepts and future trends. *Nat. Rev. Neurol.* **2011**, *71*, 15–29. [CrossRef] [PubMed]
4. Horn, N.; Tümer, Z. Menkes disease and the occipital horn syndrome. In *Connective Tissue and Its Heritable Disorders*; Royce, P.M., Steinman, B., Eds.; Wiley-Liss: New York, NY, USA, 2002; pp. 651–685.
5. Sartoris, D.J.; Luzzatti, L.; Weaver, D.D.; Macfarlane, J.D.; Hollister, D.W.; Parker, B.R. Type IX Ehlers-Danlos syndrome. A new variant with pathognomonic radiographic features. *Radiology* **1984**, *1523*, 665–670. [CrossRef] [PubMed]
6. Das, S.; Levinson, B.; Vulpe, C.; Whitney, S.; Gitschier, J.; Packman, S. Similar splicing mutations of the Menkes/mottled copper-transporting ATPase gene in occipital horn syndrome and the blotchy mouse. *Am. J. Hum. Genet.* **1995**, *563*, 570–576.
7. Byers, P.H.; Siegel, R.C.; Holbrook, K.A.; Narayanan, A.S.; Bornstein, P.; Hall, J.G. X-linked cutis laxa: Defective cross-link formation in collagen due to decreased lysyl oxidase activity. *N. Engl. J. Med.* **1980**, *3032*, 61–65. [CrossRef] [PubMed]
8. Petris, M.J.; Strausak, D.; Mercer, J.F. The Menkes copper transporter is required for the activation of tyrosinase. *Hum. Mol. Genet.* **2000**, *919*, 2845–2851. [CrossRef] [PubMed]
9. Kodama, H.; Okabe, I.; Yanagisawa, M.; Kodama, Y. Copper deficiency in the mitochondria of cultured skin fibroblasts from patients with Menkes syndrome. *J. Inherit. Metab. Dis.* **1989**, *124*, 386–389. [CrossRef]
10. Royce, P.M.; Camakaris, J.; Danks, D.M. Reduced lysyl oxidase activity in skin fibroblasts from patients with Menkes' syndrome. *Biochem. J.* **1980**, *1922*, 579–586. [CrossRef]
11. Dagenais, S.L.; Adam, A.N.; Innis, J.W.; Glover, T.W. A novel frameshift mutation in exon 23 of ATP7A (MNK) results in occipital horn syndrome and not in Menkes disease. *Am. J. Hum. Genet.* **2001**, *692*, 420–427. [CrossRef]
12. Kaler, S.G. Metabolic and molecular bases of Menkes disease and occipital horn syndrome. *Pediatr. Dev. Pathol.* **1998**, *11*, 85–98.
13. Tumer, Z. An overview and update of ATP7A mutations leading to Menkes disease and occipital horn syndrome. *Hum Mutat.* **2013**, *343*, 417–429. [CrossRef] [PubMed]
14. Hartwig, C.; Zlatic, S.A.; Wallin, M.; Vrailas-Mortimer, A.; Fahrni, C.J.; Faundez, V. Trafficking mechanisms of P-type ATPase copper transporters. *Curr. Opin. Cell Biol.* **2019**, *59*, 24–33. [CrossRef] [PubMed]
15. Polishchuk, R.; Lutsenko, S. Golgi in copper homeostasis: A view from the membrane trafficking field. *Histochem. Cell Biol.* **2013**, *1403*, 285–295. [CrossRef] [PubMed]
16. Dierick, H.A.; Ambrosini, L.; Spencer, J.; Glover, T.W.; Mercer, J.F.B. Molecular-Structure of the Menkes Disease Gene (ATP7A). *Genomics* **1995**, *28*, 462–469. [CrossRef] [PubMed]

17. Tumer, Z.; Vural, B.; Tonnesen, T.; Chelly, J.; Monaco, A.P.; Horn, N. Characterization of the exon structure of the Menkes disease gene using vectorette PCR. *Genomics* **1995**, *263*, 437–442. [CrossRef]
18. Chelly, J.; Tumer, Z.; Tonnesen, T.; Petterson, A.; Ishikawa-Brush, Y.; Tommerup, N.; Horn, N.; Monaco, A.P. Isolation of a candidate gene for Menkes disease that encodes a potential heavy metal binding protein. *Nat. Genet.* **1993**, *31*, 14–19. [CrossRef]
19. Mercer, J.F.B.; Livingston, J.; Hall, B.; Paynter, J.A.; Begy, C.; Chandrasekharappa, S.; Lockhart, P.; Grimes, A.; Bhave, M.; Siemieniak, D.; et al. Isolation of a Partial Candidate Gene for Menkes Disease by Positional Cloning. *Nat. Genet.* **1993**, *31*, 20–25. [CrossRef]
20. Vulpe, C.; Levinson, B.; Whitney, S.; Packman, S.; Gitschier, J. Isolation of a candidate gene for Menkes disease and evidence that it encodes a copper-transporting ATPase. *Nat. Genet.* **1993**, *31*, 7–13. [CrossRef]
21. Skjorringe, T.; Pedersen, P.A.; Thorborg, S.S.; Nissen, P.; Gourdon, P.; Moller, L.B. Characterization of ATP7A missense mutants suggests a correlation between intracellular trafficking and severity of Menkes disease. *Sci. Rep.-Uk* **2017**, *7*, 757. [CrossRef]
22. De Bie, P.; Muller, P.; Wijmenga, C.; Klomp, L.W. Molecular pathogenesis of Wilson and Menkes disease: Correlation of mutations with molecular defects and disease phenotypes. *J. Med. Genet.* **2007**, *4411*, 673–688. [CrossRef] [PubMed]
23. Gourdon, P.; Liu, X.Y.; Skjorringe, T.; Morth, J.P.; Moller, L.B.; Pedersen, B.P.; Nissen, P. Crystal structure of a copper-transporting PIB-type ATPase. *Nature* **2011**, *475*(7354), 59–64. [CrossRef] [PubMed]
24. Strausak, D.; La Fontaine, S.; Hill, J.; Firth, S.D.; Lockhart, P.J.; Mercer, J.F.B. The role of GMXCXXC metal binding sites in the copper-induced redistribution of the Menkes protein. *J. Biol. Chem.* **1999**, *27416*, 11170–11177. [CrossRef] [PubMed]
25. Gitschier, J.; Moffat, B.; Reilly, D.; Wood, W.I.; Fairbrother, W.J. Solution structure of the fourth metal-binding domain from the Menkes copper-transporting ATpase. *Nat. Struct. Biol.* **1998**, *51*, 47–54. [CrossRef]
26. Das, S.; Levinson, B.; Whitney, S.; Vulpe, C.; Packman, S.; Gitschier, J. Diverse mutations in patients with Menkes disease often lead to exon skipping. *Am. J. Hum. Genet.* **1994**, *555*, 883–889.
27. Kaler, S.G.; Gallo, L.K.; Proud, V.K.; Percy, A.K.; Mark, Y.; Segal, N.A.; Goldstein, D.S.; Holmes, C.S.; Gahl, W.A. Occipital horn syndrome and a mild Menkes phenotype associated with splice site mutations at the MNK locus. *Nat. Genet.* **1994**, *82*, 195–202. [CrossRef] [PubMed]
28. Ronce, N.; Moizard, M.P.; Robb, L.; Toutain, A.; Villard, L.; Moraine, C. A C2055T transition in exon 8 of the ATP7A gene is associated with exon skipping in an occipital horn syndrome family. *Am. J. Hum. Genet.* **1997**, *611*, 233–238. [CrossRef]
29. Qi, M.; Byers, P.H. Constitutive skipping of alternatively spliced exon 10 in the ATP7A gene abolishes Golgi localization of the menkes protein and produces the occipital horn syndrome. *Hum. Mol. Genet.* **1998**, *73*, 465–469. [CrossRef]
30. Borm, B.; Moller, L.B.; Hausser, I.; Emeis, M.; Baerlocher, K.; Horn, N.; Rossi, R. Variable clinical expression of an identical mutation in the ATP7A gene for Menkes disease/occipital horn syndrome in three affected males in a single family. *J. Pediatr.* **2004**, *1451*, 119–121. [CrossRef]
31. Moller, L.B.; Tumer, Z.; Lund, C.; Petersen, C.; Cole, T.; Hanusch, R.; Seidel, J.; Jensen, L.R.; Horn, N. Similar splice-site mutations of the ATP7A gene lead to different phenotypes: Classical Menkes disease or occipital horn syndrome. *Am. J. Hum. Genet.* **2000**, *664*, 1211–1220. [CrossRef]
32. Siegel, R.C. Lysyl oxidase. *Int. Rev. Connect. Tissue Res.* **1979**, *8*, 73–118. [PubMed]
33. Biaggioni, I.; Goldstein, D.S.; Atkinson, T.; Robertson, D. Dopamine-beta-hydroxylase deficiency in humans. *Neurology* **1990**, *402*, 370–373. [CrossRef] [PubMed]
34. Robertson, D.; Goldberg, M.R.; Onrot, J.; Hollister, A.S.; Wiley, R.; Thompson, J.G., Jr.; Robertson, R.M. Isolated failure of autonomic noradrenergic neurotransmission. Evidence for impaired beta-hydroxylation of dopamine. *N. Engl. J. Med.* **1986**, *31423*, 1494–1497. [CrossRef] [PubMed]
35. Gu, Y.H.; Kodama, H.; Murata, Y.; Mochizuki, D.; Yanagawa, Y.; Ushijima, H.; Shiba, T.; Lee, C.C. ATP7A gene mutations in 16 patients with Menkes disease and a patient with occipital horn syndrome. *Am. J. Med. Genet.* **2001**, *993*, 217–222. [CrossRef]
36. Wakai, S.; Ishikawa, Y.; Nagaoka, M.; Okabe, M.; Minami, R.; Hayakawa, T. Central nervous system involvement and generalized muscular atrophy in occipital horn syndrome: Ehlers-Danlos type IX. A first Japanese case. *J. Neurol. Sci.* **1993**, *1161*, 1–5. [CrossRef]

37. Proud, V.K.; Mussell, H.G.; Kaler, S.G.; Young, D.W.; Percy, A.K. Distinctive Menkes disease variant with occipital horns: Delineation of natural history and clinical phenotype. *Am. J. Med. Genet.* **1996**, *651*, 44–51. [CrossRef]
38. Levinson, B.; Conant, R.; Schnur, R.; Das, S.; Packman, S.; Gitschier, J. A repeated element in the regulatory region of the MNK gene and its deletion in a patient with occipital horn syndrome. *Hum. Mol. Genet.* **1996**, *511*, 1737–1742. [CrossRef]
39. De Paepe, A.; Loeys, B.; Devriendt, K.; Fryns, J.P. Occipital Horn syndrome in a 2-year-old boy. *Clin. Dysmorphol.* **1999**, *83*, 179–183. [CrossRef]
40. Yasmeen, S.; Lund, K.; De Paepe, A.; De Bie, S.; Heiberg, A.; Silva, J.; Martins, M.; Skjorringe, T.; Moller, L.B. Occipital horn syndrome and classical Menkes Syndrome caused by deep intronic mutations, leading to the activation of ATP7A pseudo-exon. *Eur. J. Hum. Genet (EJHG)* **2014**, *224*, 517–521. [CrossRef]
41. Tang, J.; Robertson, S.; Lem, K.E.; Godwin, S.C.; Kaler, S.G. Functional copper transport explains neurologic sparing in occipital horn syndrome. *Genet. Med.* **2006**, *811*, 711–718. [CrossRef]
42. De Gemmis, P.; Enzo, M.V.; Lorenzetto, E.; Cattelan, P.; Segat, D.; Hladnik, U. 13 novel putative mutations in ATP7A found in a cohort of 25 Italian families. *Metab. Brain Dis.* **2017**, *324*, 1173–1183. [CrossRef] [PubMed]
43. Donsante, A.; Tang, J.; Godwin, S.C.; Holmes, C.S.; Goldstein, D.S.; Bassuk, A.; Kaler, S.G. Differences in ATP7A gene expression underlie intrafamilial variability in Menkes disease/occipital horn syndrome. *J. Med. Genet.* **2007**, *448*, 492–497. [CrossRef] [PubMed]
44. Bazzocchi, A.; Femia, R.; Feraco, P.; Battista, G.; Canini, R.; Guglielmi, G. Occipital horn syndrome in a woman: Skeletal radiological findings. *Skeletal Radiol.* **2011**, *4011*, 1491–1494. [CrossRef] [PubMed]
45. Ogawa, E.; Kodama, H. Effects of disulfiram treatment in patients with Menkes disease and occipital horn syndrome. *J. Trace Elements Med. Biol.* **2012**, *26*, 102–104. [CrossRef] [PubMed]
46. Nascimento, A.; Rego Sousa, P.; Ortez, C.; Boronat, S.; Rebello, M.; Jimenez-Mallabrera, C.; Jou, C.; Rovira, C.; Colomer, J. Expanding the phenotype of ATP7A-related copper transport diseases: Response to copper treatment. *J. Inherit Metab. Dis.* **2013**, *36* (Suppl. 2), S91–S342.
47. Legros, L.; Revencu, N.; Nassogne, M.C.; Wese, F.X.; Feyaerts, A. Multiple bladder diverticula caused by occipital horn syndrome. *Arch. Pediatr.* **2015**, *2211*, 1147–1150. [CrossRef] [PubMed]
48. Louro, P.; Ramos, L.; Oliveira, R.; Henriques, M.; Pereira, E.; Kaler, S.G.; Pereira, J.; Diogo, L.; Garcia, P. Inherited Disorders of Metal Metabolism: A Rare Case of Occipital Horn Syndrome. In Proceedings of the 13th International Congress of Inborn Errors of Metabolism, Rio De Janeiro, Brazil, 5–8 September 2017.
49. Palcic, I.; Cvitkovic Roic, A.; Roic, G.; Krakar, G.; Jaklin Kekez, A. (Eds.) Bladder diverticula caused by occipital horn syndrome—A case report. In Proceedings of the 50th Anniversary meeting of the European Society of Paediatric Nephrology, Glasglow, UK, 6–7 September 2017.
50. Htut, E.P.; Offiah, A.C.; Holden, S.; Shenker, N.G. Compelling case of copper conundrum; Occipital horn syndrome. *Rheumatology* **2017**, *56* (Suppl. 2). [CrossRef]
51. Martin-Santiago, A.; Escudero-Gongora, M.M.; Giacaman, A.; Bauza, A.; Saus, C.; Roldan, J.; Rosell, J. Cutis laxa and copper transport anomalies, one gene, two ends of the spectrum. *Pediatr. Dermatol.* **2017**, *34*, S25.
52. Nuytinck, L.; Narcisi, P.; Nicholls, A.; Renard, J.P.; Pope, F.M.; De Paepe, A. Detection and characterisation of an overmodified type III collagen by analysis of non-cutaneous connective tissues in a patient with Ehlers-Danlos syndrome IV. *J. Med. Genet.* **1992**, *296*, 375–380. [CrossRef]
53. Syx, D.; De Wandele, I.; Symoens, S.; De Rycke, R.; Hougrand, O.; Voermans, N.; De Paepe, A.; Malfait, F. Bi-allelic AEBP1 mutations in two patients with Ehlers-Danlos syndrome. *Hum. Mol. Genet.* **2019**, *2811*, 1853–1864. [CrossRef]
54. Tsukahara, M.; Imaizumi, K.; Kawai, S.; Kajii, T. Occipital horn syndrome: Report of a patient and review of the literature. *Clin. Genet.* **1994**, *451*, 32–35. [CrossRef]
55. Lazoff, S.G.; Rybak, J.J.; Parker, B.R.; Luzzatti, L. Skeletal dysplasia, occipital horns, diarrhea and obstructive uropathy—A new hereditary syndrome. *Birth Defects Orig. Article Series.* **1975**, *115*, 71–74.
56. Callewaert, B.; Su, C.T.; Van Damme, T.; Vlummens, P.; Malfait, F.; Vanakker, O.; Schulz, B.; Mac Neal, M.; Davis, E.C.; Lee, J.G.; et al. Comprehensive clinical and molecular analysis of 12 families with type 1 recessive cutis laxa. *Hum Mutat.* **2013**, *341*, 111–121. [CrossRef] [PubMed]
57. Malfait, F.; De Coster, P.; Hausser, I.; van Essen, A.J.; Franck, P.; Colige, A.; Nusgens, B.; Martens, L.; De Paepe, A. The natural history, including orofacial features of three patients with Ehlers-Danlos syndrome, dermatosparaxis type (EDS type VIIC). *Am. J. Med. Genet. Part A.* **2004**, *1311*, 18–28. [CrossRef] [PubMed]

58. Wuyts, W.; Van Hul, W.; De Boulle, K.; Hendrickx, J.; Bakker, E.; Vanhoenacker, F.; Mollica, F.; Lüdecke, H.J.; Sayli, B.S.; Pazzaglia, U.E.; et al. Mutations in the EXT1 and EXT2 genes in hereditary multiple exostoses. *Am. J. Hum. Genet.* **1998**, *622*, 346–354. [CrossRef] [PubMed]
59. Jurik, A.G.; Jorgensen, P.H.; Mortensen, M.M. Whole-body MRI in assessing malignant transformation in multiple hereditary exostoses and enchondromatosis: Audit results and literature review. *Skeletal Radiol.* **2019**. [CrossRef] [PubMed]
60. Moller, L.B. Small amounts of functional ATP7A protein permit mild phenotype. *J. Trace Elem. Med. Biol.* **2015**, *31*, 173–177. [CrossRef]
61. Moller, J.V.; Juul, B.; le Maire, M. Structural organization, ion transport, and energy transduction of P-type ATPases. *Biochim. Biophys. Acta* **1996**, *12861*, 1–51. [CrossRef]
62. Paulsen, M.; Lund, C.; Akram, Z.; Winther, J.R.; Horn, N.; Moller, L.B. Evidence that translation reinitiation leads to a partially functional Menkes protein containing two copper-binding sites. *Am. J. Hum. Genet.* **2006**, *792*, 214–229. [CrossRef]
63. Uitto, J.; Li, Q.; Urban, Z. The complexity of elastic fibre biogenesis in the skin–A perspective to the clinical heterogeneity of cutis laxa. *Exp. Dermatol.* **2013**, *222*, 88–92. [CrossRef]
64. Symoens, S.; Renard, M.; Bonod-Bidaud, C.; Syx, D.; Vaganay, E.; Malfait, F.; Ricard-Blum, S.; Kessler, E.; Van Laer, L.; Coucke, P.; et al. Identification of binding partners interacting with the alpha1-N-propeptide of type V collagen. *Biochem. J.* **2011**, *4332*, 371–381. [CrossRef] [PubMed]
65. Vogel, A.; Holbrook, K.A.; Steinmann, B.; Gitzelmann, R.; Byers, P.H. Abnormal collagen fibril structure in the gravis form (type I) of Ehlers-Danlos syndrome. Laboratory investigation. *J. Techn. Methods Pathol.* **1979**, *402*, 201–206.
66. Malfait, F.; Coucke, P.; Symoens, S.; Loeys, B.; Nuytinck, L.; De Paepe, A. The molecular basis of classic Ehlers-Danlos syndrome: A comprehensive study of biochemical and molecular findings in 48 unrelated patients. *Hum. Mutat.* **2005**, *251*, 28–37. [CrossRef] [PubMed]
67. Birk, D.E. Type V collagen: Heterotypic type I/V collagen interactions in the regulation of fibril assembly. *Micron (Oxford, England: 1993)* **2001**, *323*, 223–237. [CrossRef]
68. Hausser, I.; Anton-Lamprecht, I. Differential ultrastructural aberrations of collagen fibrils in Ehlers-Danlos syndrome types I-IV as a means of diagnostics and classification. *Hum. Genet.* **1994**, *934*, 394–407. [CrossRef]

© 2019 by the authors. Licensee MDPI, Basel, Switzerland. This article is an open access article distributed under the terms and conditions of the Creative Commons Attribution (CC BY) license (http://creativecommons.org/licenses/by/4.0/).

Article

Characterization of Two Novel Intronic Variants Affecting *Splicing* in *FBN1*-Related Disorders

Carmela Fusco [1,*], Silvia Morlino [2], Lucia Micale [1], Alessandro Ferraris [2], Paola Grammatico [2] and Marco Castori [1]

1. Division of Medical Genetics, Fondazione IRCCS-Casa Sollievo della Sofferenza, 71013 San Giovanni Rotondo FG, Italy; l.micale@operapadrepio.it (L.M.); m.castori@operapadrepio.it (M.C.)
2. Laboratory of Medical Genetics, Department of Molecular Medicine, Sapienza University, San Camillo-Forlaninin Hospital, 00152 Rome, Italy; silvia_morlino@yahoo.it (S.M.); aferraris@scamilloforlanini.rm.it (A.F.); paola.grammatico@uniroma1.it (P.G.)
* Correspondence: c.fusco@operapadrepio.it; Tel.: +39-0882-416350

Received: 19 April 2019; Accepted: 7 June 2019; Published: 10 June 2019

Abstract: *FBN1* encodes fibrillin 1, a key structural component of the extracellular matrix, and its variants are associated with a wide range of hereditary connective tissues disorders, such as Marfan syndrome (MFS) and mitral valve–aorta–skeleton–skin (MASS) syndrome. Interpretations of the genomic data and possible genotype–phenotype correlations in *FBN1* are complicated by the high rate of intronic variants of unknown significance. Here, we report two unrelated individuals with the *FBN1* deep intronic variants c.6872-24T>A and c.7571-12T>A, clinically associated with MFS and MASS syndrome, respectively. The individual carrying the c.6872-24T>A variant is positive for aortic disease. Both individuals lacked ectopia lentis. In silico analysis and subsequent mRNA study by RT-PCR demonstrated the effect of the identified variant on the splicing process in both cases. The c.6872-24T>A and c.7571-12T>A variants generate the retention of intronic nucleotides and lead to the introduction of a premature stop codon. This study enlarges the mutation spectrum of *FBN1* and points out the importance of intronic sequence analysis and the need for integrative functional studies in *FBN1* diagnostics.

Keywords: fibrillin 1; Marfan syndrome; MASS syndrome; mRNA; splicing

1. Introduction

The *FBN1* gene is located in 15q21.1 and includes 65 exons which encode for fibrillin 1, a large glycoprotein constituted by 47 cysteine-rich epidermal growth factor (EGF)-like domains and seven motifs homologous to the binding protein for transforming growth factor beta (TGFβ) [1,2]. Fibrillin 1 is a major structural component of the extracellular matrix microfibrils and thus, gives stability and elasticity to many tissues [2]. Variants in *FBN1* cause a wide range of autosomal dominant heritable connective tissue disorders, including Marfan syndrome (MFS; OMIM 154700), mitral valve–aorta–skeleton–skin syndrome (MASS syndrome; OMIM 604308), Marfan lipodystrophy syndrome (OMIM 616914), isolated autosomal dominant ectopia lentis (OMIM 129600), Weill–Marchesani syndrome type 2 (OMIM 608328), acromicric dysplasia (OMIM 102370), geleophysic dysplasia type 2 (OMIM 614185), stiff skin syndrome (OMIM 184900), and autosomal dominant thoracic aortic aneurysms and dissections [3].

To date, over 1800 different *FBN1* germline variants have been identified so far (UMD-*FBN1*, http://www.umd.be/FBN1/) [4,5] and only a few genotype–phenotype correlations are available [6,7]. For example, nonsense, frameshift, and some splicing variants appear correlated with a more severe skin and skeletal phenotype, as compared to in-frame variants [8]. Other studies suggest that variants affecting or creating cysteine residues are more commonly associated with ectopia lentis, while *null* alleles typically combine with an increased rate of aortic events in young age and thoracic deformities [7]. In the suspicion of a *FBN1*-related disorder, the identification of the causative variant is not only relevant for diagnosis confirmation and genetic counseling but is also increasingly useful for personalized medicine approaches, comprising preventive drug treatment and aortic surgery planning [9].

The introduction of next generation sequencing (NGS) technologies in molecular diagnostics improved turnaround time and costs but did not significantly affect the clinical interpretation of variants of uncertain significance, such as intronic nucleotide changes which potentially affect splicing. The burden associated with uncertain results related to these variants is still high in MFS and related disorders, as point variants possibly impacting FBN1 pre-mRNA splicing seem to account for ~10% of reported molecular findings in MFS [10]. In these cases, optimal clinical interpretation should integrate molecular findings with customized studies exploring variant effects at the transcriptional or translational levels.

Here, we report two unrelated individuals with MFS and MASS syndrome, respectively, which were associated with different intronic variants in *FBN1*. Characterization of the effects of these variants on mRNA allowed us to support their pathogenicity and, therefore, to use them for patient and family management.

2. Materials and Methods

All investigated subjects signed informed consent for molecular testing and research use of biological and related clinical data. Molecular testing on cDNA and mRNA in individual 1 and 2 were carried out within the routine clinical diagnostic activities of the Division of Medical Genetics of Foundation IRCCS-Casa Sollievo della Sofferenza, San Giovanni Rotondo (Italy). The results of this work are part of a larger research project approved by our local ethics committee (approval protocol no. GTB12001).

2.1. Clinical Description: Individual 1

This was a 27-year-old man, the second child of unrelated parents. His father was 50-year-old, 191 cm tall and affected by dilatation of the aortic root and ascending aorta, which required surgery at 45 years. The mother and sister were healthy, and 172 and 161 cm tall, respectively. The proband was born at term after an uneventful pregnancy. At birth, he presented respiratory distress, which needed intensive care support for a few days. The neonatal period was otherwise normal, as well as psychomotor development, scholarship, and mentation. At 11 years, progressive scoliosis required an orthopedic corset for some years. The patient suffered from recurrent spontaneous pneumothorax with four episodes between 12 and 14 years of age. For this, he successfully underwent pleurodesis at 14 years. At 18 years, the previous diagnosis of aortic disease in the father prompted full cardiologic assessment in the proband who showed dilatation of the aortic root and mitral valve prolapse. Since then, he was under preventive pharmacologic therapy. Current treatment includes losartan 50 mg/day (single dose) and nebivolol 50 mg/day (single dose), without significant side effects. The last heart ultrasound, at 27 years, showed a stable dilatation of the aortic root (48 mm; Z score 171.13 = 4.66 SD [11]) with normal ascending aorta (34 mm), minimal insufficiency of the aortic valve, prolapse of the anterior mitral leaflet with minimal insufficiency, and mild insufficiency of the tricuspid valve. Dilatation of the aortic root (48 × 43 mm) was also confirmed by angio-computer tomography. A total spine MRI at 23 years showed bilateral dural ectasias of the lumbar spine. A recent ophthalmologic exam excluded lens dislocation and revealed mild myopia (−2 diopters on the right and −1.5 diopters on the left).

Physical examination included height 196 cm, weight 72 kg, arm span 198 cm, arm span/height ratio 1.01 (normal), bilateral positive wrist sign, bilateral negative thumb sign, apparent enophthalmos, down slanting palpebral fissures, malar hypoplasia, high-arched palate, severe scoliosis, pectus carinatum, bilateral valgus deformity of the elbow, bilateral metatarsus varus, some mildly atrophic post-surgical scars of the thorax, and striae rubrae of the back (Figure 1A–D). The combination of aortic root dilatation (Z-score > 2 SD) and 10 points of systemic features lead to the clinical suspect of MFS.

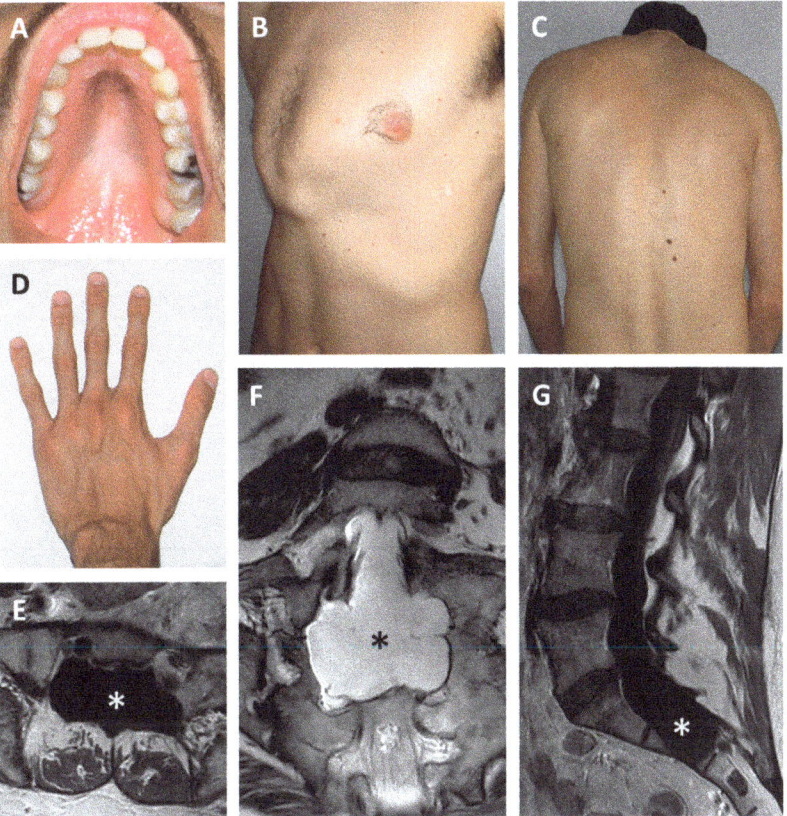

Figure 1. Clinical features. Individual 1 showing high-arched palate (**A**), pectus carinatum (**B**), scoliosis (**C**) and arachnodactyly (**D**). Lumbar spine MRI of Individual 2 demonstrating bilateral dural ectasias at axial (**E**), coronal (**F**), and sagittal (**G**) views. Asterisks indicate the dilated dural sac.

2.2. Clinical Description: Individual 2

Individual 2 was a 48-year-old, nullipara woman, the seventh child of a sibship of eleven. All brothers and sisters were healthy, but two brothers were described with tall stature (~194 cm). The mother was 165 cm tall and died at 65 years due to rupture of a cerebral aneurysm. The father was 175 tall and died at 78 years due to a stroke. Family history was negative for aortic disease. The proband was sent to clinical genetics assessment by the neurosurgeon, who first evaluated the woman for chronic back pain, diffuse spondylosis, and marked dural ectasias of the lumbar spine (Figure 1E–G). Relevant additional clinical features included moderately severe scoliosis since her teens, severe myopia (−6 diopters on the right and −8 dipters on the left), and chronic fatigue. A heart ultrasound repeatedly showed normal aortic diameters, prolapse, and mild insufficiency of the mitral valve. At 47 years, aortic root diameter was 36 mm (Z score 1.26 SD [11]). Routine ophthalmologic assessments always excluded lens dislocations. Physical examination showed height 181 cm, weight 80 kg, arm span 188 cm, arm span/height ratio 1.038 (normal), bilaterally positive wrist and thumb signs, hypermobility of fingers (but the Beighton score was 0/9), elbow limitation on both sides, enophthalmos, down slanting palpebral fissures, retrognathia, high-arched palate, striae distensae of the shoulders. The presence of 10 points of systemic features and an aortic root diameter <2 SD suggested a diagnosis of the MASS syndrome. Positive family history for tall stature and absence of aortic disease in relatives were in accordance with the diagnosis. Molecular testing in Individual 2 was carried out with a multigene panel approach including extended intronic regions around exon–intron junctions (see below).

2.3. Genomic DNA Extraction

Genomic DNA was extracted from individuals' peripheral bloods by using Bio Robot EZ1 (Qiagen, Hilden, Germany), according to standard procedures. The DNA was quantified with Nanodrop 2000 C spectrophotometer (Thermo Fisher Scientific, Waltham, MA, USA).

2.4. Genetic Testing (Sanger Sequencing): Individual 1

Molecular analysis for individual 1 was performed by Sanger sequencing. Primers were designed using the Primer 3 Output program (http://frodo.wi.mit.edu/primer3/) to amplify extended intronic sequences and exon–intron flanking sequences of *FBN1* (RefSeq NM_000138). Primers were checked both by BLAST and BLAT against the human genome to ensure specificity. Amplified products were subsequently purified and sequenced with a ready reaction kit (BigDye Terminator v1.1 Cycle, Applied Biosystems, Foster City, CA, USA). Fragments were then purified using DyeEx plates (Qiagen, Hilden, Germany) and resolved on an automated sequencer (3130xl Genetyc analyzer DNA Analyzer, ABI Prism, Foster City, CA, USA). Sequences were analyzed using the Sequencer software (Gene Codes, Ann Arbor, MI, USA). To investigate familial segregation, both parents were sequenced for the identified variant.

2.5. Genetic Testing (Next-Generation Sequencing): Individual 2

Individual 2's DNA underwent sequencing with a HaloPlex gene panel (Agilent Technologies, Santa Clara, CA, USA) designed to selectively capture known genes associated with syndromic and non-syndromic thoracic aneurysms and/or Mafanoid habitus, including: *ACTA2* (NM_001141945), *BGN* (NM_001711.5), *CBS* (NM_000071.2), *COL3A1* (NM_000090), *COL4A1* (NM_001303110), *COL5A1* (NM_000093), *COL5A2* (NM_000393), *DLG4* (NM_001365.4), *EFEMP2* (NM_016938), *ELN* (NM_000501), *EMILIN* (NM_007046.3), *FBN1* (NM_000138), *FBN2* (NM_001999), *FLNA* (NM_001110556), *FOXE3* (NM_012186), *GATA5* (NM_080473), *LOX* (NM_002317), *MAT2A* (NM_005911), *LTBP3* (NM_001130144.2), *MAT2A* (NM_005911.5), *MED12* (NM_005120.2), *MFAP5* (NM_003480), *MYH11* (NM_001040113), *MYLK* (NM_053025), *NOTCH1* (NM_017617), *PLOD1* (NM_001316320), *PRKG1* (NM_006258), *SKI* (NM_003036), *SLC2A10* (NM_030777), *SMAD2* (NM_005901.5), *SMAD3* (NM_005902), *SMAD4* (NM_005359), *TAB2* (NM_015093), *TGFB2* (NM_001135599), *TGFB3* (NM_003239), *TGFBR1* (NM_001306210), *TGFBR2* (NM_001024847), *UPF3B* (NM_080632.2), and *ZDHHC9* (NM_016032.3). Targeted fragments were then sequenced on a MiSeq platform (Illumina, San Diego, CA, USA) using a MiSeq Reagent kit V3 300 cycles flow cell. Data analysis was performed considering the frequency, impact on the encoded protein, conservation, and expression of variants using distinct tools, as appropriate ANNOVAR (http://annovar.openbioinformatics.org/en/latest/), dbSNP (https://www.ncbi.nlm.nih.gov/snp), 1000 Genomes (http://www.internationalgenome.org), ExAC (http://exac.broadinstitute.org). Selected variants are interpreted according to the American College of Medical Genetics and Genomics/Association for Molecular Pathology (ACMGG/AMP) [12]. The identified variant was confirmed by Sanger sequencing. We checked the selected variant in the UMD-*FBN1* database, a computerized database that currently contains information about the published mutations of the *FBN1* gene.

2.6. Variant Designation

Nucleotide variant nomenclature follows the format indicated in the Human Genome Variation Society (HGVS, http://www.hgvs.org) recommendations. DNA variant numbering system refers to cDNA. Nucleotide numbering uses +1 as the A of the ATG translation initiation codon in the reference sequence, with the initiation codon as codon 1.

2.7. In Silico Prediction

In silico analysis of identified intronic variants was conducted by running three independent algorithms for splice signal detection: NetGene2 (http://www.cbs.dtu.dk/services/NetGene2/), Berkeley Drosophila Genome Project (BDGP, http://www.fruitfly.org/seq_tools/splice.html), and Human Splicing Finder (HSF, http://www.umd.be/HSF3/).

2.8. RNA Extraction and Reverse Transcription–Polymerase Chain Reaction

Total RNA was extracted from lymphocytes fraction isolated from peripheral blood. Total RNA was extracted using RNeasy Mini Kit (Qiagen, Hilden, Germany), treated with DNase-RNase free (Qiagen, Hilden, Germany), quantified by Nanodrop (Thermo Fisher Scientific, Waltham, MA, USA), and reverse-transcribed using QuantiTect Reverse Transcription Kit (Qiagen, Hilden, Germany) according to the manufacturer's instructions. Primers used for DNA amplification and reverse transcription–polymerase chain reaction are reported in Table 1.

Table 1. Primers used for DNA amplification reverse transcription of *FBN1* fragments carrying the investigated variants (RefSeq NM_000138) in individual 1 and individual 2.

Individual	Primer	Sequence	Amplicon Size (pb)	Application
Individual 1	FBN1_Int56-57_F	TTTTGAGCCATGTGAACAGATT	320	DNA
	FBN1_EX57_R	AAACCCATCATTACACTCACAGG		
	FBN1_EX55_F	ATATGTGCTCAGAGAAGACCGTA	262	cDNA
	FBN1_EX57_R	AAACCCATCATTACACTCACAGG		
Individual 2	FBN1_Int61-62_F	TCCGAGTTATCCTTCTAATTTTCT	313	DNA
	FBN1_EX62_R	TATCTCATAGAGGCTGATGATGAAG		
	FBN1_EX61_F	GCAACCAAGCAACACAACTG	254	cDNA
	FBN1_EX63_R	TTACCCTCACACTCGTCCAC		

3. Results

3.1. Individual 1

Molecular testing was carried out on the Individual 1 DNA by Sanger sequencing of the *FBN1* gene, and this gave normal results. Multiple ligation-dependent probe amplification analysis for *FBN1* intragenic deletions/insertions resulted normal. Given the convincing clinical picture and positive family history, re-analyzing raw data from the Sanger sequence of extended intronic sequences was carried out and revealed a novel *FBN1* heterozygous splicing variant c.6872-24T>A, p.(?) (Figure 2A). Co-segregation with the disease was demonstrated on the father's DNA.

Computational predictions conducted using NetGene2, BDGP, and HSF revealed that the intronic c.6872-24T>A variant might influence the splicing process by differentially affecting canonical versus cryptic splice site utilization. The newly detected variant was predicted to insert 22 nucleotides (nt) of intron 56 into the mature mRNA (Figure 2B). This should give rise to frameshift and a PTC (p.(Asp2291Glyfs*9)), leading to the loss of the TGFβ binding protein homologous motifs, the last seven cysteine-rich EGF-like domains, and the FibuCTDIII-like motif.

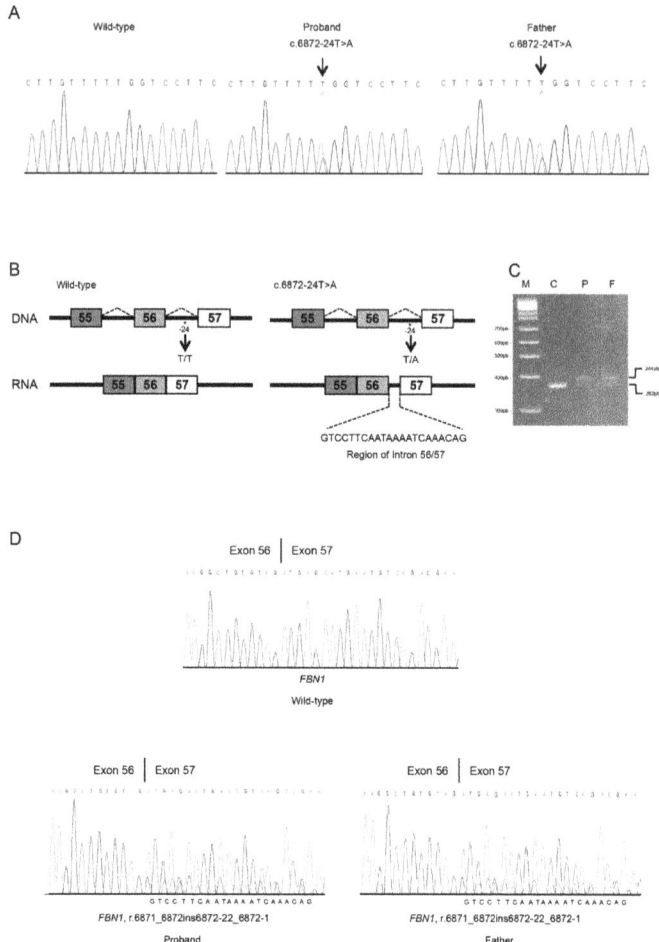

Figure 2. Molecular analyses of individual 1. (**A**) Sanger sequencing in an unaffected (unrelated) control (CTRL), the proband and his father of the *FBN1* region spanning the 56 to 57 exon–intron junction. The *FBN1* genomic variant is indicated by an arrow. (**B**) Schematic representation of *FBN1* region including 55 to 57 exons and introns. Rectangles represent exons; thin horizontal lines represent introns. The position of the identified variant is indicated with an asterisk. Wild type and abnormal transcripts generated by the c.6872-24T>A splicing variant are shown. (**C**) Total RNA of a normal (unrelated) control, the proband, and his father were used for RT-PCR of the *FBN1* transcript. The same amounts of patient's and control cDNA were PCR amplified. PCR products were analyzed by electrophoresis on 3.5% agarose gel. M, DNA marker; C, control; P, proband; F, father. Affected individuals (lane P and F) present both the wild-type mRNA (lower band) and aberrantly spliced transcript (upper band). (**D**) Sanger sequencing of the *FBN1* transcript, including exons 56 to 57 in an unaffected (unrelated) control (C), the proband (P), and his father (F). Variant position on mRNA level was annotated (r.6871_6872ins6872-22_6872-1). The intronic region was reported at the bottom of the Sanger chromatograms of all affected individuals.

The effect of the *FBN1* splice site variant c.6872-24T>A at the mRNA level was ascertained by in vitro RT-PCR amplification of *FBN1* RNA of Individual 1 and his father. The sequencing of the RT-PCR products spanning the 55 to 57 exons of *FBN1* showed the insertion of 22 nts of intron 56–57 in the *FBN1* mRNA of Individual 1 and his father (r.6871_6872ins6872-22_6872-1) (Figure 2C,D). Electropherogram showed a significant reduction of the peaks corresponding to the mutated allele. The aberrantly spliced transcript is not present in the expressed sequence tag database (EST, https://www.ncbi.nlm.nih.gov/dbEST/index.html) of *FBN1*. Following available standards for the interpretation of sequence variants [12], the *FBN1* c.6872-24T>A variant was classified as likely pathogenic and submitted to the Leiden Open Variation Database (LOVD, https://www.lovd.nl; accession number: # 00229824).

3.2. Individual 2

Targeted NGS analysis revealed the *FBN1* heterozygous c.7571-12T>A variant in intron 61 (Figure 3A). No other candidate variant was found in the remaining genes. The c.7571-12T>A variant was not reported in the UMD-*FBN1* and was absent in the proband's healthy sister. Both parents died, and the other brothers and sisters were not available for genetic testing.

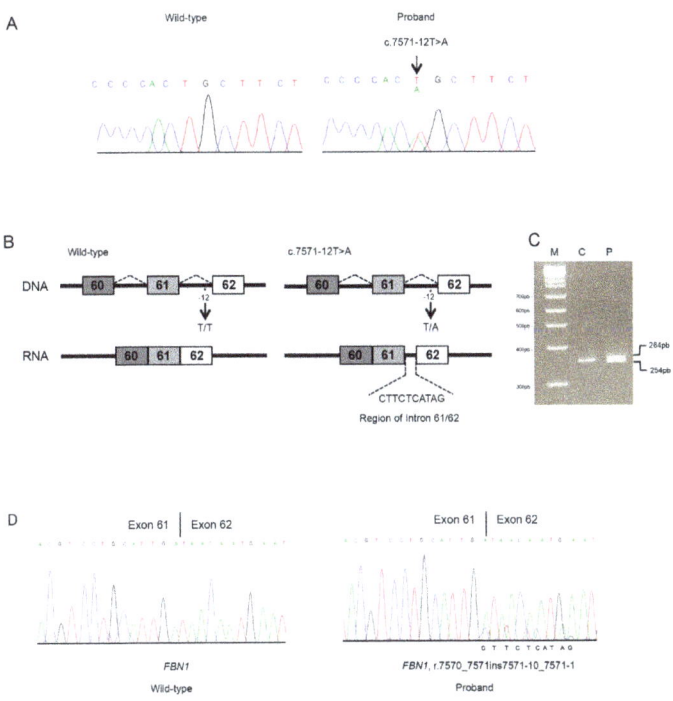

Figure 3. Molecular analyses of individual 2. (**A**) Sanger sequencing in an unaffected (unrelated) control and the proband of the *FBN1* region spanning 61 to 62 exon–intron junction. The *FBN1* genomic variant is indicated by an arrow. (**B**) Schematic representation of *FBN1* region including 61 to 62 exons and introns. Rectangles represent exons; thin horizontal lines represent introns. The position of the identified variant is indicated with an asterisk. Abnormal transcripts generated by the c.7571-12T>A splicing variant. (**C**) Total RNA of a normal (unrelated) control and the proband were used for RT-PCR of the *FBN1* transcript. The same amounts of patient and control cDNA were PCR amplified. PCR products were analyzed by electrophoresis on 3.5% agarose gel. M, DNA marker; C, control; P, the proband. M, DNA marker; C, control; P, proband; F, father. Affected individual (lane P) presents both wild-type (lower band) and aberrant transcript (upper band). (**D**) Sanger sequencing of the *FBN1* transcript, including exons 61 to 62 in an unaffected (unrelated) control (C) and the proband (P). Variant position on mRNA level was annotated (r.7570_7571ins7571-10_7571-1). The intronic region was reported at the bottom of the Sanger chromatogram of the affected individual.

The splicing predictor tools showed that the variant might alter the splicing process. In particular, all predicted with high probability that the c.7571-12T>A intronic variant activates a cryptic acceptor splice site within intron 61 that, in turn, leads to an inefficient recognition of the canonical splice site (Figure 3B) and insertion of intronic nucleotides in the *FBN1* mRNA. Sequence analysis on cDNA confirmed the c.7571-12T>A variant causes a shift from canonical toward cryptic splicing that leads to the inclusion of 10 nts of the 61 to 62 intronic sequence (r.7570_7571ins7571-10_7571-1) (Figure 3C,D) and a frameshift which generate a PTC in exon 61 (p.(Asp2524Argfs*5)). The truncated protein loses the last seven cysteine-rich EGF-like domains and the FibuCTDIII-like motif. Electropherogram showed a reduction of the peaks corresponding to the mutated allele. The sequence of the alternative mRNA was absent in the EST database. The *FBN1* c.7571-12T>A variant was therefore interpreted as likely pathogenic and submitted to LOVD (accession number: #00229832).

4. Discussion

In this work, we added two novel intronic variants to the *FBN1* mutational repertoire and demonstrated their effect at the mRNA level. Our findings highlight the impact of variants affecting non-coding DNA in MFS and related disorders, and the opportunity to re-consider the standard diagnostic approaches, at least, in specific cases. In particular, molecular studies revealed that the novel c.6872-24T>A and c.7571-12T>A *FBN1* variants induce aberrant pre-mRNA splicing by the generation of a new cryptic acceptor site that outcompetes the canonical splice site and generate exonization of intronic sequences. This, in turn, results in a frameshift with the introduction of a PTC. In both cases, PTC is predicted to produce a truncated fibrillin 1 with a loss of critical C-terminal domains of fibrillin 1 [13].

The mechanism(s) by which *FBN1* splicing variants might exert their pathogenic effect is (are) complex, and both dominant negative and loss-of-function effects have been proposed based on the predicted consequences on mRNA, and protein structure and function(s) [14–16]. In splicing variants generating out-of-frame deletions/insertions, the introduction of a PTC and the synthesis of a truncated protein are assumed. However, PTC might elicit nonsense-mediated mRNA decay (NMD), with a variable proportion of the mutated allele acting as a *null* allele. If NMD affects ~100% of the transcripts, haploinsufficiency is the leading molecular pathogenesis. Conversely, the production of a truncated fibrillin 1 due to complete NMD escape, can impact its function due to truncated protein production. In proteins with pleiotropic effects and distinct domains with different functions, such as fibrillin 1 [3], the length of the synthesized truncated protein and the differential expression of the gene in the different tissues are the leading modulator factors of its residual functions. A mixed pathogenesis should also be considered for PTC that allow only in part NMD. In splicing variants leading to in-frame deletions/insertions, the predicted consequence might mirror the dominant negative effect attributed to point variants [17].

Literature on *FBN1* intronic variants affecting non-canonical splice sites is currently limited to single reports, and the emerging genotype–phenotype correlations are purely speculative. For example, Wang et al., by reporting a small case series of familial aortic aneurysms and dissection, studied the effect of the *FBN1* c.5917+6T>C variant in HEK293 cells and concluded that skipping of exon 47 seems to associate with a higher rate of cardiovascular involvement [18]. Wypasek and coworkers reported an adult with MFS, severe cardiac involvement but normally placed lens, and the *FBN1* c.1589-9T>A affecting splicing by in silico and in vitro investigations [19]. Recently, a further MFS patient with full cardiac, ocular, and skeletal phenotype has been reported with the *FBN1* c.2678-15C>A variant affecting splicing by minigene approaches [20].

In our study, both variants are predicted to result in a truncated protein. Although we did not further investigate their effect, both transcripts are likely destined, at least in part, to NMD, as the PTC falls in more than 50 nts upstream of the last exon–exon junction within the mRNA [21]. This might be hypothesized by the differences in peak heights in the electropherograms of *FBN1* transcripts (Figures 2 and 3). This result suggests that a portion of aberrantly *FBN1*-spliced transcripts could be degraded by NMD process in both cases. *FBN1* haploinsufficiency due to NMD-mediated degradation of aberrant mRNA might be the leading candidate molecular mechanism causing disease for the novel c.6872-24T>A and c.7571-12T>A variants. However, we cannot exclude that a proportion of the mutated transcripts escapes NMD at least in blood cells. Therefore, both NMD and the synthesis of a truncated protein may result from the identified causative variants. In Individual 1 the predicted truncated protein (p.(Asp2291Glyfs*9)) losses the seven TGFβ binding protein homologous motifs, the last seven cysteine-rich EGF-like domains, and the FibuCTDIII-like motif; while in Individual 2, protein truncation should occur later (p.(Asp2524Argfs*5)) preserving the loss of TGFβ binding protein homologous motifs. Individual 1 (and his affected father) showed severe aortic disease, a feature that was not observed in Individual 2, and both individuals lack ectopia lentis.

The reasons underlying the different clinical diagnoses in these two patients are hard to define. We can speculate that in Individual 1 aortic involvement might be, at least in part, due to the loss of TGFβ binding domain, which, in turn, is preserved in the truncated protein predicted in Individual 2 (MASS syndrome), who showed normal aorta. In fact, perturbed binding with TGFβ results in the alteration of elasticity and microfibrils stability, two factors primary influencing aortic wall integrity [22]. A trend toward an increased risk for ectopia lentis was reported in cysteine-destroying or cysteine-creating variants [2,15], a mechanism not easily evoked in our two Individuals. Therefore, our observation testifies once more the difficulties in tracing consistent genotype–phenotype correlations in MFS and the need for international collaborations working towards this aim.

In this work, we highlighted the importance of including extended intronic regions in the analysis of *FBN1* for clinical purposes. This remains valid in the era of NGS and should be considered at the stage of (virtual) multigene panel design, as well as during variant prioritization and post-analytical phase. Such an approach also needs the availability of additional tests validated for diagnostics and able to explore variant effects at the mRNA or post-transcriptional levels for supporting the clinical interpretation of the genomic data. Therefore, this work points out the need for introducing functional investigations in the diagnostic workflow of at least selected disorders or gene panels, for which a high rate of intronic variants with a potential pathogenic effect is expected. The description of further individuals with splicing variants with documented effects at the various post-genomic levels seems critical for improving our knowledge on this issue.

Author Contributions: Formal analysis, C.F.; Funding acquisition, L.M.; Investigation, C.F. and L.M.; Resources, S.M., A.F. and P.G.; Supervision, M.C.; Writing—original draft, C.F., P.G. and M.C.; Writing—review and editing, C.F., S.M., L.M., A.F., P.G. and M.C.

Funding: This work was supported by the Ricerca Corrente 2018–2021 Program from the Italian Ministry of Health.

Acknowledgments: The authors thank the families for their kind availability in sharing the findings within the scientific community The funders had no role in study design, data collection, and analysis, decision to publish, or preparation of the manuscript.

Conflicts of Interest: All authors declare no conflict of interest concerning this work.

References

1. Ramirez, F.; Dietz, H.C. Fibrillin-rich microfibrils: Structural determinants of morphogenetic and homeostatic events. *J. Cell Physiol.* **2007**, *213*, 326–330. [CrossRef] [PubMed]
2. Verstraeten, A.; Alaerts, M.; Van Laer, L.; Loeys, B. Marfan Syndrome and Related Disorders: 25 Years of Gene Discovery. *Hum. Mutat.* **2016**, *37*, 524–531. [CrossRef] [PubMed]
3. Sakai, L.Y.; Keene, D.R. Fibrillin protein pleiotropy: Acromelic dysplasias. *Matrix Biol* **2018**, *80*, 6–13. [CrossRef] [PubMed]

4. Collod-Beroud, G.; Le Bourdelles, S.; Ades, L.; Ala-Kokko, L.; Booms, P.; Boxer, M.; Child, A.; Comeglio, P.; De Paepe, A.; Hyland, J.C.; et al. Update of the UMD-FBN1 mutation database and creation of an FBN1 polymorphism database. *Hum. Mutat.* **2003**, *22*, 199–208. [CrossRef] [PubMed]
5. Frederic, M.Y.; Lalande, M.; Boileau, C.; Hamroun, D.; Claustres, M.; Beroud, C.; Collod-Beroud, G. UMD-predictor, a new prediction tool for nucleotide substitution pathogenicity-application to four genes: FBN1, FBN2, TGFBR1, and TGFBR2. *Hum. Mutat.* **2009**, *30*, 952–959. [CrossRef] [PubMed]
6. Faivre, L.; Collod-Beroud, G.; Callewaert, B.; Child, A.; Binquet, C.; Gautier, E.; Loeys, B.L.; Arbustini, E.; Mayer, K.; Arslan-Kirchner, M.; et al. Clinical and mutation-type analysis from an international series of 198 probands with a pathogenic FBN1 exons 24-32 mutation. *Eur. J. Hum. Genet.* **2009**, *17*, 491–501. [CrossRef] [PubMed]
7. Aubart, M.; Gazal, S.; Arnaud, P.; Benarroch, L.; Gross, M.S.; Buratti, J.; Boland, A.; Meyer, V.; Zouali, H.; Hanna, N.; et al. Association of modifiers and other genetic factors explain Marfan syndrome clinical variability. *Eur. J. Hum. Genet.* **2018**, *26*, 1759–1772. [CrossRef]
8. Ergoren, M.C.; Turkgenc, B.; Terali, K.; Rodoplu, O.; Verstraeten, A.; Van Laer, L.; Mocan, G.; Loeys, B.; Tetik, O.; Temel, S.G. Identification and characterization of a novel FBN1 gene variant in an extended family with variable clinical phenotype of Marfan syndrome. *Connect. Tissue Res.* **2019**, *60*, 146–154. [CrossRef]
9. Wagner, A.H.; Zaradzki, M.; Arif, R.; Remes, A.; Muller, O.J.; Kallenbach, K. Marfan syndrome: A therapeutic challenge for long-term care. *Biochem. Pharmacol.* **2019**, *164*, 53–63. [CrossRef]
10. Zeyer, K.A.; Reinhardt, D.P. Engineered mutations in fibrillin-1 leading to Marfan syndrome act at the protein, cellular and organismal levels. *Mutat. Res. Rev. Mutat. Res.* **2015**, *765*, 7–18. [CrossRef]
11. Campens, L.; Renard, M.; Callewaert, B.; Coucke, P.; De Backer, J.; De Paepe, A. New insights into the molecular diagnosis and management of heritable thoracic aortic aneurysms and dissections. *Pol. Arch. Med.Wewn.* **2013**, *123*, 693–700. [CrossRef] [PubMed]
12. Richards, S.; Aziz, N.; Bale, S.; Bick, D.; Das, S.; Gastier-Foster, J.; Grody, W.W.; Hegde, M.; Lyon, E.; Spector, E.; et al. Standards and guidelines for the interpretation of sequence variants: a joint consensus recommendation of the American College of Medical Genetics and Genomics and the Association for Molecular Pathology. *Genet. Med.* **2015**, *17*, 405–424. [CrossRef] [PubMed]
13. Jensen, S.A.; Aspinall, G.; Handford, P.A. C-terminal propeptide is required for fibrillin-1 secretion and blocks premature assembly through linkage to domains cbEGF41-43. *Proc. Natl. Acad. Sci. USA* **2014**, *111*, 10155–10160. [CrossRef] [PubMed]
14. Matyas, G.; Alonso, S.; Patrignani, A.; Marti, M.; Arnold, E.; Magyar, I.; Henggeler, C.; Carrel, T.; Steinmann, B.; Berger, W. Large genomic fibrillin-1 (FBN1) gene deletions provide evidence for true haploinsufficiency in Marfan syndrome. *Hum. Genet.* **2007**, *122*, 23–32. [CrossRef] [PubMed]
15. Judge, D.P.; Biery, N.J.; Keene, D.R.; Geubtner, J.; Myers, L.; Huso, D.L.; Sakai, L.Y.; Dietz, H.C. Evidence for a critical contribution of haploinsufficiency in the complex pathogenesis of Marfan syndrome. *J. Clin. Investig.* **2004**, *114*, 172–181. [CrossRef] [PubMed]
16. Takeda, N.; Inuzuka, R.; Maemura, S.; Morita, H.; Nawata, K.; Fujita, D.; Taniguchi, Y.; Yamauchi, H.; Yagi, H.; Kato, M.; et al. Impact of Pathogenic FBN1 Variant Types on the Progression of Aortic Disease in Patients With Marfan Syndrome. *Circ. Genom. Precis. Med.* **2018**, *11*, e002058. [CrossRef]
17. Dietz, H.C.; Valle, D.; Francomano, C.A.; Kendzior, R.J., Jr.; Pyeritz, R.E.; Cutting, G.R. The skipping of constitutive exons in vivo induced by nonsense mutations. *Science* **1993**, *259*, 680–683. [CrossRef] [PubMed]
18. Wang, W.J.; Han, P.; Zheng, J.; Hu, F.Y.; Zhu, Y.; Xie, J.S.; Guo, J.; Zhang, Z.; Dong, J.; Zheng, G.Y.; et al. Exon 47 skipping of fibrillin-1 leads preferentially to cardiovascular defects in patients with thoracic aortic aneurysms and dissections. *J. Mol. Med.* **2013**, *91*, 37–47. [CrossRef]
19. Wypasek, E.; Potaczek, D.P.; Hydzik, M.; Stapor, R.; Raczkowska-Muraszko, M.; Weiss, J.; Maugeri, A.; Undas, A. Detection and a functional characterization of the novel FBN1 intronic mutation underlying Marfan syndrome: Case presentation. *Clin. Chem. Lab. Med.* **2018**, *56*, 87–91. [CrossRef]
20. Torrado, M.; Maneiro, E.; Trujillo-Quintero, J.P.; Evangelista, A.; Mikhailov, A.T.; Monserrat, L. A Novel Heterozygous Intronic Mutation in the FBN1 Gene Contributes to FBN1 RNA Missplicing Events in the Marfan Syndrome. *BioMed Res. Int.* **2018**, *2018*, 3536495. [CrossRef]

21. Isken, O.; Maquat, L.E. Quality control of eukaryotic mRNA: safeguarding cells from abnormal mRNA function. *Genes Dev.* **2007**, *21*, 1833–1856. [CrossRef] [PubMed]
22. Le Goff, C.; Mahaut, C.; Wang, L.W.; Allali, S.; Abhyankar, A.; Jensen, S.; Zylberberg, L.; Collod-Beroud, G.; Bonnet, D.; Alanay, Y.; et al. Mutations in the TGFbeta binding-protein-like domain 5 of FBN1 are responsible for acromicric and geleophysic dysplasias. *Am. J. Hum. Genet.* **2011**, *89*, 7–14. [CrossRef] [PubMed]

© 2019 by the authors. Licensee MDPI, Basel, Switzerland. This article is an open access article distributed under the terms and conditions of the Creative Commons Attribution (CC BY) license (http://creativecommons.org/licenses/by/4.0/).

Article

Genotypic Categorization of Loeys-Dietz Syndrome Based on 24 Novel Families and Literature Data

Letizia Camerota [1,†], Marco Ritelli [2,†], Anita Wischmeijer [3], Silvia Majore [4,5], Valeria Cinquina [2], Paola Fortugno [6], Rosanna Monetta [1,6], Laura Gigante [7,8], Marfan Syndrome Study Group Tor Vergata University Hospital [‡], Federica Carla Sangiuolo [7,8], Giuseppe Novelli [7,8,9], Marina Colombi [2,*] and Francesco Brancati [1,6,*]

1. Human Genetics Institute, Department of Life, Health, and Environmental Sciences, University of L'Aquila, 67100 L'Aquila, Italy; letizia.camerota@guest.univaq.it (L.C.); rosanna.monetta@graduate.univaq.it (R.M)
2. Division of Biology and Genetics, Department of Molecular and Translational Medicine, University of Brescia, 25123 Brescia, Italy; marco.ritelli@unibs.it (M.R.); valeria.cinquina1@unibs.it (V.C.)
3. Clinical Genetics Unit, Department of Pediatrics, Regional Hospital of Bolzano, 39100 Bolzano, Italy; titiaanita.wischmeijer@sabes.it
4. Medical Genetics Laboratory, Department of Molecular Medicine, Sapienza University, 00185 Rome, Italy; smajore@scamilloforlanini.rm.it
5. San Camillo-Forlanini Hospital, 00152 Rome, Italy
6. Laboratory of Molecular and Cell Biology, Istituto Dermopatico dell'Immacolata, IDI-IRCCS, 00167 Rome, Italy; p.fortugno@idi.it
7. Department of Biomedicine and Prevention, Tor Vergata University, 00133 Rome, Italy; laura.gigante84@gmail.com (L.G.); sangiuolo@med.uniroma2.it (F.C.S.); novelli@med.uniroma2.it (G.N.)
8. Medical Genetics Unit, Policlinico Tor Vergata University Hospital, 00133 Rome, Italy
9. IRCCS Neuromed Institute, 86077 Pozzilli, Italy
* Correspondence: marina.colombi@unibs.it (M.C.); francesco.brancati@univaq.it (F.B.); Tel.: +39 030 3717 240 (M.C.); +39-086-2434-716 (F.B.)
† Both authors equally contributed to this work.
‡ Members of the Marfan Syndrome Study Group Tor Vergata University Hospital: Giovanni Ruvolo (Coordinator), Fabio Bertoldo, Concettina Donzelli (Cardiovascular surgery); Patrizio Polisca (Cardiology); Giuseppe Novelli, Federica Carla Sangiuolo, Leila Baghernajad Salehi (Medical Genetics); Raffaele Mancino, Emiliano Di Carlo (Ophthalmology); Patrizio Bollero (Dentistry); Paola Cozza, Giuseppina Laganà (Pediatric Dentistry); Pasquale Farsetti, Fernando De Maio, Vincenzo De Luna, Federico Mancini (Orthopedics); Loredana Chini, Simona Graziani (Pediatrics); Roberto Floris, Massimiliano Sperandio (Radiology); Angela Infante (Counseling); Alberto De Stefano (volunteer association). Past members: Luigi Chiariello (Founder and head, Cardiovascular Surgery) and Susanna Grego (Clinical Coordinator, Cardiology).

Received: 27 August 2019; Accepted: 21 September 2019; Published: 28 September 2019

Abstract: Loeys-Dietz syndrome (LDS) is a connective tissue disorder first described in 2005 featuring aortic/arterial aneurysms, dissections, and tortuosity associated with craniofacial, osteoarticular, musculoskeletal, and cutaneous manifestations. Heterozygous mutations in 6 genes (TGFBR1/2, TGFB2/3, SMAD2/3), encoding components of the TGF-β pathway, cause LDS. Such genetic heterogeneity mirrors broad phenotypic variability with significant differences, especially in terms of the age of onset, penetrance, and severity of life-threatening vascular manifestations and multiorgan involvement, indicating the need to obtain genotype-to-phenotype correlations for personalized management and counseling. Herein, we report on a cohort of 34 LDS patients from 24 families all receiving a molecular diagnosis. Fifteen variants were novel, affecting the TGFBR1 (6), TGFBR2 (6), SMAD3 (2), and TGFB2 (1) genes. Clinical features were scored for each distinct gene and matched with literature data to strengthen genotype-phenotype correlations such as more severe vascular manifestations in TGFBR1/2-related LDS. Additional features included spontaneous pneumothorax in SMAD3-related LDS and cervical spine instability in TGFB2-related LDS. Our study broadens the clinical and molecular spectrum of LDS and indicates that a phenotypic continuum emerges as more patients are described, although genotype-phenotype correlations may still contribute to clinical management.

Keywords: hereditary connective tissue disorders; Loeys-Dietz syndrome; Ehlers-Danlos syndrome; arterial aneurysms; *TGFBR1*; *TGFBR2*; *SMAD2*; *SMAD3*; *TGFB2*; *TGFB3*

1. Introduction

Loeys–Dietz syndrome (LDS) is a rare hereditary connective tissue disorder (HCTD) with an autosomal dominant inheritance characterized by a widespread systemic involvement. The disorder was first described in 2005 and in its most typical presentation it features aortic/arterial aneurysms and/or dissections, as well as arterial tortuosity in association with variable craniofacial, osteoarticular, musculoskeletal, and cutaneous manifestations [1,2]. LDS is caused by pathogenic variants in transforming growth factor β (TGF-β) signaling pathway-related genes, i.e., *TGFBR1*, *TGFBR2*, *SMAD2*, *SMAD3*, *TGFB2*, and *TGFB3*, which alter the physiological development and function of the extracellular matrix, leading to cardiovascular and multisystem abnormalities [3]. Mutations in the genes encoding the TGF-β receptor subunits (*TGFBR1* and *TGFBR2*) were firstly identified and an initial clinical classification of LDS type I and type II was proposed based on the presence of typical craniofacial features in LDS type I [2]. Over the years, the discovery of novel disease-causative genes, eased by the advent of next-generation sequencing techniques, paved the way for novel genotype-phenotype correlations. For example, the identification of a subset of LDS patients with pathogenic variants in *SMAD3* displaying high frequency of osteoarthritis, prompted some authors to define the resulting phenotype as "aneurysms-osteoarthritis syndrome" or LDS type III [4]. McCarrick and coworkers argued that this phenotype fits within the LDS spectrum based on the analysis of large cohorts of mutated patients [5]. In particular, the authors concluded that, in the absence of formal diagnostic criteria for LDS, a pathogenic variant in any of the six LDS disease-causative genes in combination with the presence of arterial aneurysm or dissection or a positive family history should be considered sufficient for the diagnosis of LDS [5]. This stimulated the study of genotype-phenotype correlations, which are of paramount value for correct management of patients and to plan appropriate prevention programs in clinical practice. Even if wide inter- and intrafamilial variability in the distribution and severity of clinical features is observed, a more accurate gene-based categorization of LDS may have a great clinical impact in early diagnosis and management and guide treatment strategies for patients and family members with this life-threatening condition. Of note, the diagnosis of patients with LDS prompting molecular testing is not always straightforward, since the spectrum of clinical manifestations is broad and often overlaps other HCTDs, including Marfan syndrome (MFS), Shprintzen-Goldberg syndrome, some types of cutis laxa, Ehlers-Danlos syndromes (EDS) (particularly the vascular type), arterial tortuosity syndrome, congenital contractual arachnodactyly, and biglycan (*BGN*)-associated aortic aneurysm syndrome. Additionally, in some cases further genetic testing may be indicated for the differential diagnosis with other hereditary thoracic aortic disorders caused by pathogenic variants in distinct genes including *NOTCH1*, *ACTA2*, *MYH11*, *MYLK*, *PRKG1*, *MAT2A*, *FOXE3*, *MFAP5*, and *LOX* [6]. Herein, we describe 34 patients from 24 families with LDS in which we identified the genetic defect in one of the known genes; 15 variants were novel, expanding the mutational repertoire. We further strengthened genotype-phenotype correlations by fitting our data with available studies in literature and updated the gene-to-phenotype categorization of LDS.

2. Patients and Methods

2.1. Patients

Twenty-four index patients and 10 relatives with LDS, mostly of Italian origin but two from Philippines and Sri Lanka, were evaluated from 2010 to 2018 in specialized outpatient clinics for the diagnosis of HCTDs, namely: (*i*) the Ehlers-Danlos Syndrome and Inherited Connective Tissue Disorders Clinic at the University Hospital Spedali Civili of Brescia, (*ii*) the Medical Genetics Unit of the Sant'Orsola-Malpighi Hospital of Bologna, (*iii*) the Clinical Genetics Unit of the Regional Hospital of Bolzano, (*iv*) the Centre of Expertise for Marfan syndrome and Marfan-related disorders at Policlinico Tor Vergata University Hospital of Rome and (*v*) the Medical Genetics Unit, Department of Life, Health, and Environmental Sciences of the University of L'Aquila.

2.2. Molecular Investigations

Molecular analyses were performed in the laboratory of genetic testing at the Division of Biology and Genetics, Department of Molecular and Translational Medicine, University of Brescia. Mutational screening was achieved on genomic DNA purified from peripheral blood leukocytes of affected and unaffected family members by standard procedures. All exons and their intron-flanking regions of *TGFBR1* (NM_004612.4, NP_004603.1), *TGFBR2* (NM_003242.5, NP_0033233.4), *SMAD3* (NM_003242.5, NP_005893.1), and *TGFB2* (NM_001135599.3, NP_001129071.1) were PCR amplified by using optimized genomic primers (Supplementary Table S1, primers were designed by using the Primer Express software v. 3.0.1 (Thermo Fisher Scientific, South San Francisco, CA, USA) and purchased by Metabion International AG, Planneg, Germany), which were analyzed for the absence of known variants using the GnomAD database [7]. After enzymatic cleanup of the PCR products, all fragments were sequenced in both orientations using the Big Dye Terminator Cycle Sequencing kit protocol (Thermo Fisher Scientific, South San Francisco, CA, USA) followed by capillary electrophoresis on the ABI3130XL Genetic analyzer. The sequences were analyzed with the Sequencher 5.0 software (Gene Codes Corporation, Ann Arbor, MI, USA) and variants were annotated according to the Human Genome Variation Society (HGVS) nomenclature with the Alamut Visual software version 2.11 (Interactive Biosoftware, Rouen, France), which was also used for splice site prediction, since it includes four different prediction algorithms: SpliceSiteFinder-like, MaxEntScan, NNSplice, and GeneSplicer. When available, fresh blood samples were collected in patients carrying variants affecting canonical splice sites. In order to verify the effect on splicing, RT-PCR was carried out on total RNA extracted from patients' whole blood stabilized in Paxgene tubes following the manufacturer's protocol by using the Paxgene Blood RNA Extraction Kit (PreAnalytiX, Qiagen, Hilden, Germany). In particular, amplification of cDNA covering exons 2–4 of *TGFBR2* (TGFBRex2-forw: 5′-GTGGCTGTATGGTAAGAGA-3′ and TGFBR2ex4-rev: 5′-CCAGGTTGAACTCAGCTTCTG-3′) and exons 6-8 of *SMAD3* (SMAD3ex6-forw: 5′-CCTAGGGCTGCTCTCCAATG-3′ and SMAD3ex8-rev: 5′-GTGCACATTCGGGTCAACTG-3′) was performed, respectively, and followed by Sanger sequencing of the RT-PCR products. All identified variants were submitted to the Leiden Open Variation Database (LOVD) [8].

2.3. Genotype-Phenotype Analysis and Literature Review

The medical records of 34 patients with a molecularly proven diagnosis of LDS were accessed and their phenotypes were defined using the Human Phenotype Ontology (HPO) terms by a careful review of clinical notes. Phenotypic categories were created based on the number of observations in the patient's cohort and their relationship to LDS. In reporting the frequency of clinical features of our case series, the used denominator refers to the number of subjects for which information on a specific feature was available (Table 1). We reviewed the medical literature on the clinical manifestations of patients with a confirmed genetic diagnosis of LDS, i.e., patients from large cohorts harboring mutations in one of the six known disease-causative genes. The primary search was completed in the PubMed database and limited to articles in the English language literature without restrictions on the date of publication. Keywords used were "Loeys-Dietz syndrome" AND "genotype", "Loeys-Dietz syndrome" AND "phenotype" as well as "Loeys-Dietz syndrome" AND "review." A secondary search was performed to identify pertinent articles cited in those selected in the primary search. We tried to identify all papers for which genotype-phenotype correlation data were available, focusing on reviews and large cohort studies.

2.4. Ethical Compliance

This study was approved by the relevant Ethical Authorities of the: Policlinico Tor Vergata University Hospital (Progetto di Ricerca PGR00229, R.S. 204/16); University of L'Aquila, Medical Genetics Section, Department of Life, Health, and Environmental Sciences (PGR00229, prot. 20251); and local Ethical Committees. This study was performed in agreement with the principles of the 1964 Helsinki Declaration. All subjects (or their legally authorized representative) enrolled into the genetic study provided written informed consent.

Table 1. Comparison of the clinical features observed in previously reported and LDS patients described in this cohort for each distinct gene.

Clinical Features	TGFBR1 Lit.	TGFBR1 This Cohort n = 12 (%)	TGFBR2 Lit.	TGFBR2 This Cohort n = 12 (%)	SMAD3 Lit.	SMAD3 This Cohort n = 9 (%)	TGFB2 Lit.	TGFB2 This Cohort n = 1	SMAD2 Lit.	TGFB3 Lit.
Hypertelorism	++++	10/12 (84)	++++	6/12 (50)	++	4/9 (44)	++	1/1	+	++
Strabismus	+	1/12 (8)	+	1/12 (8)	+	0/9	+	0/1	–	–
Malar hypoplasia	+++	8/12 (67)	+++	9/12 (75)	++	9/9 (100)	++	1/1	++++	++
Bifid uvula/Cleft palate	++++	3/12 (25)	++++	5/12 (42)	++	2/9 (22)	+	1/1	–	++
Dolichocephaly	+++	11/12 (92)	+++	7/12 (58)	+	3/9 (33)	–	0/1	++++	–
Hernia	+++	4/12 (33)	+++	6/12 (50)	++	5/9 (55)	++	1/1	++++	++
Striae	++	5/12 (42)	++	3/12 (25)	++	3/8 (37)	++	1/1	++	+
Pectus deformity	+++	5/12 (42)	+++	7/12 (58)	++	6/9 (66)	++	1/1	++	+++
Scoliosis	+++	10/12 (84)	+++	8/12 (67)	++	3/9 (33)	++	1/1	++	+++
Arachnodactyly	+++	5/12 (42)	+++	6/12 (50)	++	1/9 (11)	++	1/1	++	++
Talipes equinovarus	++	1/12 (8)	++	5/12 (42)	+	1/9 (11)	+	0/1	–	++
Osteoarthritis	++	0/11	++	0/10	++	3/6 (50)	+	0/1	++++	++
Cervical spine malformation/instability	+	1/11 (9)	+	2/9 (22)	+	0/3	–	1/1	–	+
Dural ectasia	++	1/11 (9)	++	3/8 (37)	+++	1/4 (25)	++	1/1	+	–
Mitral valve prolapse or insufficiency	++	5/12 (42)	++	7/10 (70)	++	5/9 (55)	++	1/1	++	++
Arterial tortuosity	++++	3/11 (27)	++++	5/11 (45)	++	1/6 (17)	++	1/1	+	+
Aortic root aneurysm	++++	12/12 (100)	++++	9/12 (75)	+++	7/9 (77)	+++	0/1	++++	++
Arterial aneurysms	+++	5/12 (42)	+++	5/11 (45)	+	2/9 (22)	+	0/1	+	+
Aortic dissection	++++	3/12 (25)	++++	2/12 (17)	++	1/9 (11)	+	0/1	–	++

Lit: Literature. Frequencies of clinical feature associated with LDS were scored as: – absent/infrequent; + <25%; ++ 25–50%; +++ 50–75%; ++++ >75%.

3. Results

3.1. Demographic Data and Genotype-Phenotype Analysis of LDS Patient's Cohort

We report the clinical and genetic features of 34 individuals from 24 families. The female-to-male ratio of affected probands was 1/1. The mean age at diagnosis of the 24 unrelated probands was 26 years (range 1–51 years). Thirteen of them had a positive family history, while 11 were sporadic. Ten individuals were diagnosed upon familial segregation. The detailed clinical and molecular features of each affected individual are summarized in Supplementary Table S2. The overall frequency of selected clinical features observed in our cohort, categorized for each distinct gene as compared to in the literature, is shown in Table 1. An overview of characteristic features observed in the patients of our cohort is presented in Figure 1.

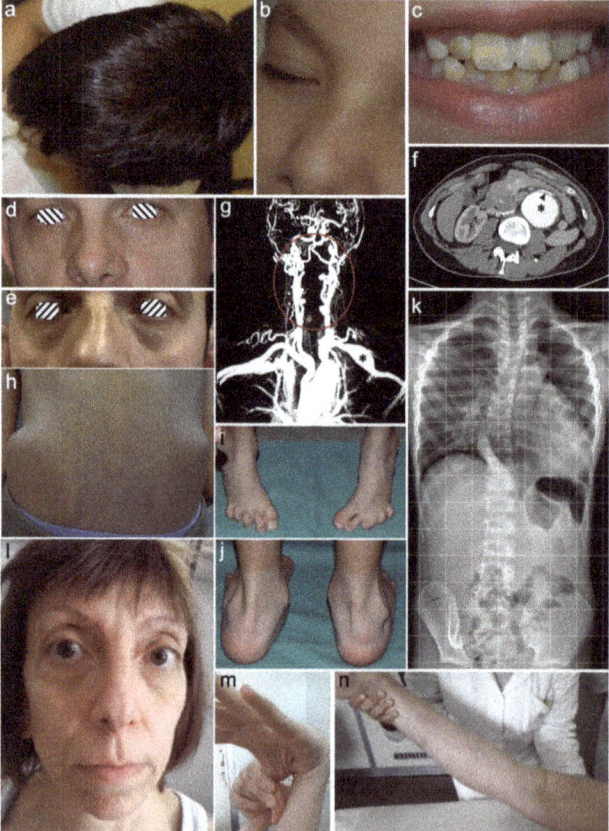

Figure 1. Clinical and instrumental findings observed in our Loeys-Dietz syndrome (LDS) patients.

(**a**–**c**) Proband of family 1 (p.Asp400Gly,*TGFBR1*), aged 7 years manifesting dolichocephaly (**a**), milia (**b**) and enamel defect of permanent dentition. (**d**,**e**) Family 4 (p.Gly271Asp, *TGFBR1*), proband (**d**) and his brother (**e**) outlining intrafamilial variability of palpebral fissures, horizontal in the proband (**d**) and downslanted in his brother (**e**) in the presence of severe milia in both. (**f**) CT-angiography of the abdomen in the proband of family 7 (p.Gly353Arg, *TGFBR1*), aged 23 years, showing aneurysm of the abdominal aorta (asterisk) with intimal blister flap (arrow head). (**g**) Proband of family 15 (p.Asp522Asn, *TGFBR2*), aged 36 years: CT-angiography of head and neck showing fusiform aneurysm of the left subclavian artery (asterisk), middle cerebral artery stenosis (arrow heads), diffuse tortuosity of vertebral arteries (area within the circle). (**h**–**k**) Proband of family 16 (p.Asp446Asn, *TGFBR2*), aged 7 years, displays asymmetry of iliac crest, thin and translucent skin with highly visible subcutaneous venous reticulum (**h**); Limited function of the feet after corrective surgery for bilateral clubfoot (**i**); hind foot deformity with bilateral pronated valgus pes planus (**j**); X-ray frontal view performed at 6 years shows the displaced heart in the left thorax due to severe pectus excavatum and scoliosis with lumbar left rotation and asymmetric iliac wings (**k**). (**l**–**n**), Proband of family 24 (p.Phe160Leufs*14, *TGFB2*), aged 48 years featuring facial asymmetry and dysmorphism (highly arched eyebrows, hypertelorism, bilateral exophthalmos, bifid nasal tip, long philtrum with thin upper lip, micrognathia) in addition to premature aging appearance (**l**); Joint laxity of the thumb (**m**) and the knee (**n**); Note the thin and translucent skin with visible subcutaneous veins.

3.2. Molecular Findings

In this study, 24 different variants in four distinct LDS-related genes were identified in the proband of each family by direct sequencing (Table 2). All variants were deposited in LOVD. The majority of variants affected the *TGFBR1* (9/24) and *TGFBR2* (10/24) genes, four families showed variants in *SMAD3*, and 1 patient harbored a *TGFB2* variant. No variants were identified in *SMAD2* and *TGFB3*. Fifteen variants affecting *TGFBR1*, *TGFBR2*, *SMAD3*, and *TGFB2* were novel, while nine in *TGFBR1*, *TGFBR2*, *SMAD3* were previously reported either in the literature or in the ClinVar database (Supplementary Table S2 and Table 2).

Among novel variants, 10 missense substitutions and one in-frame deletion of two highly conserved amino acid residues p.(Lys232_Ile233del) were evaluated for their putative pathogenicity through different *in silico* prediction algorithms, all predicted them as being high impacting variants. Given that they are: (i) located in critical and well-established functional domains without benign variation; (ii) absent in publicly available population databases, (iii) predicted as deleterious my multiple lines of computational evidence and based on the matching of the clinical phenotypes, all these variants were classified as pathogenic (class 5) according to the American College of Medical Genetics and Genomics (ACMG) guidelines [9].

Likewise, the novel c.480del variant in *TGFB2* was classified as pathogenic (class 5), since it leads to frameshift and formation of a premature termination codon p.(Phe160Leufs*14) that likely activates nonsense-mediated mRNA decay (NMD).

For two out of three variants affecting splice sites, we obtained fresh blood samples and demonstrated an effect on the splicing process, corroborating their pathogenicity. In particular, the c.263+6C>T variant in *TGFBR2* (rs758501054, Minor allele frequency in ExAC, MAF = 0.000008) was classified as likely pathogenic (class 4) according to the ACMG, since it creates a new splice donor site 6 bases downstream of the wild-type donor with retention of 4 nucleotides of intron 3, formation of a stop codon (p.Arg114*) and activation of NMD. The c.1009+1G>A variant in *SMAD3* was classified as pathogenic (class 5), since it abolishes the canonical splice donor site of exon 7 causing in-frame exon skipping (p.Arg292_Gly337del). Lastly, although RNA was not available to investigate its effect, the c.862_871+8del pathogenic variant (class 5) variant in *SMAD3*, thought formally a frameshift variant, likely leads to abnormal splicing as well (i.e., in-frame exon 6 skipping), as the canonical splice donor site is lost.

Table 2. List of variants identified in this study in each gene and molecular details.

Gender	Age at Diagnosis	Family History	Origin	Gene	HGVS	Protein	dbSNP	Patient ID (LOVD)	Variant ID (LOVD)
M	7 years	−	Italy	TGFBR1	c.1199A>G	p.(Asp400Gly)	rs121918711	#00245208	#0000498906
F	31 years	+	Italy	TGFBR1	c.1120G>A	p.(Gly374Arg) §		#00245211	#0000498909
F	29 years	−	Italy	TGFBR1	c.1052A>T	p.(Asp351Val) §		#00245212	#0000498911
M	29 years	+	Italy	TGFBR1	c.812G>A	p.(Gly271Asp) §		#00245213	#0000498912
M	47 years	+	Italy	TGFBR1	c.705_707del	p.(Ser236del)	rs863223830	#00245343	#0000499180
F	17 years	−	Philippines	TGFBR1	c.650G>T	p.(Gly217Val) §		#00245345	#0000499182
M	23 years	−	Italy	TGFBR1	c.1057G>C	p.(Gly353Arg) §		#00245346	#0000499183
M	17 years	−	Italy	TGFBR1	c.1460G>A	p.(Arg487Gln)	rs113605875	#00245347	#0000499184
M	43 years	+	Italy	TGFBR1	c.693_699delinsC	p.(Lys232_Ile233del) §		#00245348	#0000499185
F	51 years	+	Italy	TGFBR2	c.1609C>T	p.(Arg537Cys)	rs104893809	#00245350	#0000499187
F	3 years	−	Sri Lanka	TGFBR2	c.1582C>T	p.(Arg528Cys)	rs104893810	#00245351	#0000499232
F	3 years	−	Italy	TGFBR2	c.1598G>T	p.(Cys533Phe) §		#00245396	#0000499233
M	9 years	−	Italy	TGFBR2	c.1336G>T	p.(Asp446Tyr) §		#00245398	#0000499234
F	45 years	+	Italy	TGFBR2	c.263+6C>T	r.263_264insguaa * p.(Arg114 *) §	rs758501054	#00245408	#0000499245
F	36 years	+	Italy	TGFBR2	c.1564G>A	p.(Asp522Asn)	rs863223854	#00245409	#0000499246
M	1 year	−	Italy	TGFBR2	c.1336G>A	p.(Asp446Asn)	rs886039551	#00245410	#0000499247
M	37 years	+	Italy	TGFBR2	c.1187G>A	p.(Cys396Tyr) §		#00245411	#0000499248
F	3 years	−	Italy	TGFBR2	c.1184T>C	p.(Leu395Pro) §		#00245412	#0000499249
M	12 years	−	Italy	TGFBR2	c.1270T>G	p.(Tyr424Asp) §		#00245413	#0000499250
F	31 years	+	Italy	SMAD3	c.1247C>T	p.(Ser416Phe)		#00245414	#0000499251
M	13 years	+	Italy	SMAD3	c.1009+1G>A	r.872_1009del * p.(Arg292_Gly337del) §		#00245415	#0000499252
F	41 years	+	Italy	SMAD3	c.803G>A	p.(Arg268His)	rs863223740	#00245416	#0000499253
M	23 years	+	Italy	SMAD3	c.862_871+8del	p.(Arg288Glufs*50) §		#00245417	#0000499254
F	48 years	+	Italy	TGFB2	c.480del	p.(Phe160Leufs*14) §		#00245418	#0000499255

*: demonstrated by RT-PCR; §: newly reported variants.

4. Discussion

Since the relatively recent description of LDS in 2005, six disease-causative genes have been identified, namely *TGFBR1* (MIM #190181) [1], *TGFBR2* (MIM#190182) [1], *SMAD2* (MIM #601366) [10], *SMAD3* (MIM #603109) [4], *TGFB2* (MIM #190220) [11], and *TGFB3* (MIM#190230) [12]. This advised the subdivision of LDS into multiple classes based on the causative gene (LDS1-5) providing a general indication of the spectrum of disease severity, from most to least severe form: LDS1=LDS2>LDS3>LDS4>LDS5 [6]. Still, there are not enough data on LDS caused by heterozygous mutations in *SMAD2* to place this form (LDS6) in this spectrum [13]. Recently, it has been suggested that mutations in genes different from *TGFBR1/2* may give rise to a phenotypic continuum hard to categorize in clear-cut genotype-phenotype correlations [3,6]. In this study, we performed a retrospective multi-center study on clinical and mutational analyses in 24 novel LDS families with 34 patients. Furthermore, a systematic overview of LDS patient's features for each mutated gene was conducted to verify if observed genotype-phenotype correlations fitted current knowledge.

Overall, we noticed significant overlaps for *TGFBR1/2*, *SMAD3*, and *TGFB2* patients (Table 1), while emerging genotype-phenotype correlations are described below, subdivided for each gene.

4.1. TGFBR1/2 Genes

According to previous studies, most of the identified mutations affected *TGFBR1* (9/24) and *TGFBR2* (10/24) genes. While nearly all mutated patients displayed ascending/aortic root aneurysm, severe arterial involvement was recorded in half. Early aortic dissection was more rarely reported together with arterial tortuosity. Interestingly, we noticed a high number of different arteries being affected. These findings highlight the need for extended arterial imaging, as was already recommended [5]. Valve abnormalities, in particular mitral valve prolapse and insufficiency, affected nearly half of the patients; isolated mitral valve prolapse was also present in seven out of 10 *TGFBR2* patients. As previously described, craniofacial features may be considered to be characteristics of *TGFBR1/2*-LDS including dolichocephaly, hypertelorism, malar hypoplasia and highly arched palate, while abnormal palate/bifid uvula was not as frequently found as described in the literature. Recurrent skeletal anomalies included scoliosis, pes planus, long slender fingers, marfanoid habitus, and pectus deformity. Hernias were of different types (diaphragmatic, inguinal, or umbilical) and together with joint laxity were registered in half of patients. Cutaneous findings appeared extremely frequent including translucent skin, easy bruising and striae; together with facial milia, described in half mutated patients, skin features appeared as extremely useful handle for diagnosis [14]. Overall, our data confirm broad overlap between LDS phenotypes caused by *TGFBR1/2* pathogenic variants, as previously assessed [6]. Also, in four families the presence of significant craniofacial malformations correlated with a higher risk of vascular manifestations [5,6]. Finally, two children had tooth abnormalities, in particular enamel defects, a possibly underestimated features in LDS in line with the role of TGF-β signaling in dental and enamel formation [15].

4.2. SMAD3 Gene

Among our three *SMAD3*-mutated families, early osteoarthritis, considered a hallmark of this subset of patients, was described only in one, although it segregated in the proband and his father. In line with our data, despite early studies reported osteoarthritis in all *SMAD3* patients [4], others noticed a lower incidence [16]. Interestingly, this family presented very mild cardiovascular anomalies, with only mitral valve prolapse registered in the proband. Generally, *SMAD3*-LDS patients displayed a less severe degree of cardiovascular complications. In fact, despite all the presence of thoracic aortic (and renal arteries) aneurysms, dissection was reported only once, as well as arterial tortuosity and varices. Additional clinical findings overlapped with those recorded in other LDS patients, including recurrent craniofacial dysmorphism such as hypertelorism, malar hypoplasia, micrognathia and highly arched palate, cleft palate and/or uvula, cutaneous features with translucent skin, striae and facial milia, hernias and skeletal manifestation like pes planus, pectus deformity, marfanoid habitus, and scoliosis. It is worth noting that one subject had spontaneous pneumothorax, recorded in a single *SMAD3* patient in the literature [3]. The causes of spontaneous pneumothorax in LDS have not been clearly elucidated but its association with LDS independently from the genetic subtype as well as the occurrence in other HCTD such as Marfan syndrome argues in favor of a common role for TGF-beta signaling. Several lines of evidence based on mouse models, knockout for genes encoding members of the TGF-beta pathway, support this observation (reviewed in reference [17]). In particular, perturbed TGF-beta signaling was implicated in abnormal pulmonary alveolarization and alveolar destruction, leading with age to emphysema and spontaneous lung rupture (i.e., primary spontaneous pneumothorax).

4.3. TGFB2 Gene

We identified only one patient with a *TGFB2* pathogenic variant, whose cardiovascular features consisted in mitral valve prolapse and insufficiency, mild arterial tortuosity and varicose veins, in the absence of aneurysms. Additional features included bifid uvula, striae, translucent and hyperextensible skin, facial milia, hernias, joint hypermobility, pectus deformity, scoliosis, and pes planus observed in other LDS genetic subtypes. In line with previous observations, this patient presented the mildest end of LDS phenotypic spectrum [18]. However, it should be noted that phenotypic variability ranging from mild to severe expression was also recorded in *TGFB2*-LDS [19]. Further broadening the clinical picture of this genetic subtype, she also showed cervical spine instability, which was not previously registered in association with *TGFB2*, although it has been commonly encountered in LDS [3]. Interestingly, in Tgfb2 heterozygous knockout mice severe skeletal defects are also observed including skull base and vertebral malformations, in line with the human phenotype [20]. These data corroborate the importance to investigate and diagnose malformation of cervical spine and/or instability, which should be carefully monitored to prevent severe complications in these patients.

4.4. Other Genes

No mutations were identified in the latest described LDS genes, namely *SMAD2* and *TGFB3*. Still, limited numbers are available in literature about genotype-phenotype correlation. To date, 9 (likely) pathogenic variants in *SMAD2* have now been described in 15 subjects displaying a broad range of features, including aneurysms, tortuosity of the entire arterial tree, and coronary artery dissections, even in the absence of prominent connective tissue characteristics [13]. Concerning *TGFB3*, 15 different variants were reported in 56 individuals presenting with phenotypic overlap between LDS and MFS [12].

5. Conclusions

In this work, we broadened the clinical and molecular spectrum of LDS, corroborated, and expanded previously delineated genotype-phenotype correlations, paving the way for a gene-based classification of different disease subtypes. Larger cohort screenings are needed to unravel the thorough clinical and molecular repertoires in LDS to accurately establish diagnostic criteria, define genotype-phenotype correlations, and collect natural history data for clinical prognostication.

Supplementary Materials: The following are available online at http://www.mdpi.com/2073-4425/10/10/764/s1, Table S1: Primers, Table S2: Clinical and molecular features of LDS patients described in this cohort.

Author Contributions: All authors fulfill the criteria for authorship. F.B. and M.C. conceived the study and wrote the manuscript drafted by L.C. and M.R. who organized data contents, reviewed the literature and performed the tables; F.B. and L.C. prepared the figure; F.B., L.C. and L.G. performed clinical evaluation and follow-up of families 1, 2, 3, 4, 5, 10, 20; F.C.S. and G.N. performed genetic counseling and reviewed molecular data for families 1, 2, 3, 4, 5, 10, 20; A.W. performed clinical evaluation, follow-up and genetic counseling for families 6, 8, 16, 17, 18, 19, 21, 22, 23; S.M. performed clinical evaluation and follow-up of families 7 and 15; M.C. and M.R. performed genetic counseling and follow-up for families 9,11, 12, 13, 14, 22, 24; M.C., M.R. and V.C. carried out molecular investigations and RNA studies; R.M. and P.F. reviewed, analyzed and interpreted molecular data; L.C. submitted the study. All authors discussed, read, and approved the final manuscript.

Funding: F.B. receives funding from the University of L'Aquila (Fondi RIA 2019 and Fondi Ricerca Premiale 2017), the Italian Ministry of Health (Ricerca Finalizzata GR-2013-02356227) and the Undiagnosed Disease Network Italy established at Istituto Superiore di Sanità, Italy (Farmindustria, contributo incondizionato FAC. X7E).

Acknowledgments: The authors wish to thank all patients for their cooperation during the diagnostic process. M.R, V.C., and M.C. thank the Fazzo Cusan family for its generous support. We thank Domenica Taruscio and Marco Salvatore from the Istituto Superiore di Sanità (Italy) for their constant support.

Conflicts of Interest: All authors declare that there is no conflict of interest concerning this work.

References

1. Loeys, B.L.; Chen, J.; Neptune, E.R.; Judge, D.P.; Podowski, M.; Holm, T.; Meyers, J.; Leitch, C.C.; Katsanis, N.; Sharifi, N.; et al. A syndrome of altered cardiovascular, craniofacial, neurocognitive and skeletal development caused by mutations in *TGFBR1* or *TGFBR2*. *Nat. Genet.* **2005**, *37*, 275–281. [CrossRef] [PubMed]
2. Loeys, B.L.; Schwarze, U.; Holm, T.; Callewaert, B.L.; Thomas, G.H.; Pannu, H.; De Backer, J.F.; Oswald, G.L.; Symoens, S.; Manouvrier, S.; et al. Aneurysm syndromes caused by mutations in the TGF-beta receptor. *N. Engl. J. Med.* **2006**, *355*, 788–798. [CrossRef] [PubMed]
3. Schepers, D.; Tortora, G.; Morisaki, H.; MacCarrick, G.; Lindsay, M.; Liang, D.; Mehta, S.G.; Hague, J.; Verhagen, J.; van de Laar, I.; et al. A mutation update on the LDS-associated genes *TGFB2/3* and *SMAD2/3*. *Hum. Mutat.* **2018**, *39*, 621–634. [CrossRef] [PubMed]
4. Van de Laar, I.M.; Oldenburg, R.A.; Pals, G.; Roos-Hesselink, J.; de Graaf, B.; Verhagen, J.; Hoedemaekrs, Y.; Willemsen, R.; Severijnen, L.; Venselaar, H.; et al. Mutations in *SMAD3* cause a syndromic form of aortic aneurysms and dissections with early-onset osteoarthritis. *Nat. Genet.* **2011**, *43*, 121–126. [CrossRef] [PubMed]
5. MacCarrick, G.; Black, J.H., 3rd; Bowdin, S.; El-Hamamsy, I.; Frischmeyer-Guerrerio, P.A.; Guerrerio, A.L.; Sponseller, P.D.; Loeys, B.; Dietz, H.C., 3rd. Loeys-Dietz syndrome: A primer for diagnosis and management. *Genet. Med.* **2014**, *16*, 576–587. [CrossRef] [PubMed]
6. Loeys, B.L.; Dietz, H.C. Loeys-Dietz Syndrome. In *GeneReviews®*; University of Washington: Seattle, WA, USA, 2008. Available online: https://www.ncbi.nlm.nih.gov/books/NBK1133/ (accessed on 21 August 2019).
7. The Genome Aggregation Database (gnomAD) v2.1.1. Available online: https://gnomad.broadinstitute.org (accessed on 21 August 2019).
8. Leiden Open Variation Database (LOVD). Available online: https://databases.lovd.nl/shared/genes (accessed on 21 August 2019).
9. Richards, S.; Aziz, N.; Bale, S.; Bick, D.; Das, S.; Gastier-Foster, J.; Grody, W.W.; Hegde, M.; Lyon, E.; Spector, E.; et al. ACMG Laboratory Quality Assurance Committee. Standards and guidelines for the interpretation of sequence variants: A joint consensus recommendation of the American College of Medical Genetics and Genomics and the Association for Molecular Pathology. *Genet. Med.* **2015**, *17*, 405–424. [CrossRef] [PubMed]
10. Micha, D.; Guo, D.C.; Hilhorst-Hofstee, Y.; van Kooten, F.; Atmaja, D.; Overwater, E.; Cayami, F.K.; Regalado, E.S.; van Uffelen, R.; Venselaar, H.; et al. SMAD2 mutations are associated with arterial aneurysms and dissections. *Hum. Mutat.* **2015**, *36*, 1145–1149. [CrossRef] [PubMed]
11. Lindsay, M.E.; Schepers, D.; Bolar, N.A.; Doyle, J.J.; Gallo, E.; Fert-Bober, J.; Kempers, M.J.; Fishman, E.K.; Chen, Y.; Myers, L.; et al. Loss-of-function mutations in *TGFB2* cause a syndromic presentation of thoracic aortic aneurysm. *Nat. Genet.* **2012**, *44*, 922–927. [CrossRef] [PubMed]
12. Bertoli-Avella, A.M.; Gillis, E.; Morisaki, H.; Verhagen, J.M.A.; de Graaf, B.M.; van de Beek, G.; Gallo, E.; Kruithof, B.P.T.; Venselaar, H.; Myers, L.A.; et al. Mutations in a TGF-β ligand, *TGFB3*, cause syndromic aortic aneurysms and dissections. *J. Am. Coll. Cardiol.* **2015**, *65*, 1324–1336. [CrossRef] [PubMed]
13. Cannaerts, E.; Kempers, M.; Maugeri, A.; Marcelis, C.; Gardeitchik, T.; Richer, J.; Micha, D.; Beauchesne, L.; Timmermans, J.; Vermeersch, P.; et al. Novel pathogenic *SMAD2* variants in five families with arterial aneurysm and dissection: Further delineation of the phenotype. *J. Med. Genet.* **2019**, *56*, 220–227. [CrossRef] [PubMed]
14. Lloyd, B.M.; Braverman, A.C.; Anadkat, M.J. Multiple facial milia in patients with Loeys-Dietz syndrome. *Arch. Dermatol.* **2011**, *147*, 223–226. [CrossRef] [PubMed]
15. Morkmued, S.; Hemmerle, J.; Mathieu, E.; Laugel-Haushalter, V.; Dabovic, B.; Rifkin, D.B.; Dollé, P.; Niederreither, K.; Bloch-Zupan, A. Enamel and dental anomalies in latent-transforming growth factor beta-binding protein 3 mutant mice. *Eur. J. Oral Sci.* **2017**, *125*, 8–17. [CrossRef] [PubMed]

16. Wischmeijer, A.; Van Laer, L.; Tortora, G.; Bolar, N.A.; Van Camp, G.; Fransen, E.; Peeters, N.; Di Bartolomeo, R.; Pacini, D.; Gargiulo, G.; et al. Thoracic aortic aneurysm in infancy in aneurysms-osteoarthritis syndrome due to a novel *SMAD3* mutation: Further delineation of the phenotype. *Am. J. Med. Genet. A* **2013**, *161*, 1028–1035. [CrossRef] [PubMed]
17. Saito, A.; Horie, M.; Nagase, T. TGF-β Signaling in Lung Health and Disease. *Int. J. Mol. Sci.* **2018**, *19*, 2460. [CrossRef] [PubMed]
18. Ritelli, M.; Chiarelli, N.; Dordoni, C.; Quinzani, S.; Venturini, M.; Maroldi, R.; Calzavara-Pinton, P.; Colombi, M. Further delineation of Loeys-Dietz syndrome type 4 in a family with mild vascular involvement and a *TGFB2* splicing mutation. *BMC Med. Genet.* **2014**, *15*, 91. [CrossRef] [PubMed]
19. Mazzella, J.M.; Frank, M.; Collignon, P.; Langeois, M.; Legrand, A.; Jeunemaitre, X.; Albuisson, J. Phenotypic variability and diffuse arterial lesions in a family with Loeys-Dietz syndrome type 4. *Clin. Genet.* **2017**, *91*, 458–462. [CrossRef] [PubMed]
20. Sanford, L.P.; Ormsby, I.; Gittenberger-de Groot, A.C.; Sariola, H.; Friedman, R.; Boivin, G.P.; Cardell, E.L.; Doetschman, T. TGFbeta2 knockout mice have multiple developmental defects that are non-overlapping with other TGFbeta knockout phenotypes. *Development* **1997**, *124*, 2659–2670. [PubMed]

© 2019 by the authors. Licensee MDPI, Basel, Switzerland. This article is an open access article distributed under the terms and conditions of the Creative Commons Attribution (CC BY) license (http://creativecommons.org/licenses/by/4.0/).

MDPI
St. Alban-Anlage 66
4052 Basel
Switzerland
Tel. +41 61 683 77 34
Fax +41 61 302 89 18
www.mdpi.com

Genes Editorial Office
E-mail: genes@mdpi.com
www.mdpi.com/journal/genes

www.ingramcontent.com/pod-product-compliance
Lightning Source LLC
LaVergne TN
LVHW070400100526
838202LV00014B/1357